ENGINEERING HERITAGE

Uniform with this Volume

ENGINEERING HERITAGE (VOLUME I)

MECHANICAL ENGINEERING

ENGINEERING MATERIALS AND METHODS

MANAGEMENT FOR ENGINEERS

THE ENGINEER AND SOCIETY

ENGINEERING HERITAGE

Volume Two

Edited by
E. G. SEMLER, BSc, CEng, FIMechE

The Institution of Mechanical Engineers
LONDON

First published by Heinemann Educational Books Ltd
on behalf of The Institution of Mechanical Engineers
1966
Reprinted by The Institution of Mechanical Engineers
1972

ISBN 0-85298-083-3

First published as a series of articles in *The Chartered Mechanical Engineer* between 1962 and 1966. Reprinted by photolitho in this form by J. W. Arrowsmith Ltd., Bristol, on cartridge paper and bound by G. and J. Kitcat Ltd, London, England

Made and printed in Great Britain

CONTENTS

Getting down to the Roots

by E. G. Semler, BSc, CEng, FIMechE

When we published the first volume of *Engineering Heritage* its reception exceeded our highest expectations and in fact within less than a year we had to reprint it twice. It is particularly pleasing that a substantial part of the printing was sold to the lay public.

This is no accident, for the increasing interest by the layman in engineering—the basis of our technological civilisation—is matched only by the growing desire of engineers themselves to learn more about the history of their craft and about the great masters who have practised it in the past. It was to cater for this that the original articles in *The Chartered Mechanical Engineer* were published and later reprinted as Volume I of *Engineering Heritage*. However, the mine of history was by no means exhausted and as a result of the popularity of both the 'Evolution of Engineering' and 'The Great Masters' series of articles, it was decided to continue them for several more years and publish a companion volume, based on the same general principles. In other words, the response has been such that we need not apologize for giving you the mixture as before!

While a knowledge of Volume I will be helpful to an understanding of the present work, it is by no means essential: the two can be read quite independently and are related only by their common subject of mechanical engineering. Chronologically they outline the whole history from the dawn of civilisation to the beginning of this century; though, naturally, the emphasis is on the period beginning with the industrial revolution when mechanical engineering in the modern sense came of age.

One is tempted to go further and speak of birth, rather than coming-of-age: this temptation should be resisted by any serious student of the subject. While practically nothing definite is known about the mechanical arts during the middle ages and little enough in classical times, the theory that all the sophisticated techniques, whose existence made the industrial revolution possible, were invented by a few brilliant 18th century craftsmen is too great a strain on, at any rate, an engineer's credulity. Just as the idea of the steam engine can—albeit with increasing vagueness—be traced into the remotest antiquity, so the age-old existance of basic mechanical technologies, plus a fair amount of rudimentary theoretical knowledge, must be surmised. It is not for nothing that the coat of arms of the Institution of Mechanical Engineers includes both Vulcan and Archimedes.

Unfortunately throughout long ages such practical knowledge was obscured by the jealous secretiveness of the craft guilds, as well as by an academic lack of interest in mundane matters pertaining to human comfort, which began with Plato, who has a lot to answer for as far as technology is concerned.

The position of the engineers themselves was perhaps somewhat different: whether or not there were any great mechanical inventors before the 18th century we cannot say. Agricola does not mention any and while some of the engineers of the Italian Renaissance (including Leonardo) have left us with numerous mechanical doodles, they certainly did not attempt to translate most of them into machinery. Nevertheless, they were interested in linkages and mechanisms; books such as Agricola's *De Re Metallica* prove that a large amount of our everyday mechanical engineering was already established usage among the German miners of the 16th century.

But it was not until the 18th century that a number of undisputed masters began to be distinguishable in the general panorama of mechanical engineering history: these were the makers of the industrial revolution which can now be seen as the birthpangs of the modern technological era. Curiously enough, the procession of great masters begins to thin out at the beginning of the current century. Is this because the more recent engineers are too close to us in time to be seen in full stature? or, more likely, because the Limited Company and corporate research have begun to take over what the rugged individualists of an earlier age had started? Whatever the reason, although the technological drama has accelerated and intensified enormously, none of its protagonists quite measure up to a Smeaton, Fulton or Armstrong.

Could it be that the century of the great masters was merely a short interlude between the long ages of craft secrecy and rule-of-thumb working in the past and the present era of commercial security covering the work of research and design teams?

It is questions such as these that the thoughtful reader of this book must ask himself as he studies the historical background and gets down to the roots of Britain's basic technology; and the subject can hardly fail to gain in interest from the knowledge that on the answer may well depend the economic future of this country.

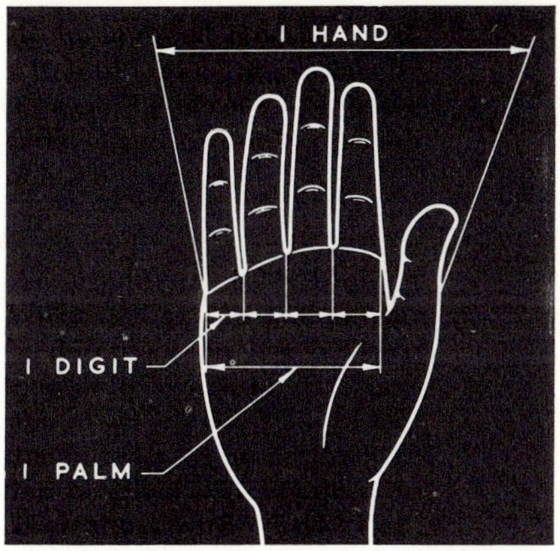

Fig. 1. Until fairly recent times, length measurements in most countries were based upon average-sized human limbs and digits

1 palm=4 digits=3 inches (approx)
1 hand=4 inches (approx)

For many centuries length measuring techniques were of interest to civil engineers only, because such mechanisms as required accurate fitting were made and matched individually.
With the advent of precision tools in the 19th century, and the desire for interchangeable parts in small arms manufacture, came the need for accurate dimensional control of engineering components which opened the way to modern production methods.
The search for permanency has led to a return to natural standards, based on the wavelength of monochromatic light.

The Measurement of Length

by L. W. Nickols, B.Sc(Eng.), M.I.Prod.E. (Member)

The earliest known measurements of length were made many thousands of years ago to meet the requirements of land surveying and civil engineering. It was not until the industrial revolution that length measurements for mechanical engineering purposes became of importance. Nevertheless, the following account goes back to the earliest civilisations in order to trace the way in which techniques of measurement have evolved through the ages.

From the industrial revolution onwards the development of better methods of measurement has proceeded hand in hand with the development of better methods of manufacture. In many cases the higher performances demanded of mechanisms necessitated their manufacture to smaller tolerances, and this could only be achieved by more accurate methods of manufacture; full advantage of these methods could not be taken until it was possible to measure the parts to a greater accuracy.

Over the years there has been a progressive increase in the accuracy of measurement and this trend still continues today.

All measurements have to be made in terms of suitable units, by comparison with some standard. The standard for length today may vary from the mechanic's graduated rule for coarse measurements, to the wavelength of a certain type of light for the most refined techniques. The accuracy achieved depends on the accuracy of the standard, as well as on the method of comparison.

The evolution of standards is thus closely bound up with the evolution of measuring equipment. Standards of length vary widely in type and accuracy. Primary, national, or international standards represent the appropriate unit of length to the highest degree of accuracy.

In some cases the primary standard is the means of defining the unit; in other cases the primary standard, in the form of a material representation of the unit, may have a certain known error.

The earliest standards

When the earliest known civilisations became established, length was measured to mark out areas of land for cultivation and taxation. The wealth of ancient Egypt was based largely on the annual flooding of the Nile valley which deposited large quantities of rich alluvial soil. During the floods, existing landmarks were obliterated and it became necessary to mark out afresh each year. Similar needs arose in the valleys of the Euphrates and Tigris where the Sumerian civilisation was contemporary with that of Egypt. A further stimulus to length measurement was the erection of buildings and monuments.

The earliest units of length were based on the lengths of limbs or parts of limbs. The main unit used in early Egyptian and Sumerian times was the cubit, based on the length of the forearm from the elbow to the tip of the middle finger.

The cubit is mentioned in the Bible as early as the book of Genesis (6.15) in connection with the building of Noah's Ark, the length of which was specified as "three hundred cubits, the breadth of it fifty cubits and the height of it thirty cubits". Such an ark, if immersed to one-third its depth, would have displaced about 14 000 tons of water!

Smaller units were based on the lengths of certain parts of the hand and foot, some of which are shown in Fig. 1. Larger units were the fathom, corresponding to the maximum distance between the finger tips of a man's outstretched arms, and the reed which was two fathoms long These natural units had the disadvantage that they varied from person to person and, as civilisation developed, attempts were made to standardise their lengths and to adopt simple relationships between the various units, shown in Fig. 1.

As early as 3000 BC it became necessary to replace natural units by standard cubit rods, sub-divided into the smaller units, and marked with the King's name. These were kept in the principal temples and copies were used as working standards. Evidence exists from measurements made on ancient Egyptian monuments, and from rods found in tombs of the period, that several sizes of cubit were used; in Egypt the short cubit of about 17·68 British inches, and a longer royal cubit of about 20·62 British inches, were the commonest. The royal cubit introduced later was one palm longer than the short cubit. The reason for its use is still obscure.

One of the most interesting royal cubit standards still in existence, dates to about 1300 BC and is now in the Egyptian Museum at Turin. It is of pentagonal section, made of wood, and fully graduated. The markings show that the royal cubit was sub-divided into seven palms, each containing four digits. Some of the digits are sub-divided into parts varying from two to sixteen in number, indicating measurements to about 0·04 in. As would be expected from primitive craftmanship, the graduations on this cubit are far from perfect.

Another royal cubit, dating from about 1500 BC is a working standard. It was found in the tomb of Kha who was superintendent of works of the necropolis at Thebes in Upper Egypt. This cubit rod is made of wood and is hinged at the centre like a carpenter's rule so that it may be folded for convenience: it is graduated into seven palms and the seventh palm is divided into four digits.

The length of the cubit was also standardised in Sumeria and afterwards in Assyria and working standards of length were used in the Indian and Chinese civilisations of the same period. Of particular interest to us is the Northern cubit of about 26·5 in which may originally have been based on one of the Assyrian measures. It was used in Egypt and later came into use in Europe as the Teutonic tribes gradually advanced westwards. It was introduced into England by the Saxons from about AD 400 onwards. The Northern cubit was sub-divided into two feet, each of about 13·2 British inches.

There is evidence that certain working standards were longer than one cubit. Ezekiel (40.3) mentions the use of a measuring reed in conjunction with a line of flax, the latter being used, presumably, to define a straight line along which to measure. Later (41.8) it refers to a "full reed of six great (Assyrian) cubits".

Even down to Tudor times similar methods were used to measure distances in surveying. For example, Leigh (AD 1577) mentions in his treatise *The most Profitable and Commendable Science of Surveying*, the use of the line and the rod, the line being four perches in length and having a knot at each perch interval.

Despite the early Egyptians' crude measuring equipment, they could build to a surprising accuracy. For example measurements of the square base of the Khufa (Cheops) Pyramid at Giza have shown that the lengths of the four 756 ft long sides were equal to within 8 in. Bearing in mind that the line of flax must have been liable to stretch in use, this is a commendable achievement. Much more remarkable is the fact that the angles at the corners of the square base do not depart from 90 degrees by more than about 3½ minutes of arc, and the sides of the base do not depart from the points of the compass by more than 5½ minutes of arc. Such accuracies are explicable only by a knowledge of geometry and line-of-sight surveying techniques.

The perimeter of the base is 3023·16 ft, which is exactly one-half the meridian distance corresponding to 1 min of arc of latitude at the Equator. This has been taken as proof that the royal cubit was derived from the measurement of the Earth but there is no other evidence to support this theory. One wonders whether it was known to the Frenchmen who adopted a similar principle in 1793.

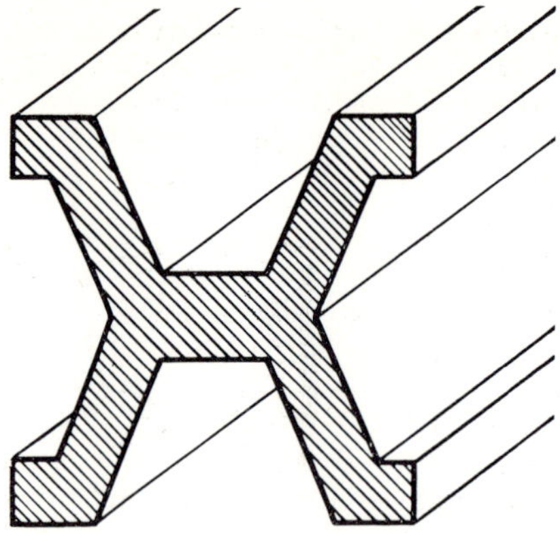

Fig. 2. The international standard metre, set up in 1870, was a platinum bar of this cross section. The idea was to combine stiffness with an exposed neutral plane on which the definition marks could be engraved

Information about the length measuring equipment of the Egyptians is scanty. It is probable that they relied on the use of simple equipment such as graduated sticks, knotted lines, simple callipers, plumb bobs and squares.

There is little evidence that the length measuring equipment used by the Greeks and Romans showed any major advance in principle over that used by the early Egyptians, although the workmanship was probably better. The Chinese of about the same period would appear to have been at least as advanced as the Greeks and Romans in their measuring techniques: the Science Museum has, for example, a model of a Chinese slider calliper which is remarkably similar to the frame of a modern vernier calliper gauge.

The Greeks, and afterwards the Romans, adopted different cubits of about 18 British inches but the need for a shorter primary unit resulted in the introduction of a new standard, generally two-thirds or three-fifths of the cubit, or approximately the length of the human foot. The foot was alternatively divided into sixteen digits (finger breadths) or twelve inches (thumb breadths). The Northern foot of 13·2 modern inches remained in common use in England from about AD 400 until it was superseded at the beginning of the 14th century.

The British yard

So far as is known, the first material primary standard in this country was the standard yard introduced by Edward I in AD 1305. This was made of iron and was called an *ulna*. This new yard had to fit into the existing system of measurement and its length was so chosen that the then existing rod of 15 Northern feet became the rod of 5½ yards, i.e. 16½ modern feet. The statute in which the use of the ulna was legalised, ordained "that three grains of barley, dry and round, make an inch, twelve inches make a foot, three feet make an ulna". The use of grains of barley with their wide variation in size indicates the low accuracy required. Later yard standards, which are still in existence, were introduced by Henry VII in 1497 and by Elizabeth I in 1598; these were end-standards, i.e. the yard was defined by the distance between the parallel end-faces, and they were shorter than the present yard by 0·04 and 0·01 in, respectively.

The Elizabethan yard continued to be the British primary standard of length until AD 1824, despite the fact that at some time it had been broken into two parts and had been joined together again. Copies were made from time to time and kept at the Tower of London. As time went on it was realised that the standard yard's sub-divisions were insufficiently accurate for scientific purposes and, in 1742, the Royal Society requested George Graham, a well-known instrument-maker, to manufacture two brass bars, on each of which were engraved parallel lines, representing the lengths of the Elizabethan and Tower yards. The preference for a line-standard was due to the improvement of line engraving techniques.

In 1758 a Royal Commission was set up under Lord Carysfort for the purpose of preparing a new primary standard. The task was entrusted to John Bird who completed it in 1760. The Bird yard was a bronze bar in which two gold plugs were mounted, each bearing a fine dot; the distance between the dots defined the length. A Bill introduced in 1760 to legalise this new primary standard was never passed into law because of the death of George II, and the Bird yard was legalised only in 1824, ten years before it was destroyed in the fire at the Houses of Parliament!

The distance between the dots on the Bird yard had to be transferred to secondary standards by means of trammels, a far from accurate method. During the 64 years between its completion and its legal adoption, considerable improvements took place in the methods of calibrating scales. Ramsden had invented his dividing engines and microscopes had come into use. It is surprising, therefore, that advantage was not taken of these improvements to produce a better standard. However, the destruction of the Bird yard led to the construction of a true line standard, legally adopted in 1856. The Imperial Standard Yard,* is a bronze bar of square section with two gold plugs on which are engraved the lines that define the length. The plugs are inset in the neutral plane of the bar so as to minimise the effect of flexure.

International efforts

In France, the chaotic situation arising from the use of a number of different standards led to the introduction of the metre. In an attempt to overcome the disadvantages of a material standard, a laudable effort was made to define the metre in terms of something more constant and durable; the ten-millionth part of the Earth's polar quadrant passing through Paris.

The practical embodiment of this definition was in the form of a platinum end-standard called the *metre des archives*. Unfortunately the geographical definition was found to be inconvenient and the *metre des archives* was adopted as the primary standard of length in France in 1799.

In 1870 an international commission was set up to consider, *inter alia*, the establishment of an international standard of length based on the *metre des archives*. It was later decided to make it a line-standard of the winged X–section shown in Fig. 2, designed to provide maximum rigidity and an exposed neutral plane on which the lines defining the metre were engraved. This platinum-iridium international prototype metre was approved by the international conference of 1889.

Another country which experienced trouble in establishing a primary standard of length was the United States. No legal primary standard existed for many years. The U.S. Treasury established, in 1832, an unofficial yard standard defined as the distance between the 27th and 63rd inch graduations on an 82-inch brass scale made by the British instrument-maker Edward Troughton. The Troughton scale remained the unofficial standard until, in 1857, it was supplanted by two copies of the Imperial Standard Yard. In 1893 an order was issued defining the U.S. yard in terms of the metre, the use of which had been authorised by Congress in 1866.

One further development must be mentioned. For many years the desirability of defining standards of length in terms of a natural standard had been recognised. In 1827, J. Babinet suggested using a wavelength of light. Such a standard is indestructible and is easily reproducible, but was not practicable until suitable sources of monochromatic light had been developed. It was not until 1892–93 that Michelson and Benoit, at the International Bureau of Weights and Measures, made the first direct measurement of the metre in terms of the wavelength of the Cadmium red radiation. After the end of the first world war the measurement of engineers' end-standards of length by interferometry became a common procedure.

Due to the expansion of metals with increase of temperature, the necessity to standardise the temperature became apparent many years ago. The *metre des archives* was the first primary standard to be related to temperature, 0°C being chosen. Britain used 62°F and the U.S.A. 20°C, so that considerable confusion arose, until, in 1931, the International Committee of Weights and Measures adopted 20°C.

Vernier and micrometer

Up to the middle of the 18th century there were few developments in the methods of length measurement. Chief among these were the invention, by Pierre Vernier (1580–1673) of the device bearing his name; and of the micrometer by Gascoigne in 1638. Full use could not be made of either of these until better methods of engraving scales and making micrometer screws had been developed. Gascoigne was an astronomer and invented a micrometer-operated calliper to measure the sizes of the images of stars as seen in his telescope. It is not known how he obtained a sufficiently accurate screw thread. Unfortunately his promising career was cut short at the age of 24 when he was killed at the battle of

* Now known as the United Kingdom Primary Standard of the yard Weights and Measures Act, 1963).

Fig. 3. James Watt's micrometer was the forerunner of today's bench instruments. The U-frame was made of brass and workpieces were inserted between two radiused steel anvils

Marston Moor during the Civil War. His invention was the progenitor of the micrometer-microscope which later became essential in the accurate calibration of linear scales.

Verniers were used by instrument-makers by the middle of the 18th century; the Science Museum in London contains 90-in and 23-in scales used by Bird (1750) which are fitted with verniers reading to 0·001 in. However, for general workshop use, the manufacture of scales and verniers had to await the invention of the linear ruling engine by Ramsden in 1775, because prior to this the ruling of accurate scales was a slow and painstaking task.

The next major advance was the famous micrometer made by James Watt in 1772 and shown in Fig. 3. This was the forerunner of today's bench instruments. The U-shaped frame is made of brass and has a radiused iron anvil fixed to one limb. Another radiused anvil is mounted in a slide on the other limb and is moved by a screw engaging a rack, cut on the lower edge of the slide. Fractions of a revolution of the screw are read off a dial and the number of complete revolutions is indicated by a pointer reading on the smaller dial. The pitch of the screw is about 19 turns per inch and each of the 51 divisions on the right-hand dial is apparently intended to represent 0·001 in. This curious number of divisions indicates the difficulty at that time of producing a screw of accurate pitch.

Henry Maudslay was the next engineer to make a significant contribution to methods of length measurement. Although not the first to introduce the screw-cutting lathe to engineering, he developed it to the form we know today. He used his knowledge of cutting accurate threads to make, in 1805, a bench micrometer with flat parallel anvils and a 100 t.p.i. micrometer screw. This could be read to the nearest 0·0001 in and was called the *Lord Chancellor* because it was the final arbiter in any dispute.

Interchangeability

Despite Maudslay's work in improving accuracy of measurement, it would appear that, at the time of his death, most mechanics still measured workpieces by means of callipers and a graduated rule. At this time also, the parts of engineering mechanisms had to be individually fitted together by hand in order to obtain the desired degree of clearance; no two similar parts were dimensionally interchangeable, except by chance, and output was low.

A major advance in the field of interchangeability took place in the U.S.A. in the 1830's when Eli Whitney introduced, for the large-scale manufacture of muskets, what became known as the American system of manufacture. This was based on using a master gauge for each critical dimension of every component. The actual sizes of the master gauges were not necessarily in strict accordance with the sizes specified on the component drawing in terms of the standard yard; there was no easy way of checking this at that time but, provided all the components were made to the same gauges, no assembly troubles were likely to arise.

The American system was introduced in England in 1854 for the manufacture of rifles at the Royal Small Arms Factory at Enfield. It had the disadvantage that interchangeability was confined to the one organisation that had the set of master gauges and subcontracting to other firms was difficult. Also, for small-scale production the scheme was not economical because of its high initial cost.

To overcome these difficulties, Joseph Whitworth introduced his system of gauges which were end-standards in the form of bars with flat parallel ends; they were usually made in

Fig. 4. One of Whitworth's measuring machines was reputed to be sensitive to a millionth of an inch but the machines used in industry, such as this one, were considerably less accurate

nominal sizes from 1 to 12 in by 1 in increments and then, by 6 in increments, up to 36 in. In conjunction with a measuring machine designed by himself, these gauges could be used on a wide range of workpieces. The machine was basically a bench micrometer used to compare the size of the workpiece with the standard gauge.

One of the Whitworth machines was reputed to be sensitive to 0·000001 in, but those generally used (one of which is shown in Fig. 4), were less accurate. Readings repeatable to 0·00001 in were obtained by means of a vernier. The position of the tailstock along the bed of the machine was adjustable. The standard was placed on suitable supports with its left-hand end touching the flat anvil of the tailstock. The micrometer-operated anvil was then advanced until the gravity piece, a short plug of $\frac{3}{8}$ in diameter, just refused to pass under its own weight between the anvil and the right-hand end of the standard. This enabled all readings to be made under a constant force.

Whitworth also introduced standard plug and ring gauges in a series of nominal sizes. The work had to be made to fit these gauges closely; the concept of specified manufacturing tolerances, based on 'not-go', as well as 'go', gauges, had not yet been developed; interchangeable assembly had been achieved but with difficulty, and at considerable cost.

Whitworth used end-standards rather than graduated scales and he told the Institution in 1856:

> We have in this mode of measurement all the accuracy we can desire; and we find in practice in the workshop that it is easier to work to the ten-thousandth of an inch from standards of end-measurement than to one hundredth of an inch from lines on a two foot rule. In all cases of fitting, end-measure of length should be used instead of lines.

In other countries appreciable progress in measuring equipment had been made. In France, Palmer produced the forerunner of the modern hand micrometer in 1848. This was very similar to the modern version but had the screw exposed. In the U.S.A. Brown and Sharpe began to manufacture hand micrometers of a type similar to Palmer's in 1867, and in 1885 the same firm introduced the modern concealed screw design. J. R. Brown and Sharpe also invented the vernier calliper in 1851, and this has remained virtually unchanged in principle.

Another useful development was the dial gauge; as one would expect from its construction, it originated in the watch-making industry and thence its use spread into the field of general engineering. The manufacture of dial gauges began in the U.S.A. about 1890.

Slip gauges

The modern mass-production of interchangeable parts to specified limits of size was greatly facilitated by the invention of slip gauges by C. E. Johansson in 1896. Johansson was an inspector at a rifle factory in Eskilstuna, Sweden. In those days every dimension on a rifle had its master gauge in the form of a steel block made to the desired size. The manufacture of a rifle entailed the provision of a very large number of these blocks. Johansson had the ingenious idea of providing a limited number of blocks in a series of preferred sizes, which could be combined together to build up standards covering a wide range in very small increments. For example, from 81 blocks, combinations can be built up in 0·0001 in increments, ranging between 0·2 in and above 18 in.

The blocks are made of hardened and stabilised rectangular steel section; the two end-faces which define the length of each block are so accurately lapped flat and parallel that, when they are brought together in sliding contact, they adhere due to molecular cohesion.

Johansson set up in business making these slip gauges in his home workshop, using improvised equipment. For a number of years industry was slow to grasp the advantages of his slip gauges but when World War 1 broke out, vast quantities of armaments had to be made by thousands of firms of all sizes, and components made by any firm had to fit those made by others.

To guarantee interchangeable assembly without the need for hand-fitting, all parts were made to specified limits of size, controlled by limit gauges; the limit gauges could best be checked against slip gauges. Johansson's manufacturing process was a secret and the import of slip gauges from Sweden became difficult. For this reason, the National Physical Laboratory developed a British method of making slip gauges to the required accuracy which led to their manufacture in this country.

Johansson-type slip gauges were convenient end-standards for lengths up to about 5 in but, above that, particularly when used horizontally, they were not so convenient. A. J. C. Brookes developed length bars of circular section at the NPL in 1919 which were made in nominal lengths up to 36 in. A set of these can be used to build up combinations by increments of 1 in, and slip gauges can be 'wrung' on to one end-face to make up the desired length. For workshop use these length-bars are assembled by means of studs.

Comparators

The manufacture of very precise end-standards brought another problem in its train. Whilst dial gauges and similar devices were used to measure the differences in size between workpieces and gauges to a reasonable accuracy, no comparators existed in industry by means of which the sizes of slip gauges could be checked against standards, to an accuracy of 1 microinch. The answer to this problem was supplied in 1918 by E. M. Eden who, in cooperation with F. H. Rolt, designed the well known Eden-Rolt millionth comparator at the NPL; the original instrument, which is now in the Science Museum, was constructed during one weekend and the comparator is still made today, substantially in its original form.

The Eden-Rolt millionth comparator is a classic example of the simplicity and elegance in design associated with Eden's instruments. The high magnification (×20 000) was obtained by a combination of a mechanical amplification system in the first stage, followed by optical magnification in the second stage, as shown in Fig. 5. The mechanical system is of particular interest because it pioneered the use of spring-strip suspensions for the moving parts in order to eliminate

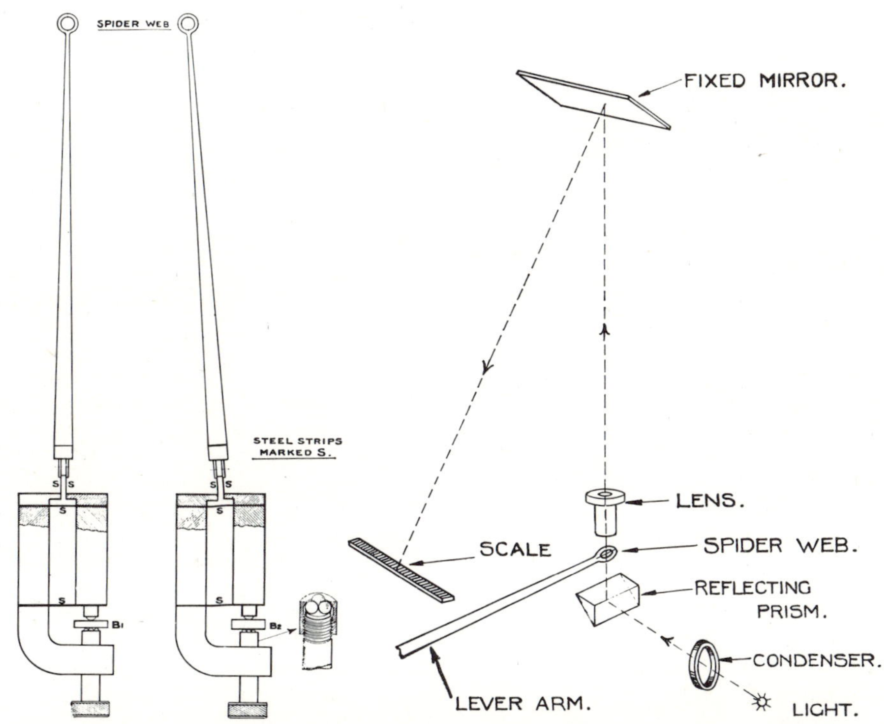

Fig. 5. In the Eden-Rolt millionth comparator a magnification of 20 000 was obtained by a combination of mechanical amplification (left) and an optical system in the second stage

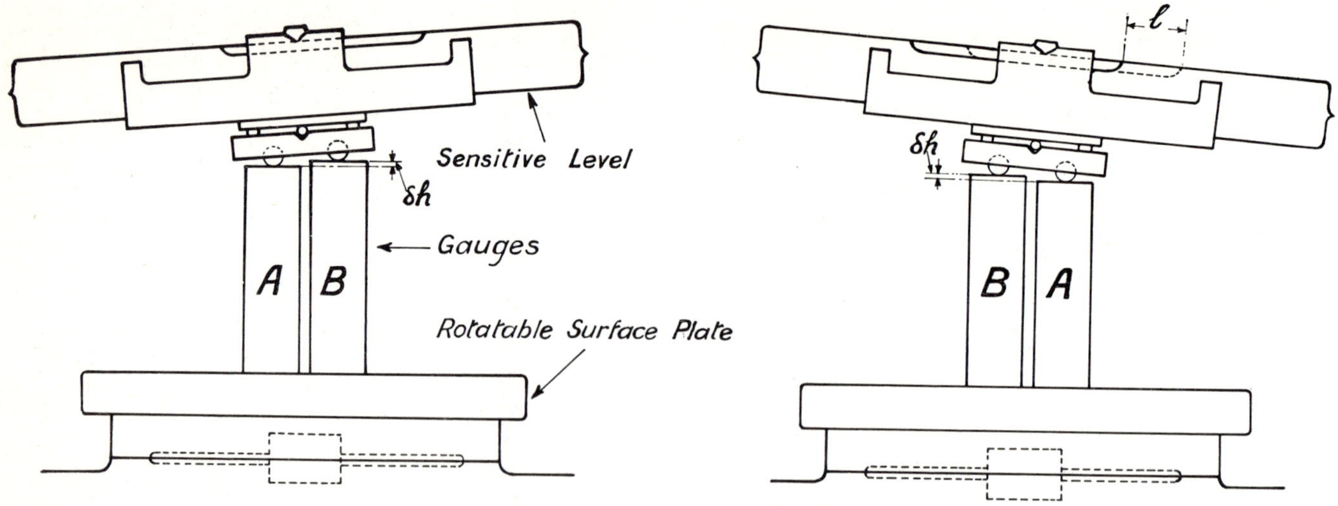

Fig. 6. The principle of the Brookes level comparator: after revolving the base plate through 180° so that the gauges are reversed with respect to the level, the movement 'l' of the bubble is a measure of twice the difference between the two gauges

the effect of friction and play caused by the types of pivots formerly used. The slip gauges to be compared (B1 and B2) are inserted in turn between a fixed 3-ball anvil and a movable single ball, connected by means of a spring-strip linkage to an arm having a spider at its end. The movement of the spider is then amplified further by means of a conventional optical lever system.

Brookes was faced with a similar problem to that of Eden and Rolt in the comparison of his length-bars to an accuracy of a few millionths of an inch: he adopted a very different but equally elegant solution. In 1918, he designed his level comparator, shown in Fig. 6, which makes use of a flat plate, the upper surface of which can be rotated in its own plane. The length-bars to be compared are wrung side-by-side on to this plate. A sensitive bubble-tube is mounted on two ball-feet, each of which contacts one exposed end face of the two bars. The bubble position is taken in position 1, the base plate is rotated through 180 degrees to position 2 and the bubble reading repeated. The radius of the bubble tube and the separation of the ball-feet are known and, from the movement of the bubble between the two positions, the difference in length is easily determined. Since the bars are not handled during the measurement, errors caused by temperature changes arising from handling are eliminated.

Optical devices

One of the earliest uses of a measuring microscope for engineering purposes was in the instrument shown in Fig. 7, made in 1903 for the committee concerned with the B.A. system of screw threads. This instrument was the forerunner of the modern toolmaker's microscope. The screw to be examined was held in a chuck, mounted on a table, the co-ordinate position of which was set by means of micrometer screws. The thread was viewed through a fixed microscope with crossed wires in its eyepiece; by bringing various parts of the screw into line with these crossed wires it was possible to measure the screw diameters and pitch.

During the 1914–18 war this measuring microscope gave Eden at the NPL the idea of projecting the magnified image of a gauge or workpiece on to a large screen, where it could be viewed more conveniently and could be compared with a master template of the required form. He experienced considerable difficulty, at first, in obtaining a suitable lens system which would produce a distortion-free image over a large field. But eventually he was successful. His first projector, which had a field of 7 in and was used for screw thread measurement, is shown in Fig. 8. It was mounted on an old lathe bed; the screw to be measured was illuminated by a lamp and condenser lens at one end of the bed and the projected image, after passing through a hole in a screen, was reflected downward on to a horizontal table.

During the early 1920's much attention in this and other countries was devoted to developing methods of measuring the lengths of slip gauges in terms of wavelengths of light. The end-faces of slip gauges and length-bars were flat and

Fig. 7. This is one of the earliest measuring microscopes which was used for engineering purposes

well finished enough to act as reflecting faces in an interferometer. This is now a standard procedure.

Conclusion

From the very early days of mechanical engineering there has been a demand for improved performances of mechanisms. This has been met partly by better designs and materials but it has, in most cases, entailed more accurate manufacture. In order to take advantage of improved machine tools and manufacturing methods, techniques had to be developed for measuring to a greater accuracy.

Up to the beginning of the 19th century the various parts of any mechanisms were hand-fitted together—even bolts and nuts were selectively fitted—so that the need for accurate measurement was rare: the machines were made in small numbers and the specification of manufacturing tolerances was unknown. It was not until Eli Whitney introduced the first practicable scheme for manufacturing interchangeable components in relatively large numbers that the need for more accurate measurement in engineering became evident; and not until Johansson invented his slip gauges in 1896 did the mass production of truly interchangeable parts become feasible to specified and fine limits of size. From this time onwards progress in the development of measuring equipment has been rapid and has played a major part in the establishment of the engineering industry.

Acknowledgements

The diagrams in this article, which are Crown Copyright, are reproduced by permission of the Director of the National Physical Laboratory.

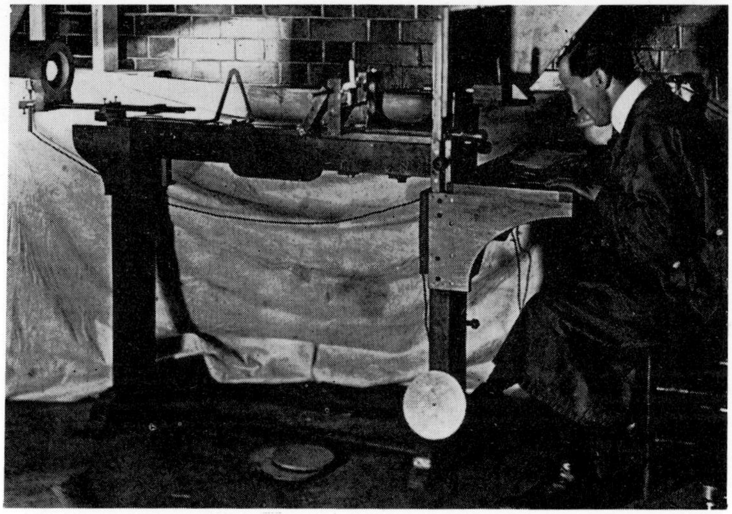

Fig. 8. The first optical projector used at the NPL for screw thread measurement had a field of 7 in

REFERENCES FOR FURTHER READING

1. EDWARDS, I. E. S., *The Pyramids of Egypt*, Max Parish, London, 1961
2. BERRIMAN, A. E., *Historical Metrology*. J. M. Dent and Sons Ltd., London, 1953
3. SINGER, C. et al, *A History of Technology*, Clarendon Press, Oxford, 1958.
4. ROLT, F. H., *The Development of Engineering Metrology*, 1952 Sir Alfred Herbert Paper, Proc. Instn. Prod. E.

THE GREAT MASTERS

First and last a mechanical engineer, it was through his Eddystone lighthouse and other massive achievements, that Smeaton became best known as a civil engineer. Nevertheless, neither the Eddystone nor much of his other work would have been possible, had he lacked deep intuition in mechanical matters. Intuition it must have been since he received practically no engineering training.

John Smeaton—Consulting Engineer

by Gerald Bowman*

[Courtesy: National Portrait Gallery]
Portrait by Romney, after Rhodes, from the frontispiece to Smiles' 'Life'

John Smeaton was born at Austhorpe Lodge in the village of Whitkirk near Leeds on 8th June 1724, the eldest son of a prosperous 'attourney', William Smeaton, and Mary Stones.

There is no doubt that his childhood was solitary because a brother born three years later died at the age of five and a sister who followed in 1732 died after only a year. John, however, was a self-sufficient youngster who had no interest in the toys that were normal to his age. His only toys were whatever tools he could get hold of and his besetting interest was in altering or making anew things which normal children take for granted.

Infant toolmaker

At the age of six he was discovered perilously clinging to the roof of one of his father's barns and there trying to attach a windmill of his own making. As a result of this—and, no doubt, after suitable admonishments—his father gave him the use of a lean-to in the grounds and bought him a set of the tools for which he craved. Young Smeaton's reaction was to make with them other tools for purposes he had in mind. Not long afterward, having begged a length of bored piping from some workmen, he fashioned a water-pump that worked as efficiently as the one in his father's kitchen.

Austhorpe Lodge in those days was a pleasant enough place in which an intelligent child could grow up. It had been built by John's grandfather in 1698 and was both charmingly rambling and extensive. The work of maintenance was fairly constant and young John's great pleasure was to stand about any gang of workmen and plague them with logical questions that, usually, were more than their foreman could answer.

By the time he was sixteen he had grown into an eager, energetic youth, constantly absorbed in his projects. His workshop was now extended and had been much better equipped, entirely by his own hands. At this time a report upon him by John Holmes, his lifelong friend and later his partner in a water-supply project, is vivid and informative. Holmes, slightly older, was already being trained in mechanics and was afterwards to become a well-known watchmaker and instrument maker in the Strand. He wrote of his visit to Austhorpe in 1742, when Smeaton was less than eighteen:

> I could not but view his works with astonishment. He forged his iron and steel and melted his metal. He had tools of every sort for working in wood, ivory and metals. He had made a lathe by which he cut a perpetual screw in brass, a thing little known at that day.

Holmes' visit lasted a month and he introduced Smeaton to his master, Henry Hindley of York, "a great lover of mechanics". On the occasions the young men visited him together, Smeaton kept their host talking until daylight. It is clear that Hindley was both indulgent and immediately aware of young Smeaton's developing genius. This month, and a shorter period much later, when he joined a 'philosophical instrument maker' were the only periods of mechanical training, if training they can be called, that he ever received.

In that day and age, however, engineers were rare creatures and for the most part they earned little more than daily wages. For this reason Smeaton the elder, although interested and amused by his son's 'play', regarded it as that and nothing else. When John turned sixteen, therefore, his father took him from Leeds Grammar School where he had received his formal education and put him to work in the family law firm. This was a well established and highly profitable one which offered his son comfort for life without

* *Technical author and journalist.*

too much effort. Any other occupation seemed out of the question.

John Smeaton appears to have accepted his fate and, for the next two years, worked diligently in his father's office, travelling to and from Leeds daily. But still every available spare moment was spent in his workshop and, having been to see a Newcomen atmospheric engine installed in the nearby Garforth coal mines, he set to work and built a model with which he proceeded to empty one of his father's fishponds. Although this feat perhaps lacked tact, the achievement would have done credit to an able senior apprentice of today, with the greatly improved tools now available.

The upshot was that Smeaton senior decided it was time John took life really seriously. He was therefore packed off to London to continue legal training at Westminster Hall, and to live in lodgings where he could no longer 'waste his time' with mechanical ideas. It is probable that he, himself, felt that this was fair enough and only to be expected.

For the next two years he worked obediently. But by now his magnificent mind was beginning to develop and, while he could not use tools, he read everything about mechanical and scientific matters that could be found; and, naturally, in London there was more to hand than in a Yorkshire village.

The result was inevitable. Soon after he was twenty, Smeaton wrote home to say that he could not bring himself to continue in the Law although he had honestly tried. He begged his father to release him and allow him to follow his natural bent. It says much for the elder Smeaton that, after natural hesitation, after pointing out that little more than a working man's future lay in mechanics, he finally agreed. What was more, he made his son a generous allowance which was continued for the rest of his life. Smeaton was thereby enabled to start his career on a fortunate basis, in sharp contradistinction to his slightly younger colleague, Brindley, who had to fight his way to recognition through grinding poverty and parental neglect. The two were destined to become close friends in later years, and leaders in their profession.

Instrument making

From that time onward Smeaton followed his heart's desire which was to frequent the company of men learned in mechanical matters. For a period, he joined an instrument maker and, as a result, met members of the Royal Society.

He began attending the meetings, and offering papers to the Society upon matters with which he had been experimenting. The first, in 1750, was upon improvements to the mariner's compass which he had carried out in conjunction with a Dr Gowan Knight. In that same year he started business on his own account as a methematical instrument maker but it is not clear that this business ever succeeded commercially and he soon dropped it.

A year later his communication to the Royal Society concerned a special log he had tested on the sloop of war *Fortune*. This was followed by a paper on improvements to air pumps, an old love. The pump he constructed and exhibited was afterwards used by the great Dr Priestly who obtained with it "a rarification of 400 times".

In 1753 he was elected a Fellow of the Royal Society when only 29. He decided to teach himself French since there were, he was told, far better technical works in that language than in English. There is a clear picture of him, given by his friend, Holmes, at this period. He was apparently of middle height, powerfully built, with a face strongly modelled, if not classically handsome. There was a touch of brusqueness about his way of speech, largely because he knew precisely what he meant and how to express it without misunderstanding. Moreover, neither then nor at any other time did he learn to suffer a fool gladly. Yet he had genuine depth of nature and could be a loyal friend. With Holmes he never had a difference, in a close collaboration of over 50 years.

In 1752 and 1753 Smeaton began a series of experiments which, five years later, resulted in his most important paper to the Royal Society which was to gain him the Copley Gold Medal.

On a visit to the Low Countries—to Rotterdam, Delft, the Hague, Amsterdam and Helder—he made copious and clear notes which were to prove invaluable in the canal-building he undertook later on. What impressed him were the docks at Amsterdam. At that time there were no docks in London. The ships were generally small and could lie on the mud of the river at low tide. Although there were plenty of quays in the Thames, unloading was mostly by barges.

Lighting of the coasts and danger-spots was just as primitive. Speculators were allowed to put up lights—for the most part blazing pitch-pots—and collect dues from shipping using the nearby waters.

The Eddystone Lighthouse

The Eddystone Reef, 14 miles from Plymouth, lay directly in the path of Channel shipping. On it a fantastic wooden lighthouse had first been built in 1696 as a speculation by one Henry Wynstanley, who died in it some seven years later when it was swept away in a gale. Another was put up by a London silk-mercer and destroyed by fire 60 years later in 1756. The leaseholder at the time, a Mr Robert Weston, asked the President of the Royal Society to recommend an engineer to undertake the rebuilding. Macclesfield suggested John Smeaton who, at 31, was acknowledged by the members to be a master of engineering.

The world was then far smaller and the Royal Society almost the only meeting place of men of education and scientific ability. Although communications were primitive, the name of a man of real accomplishment would,

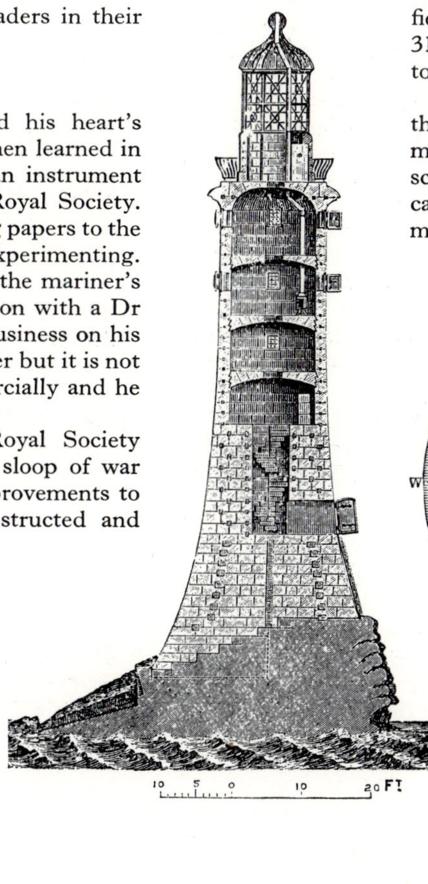

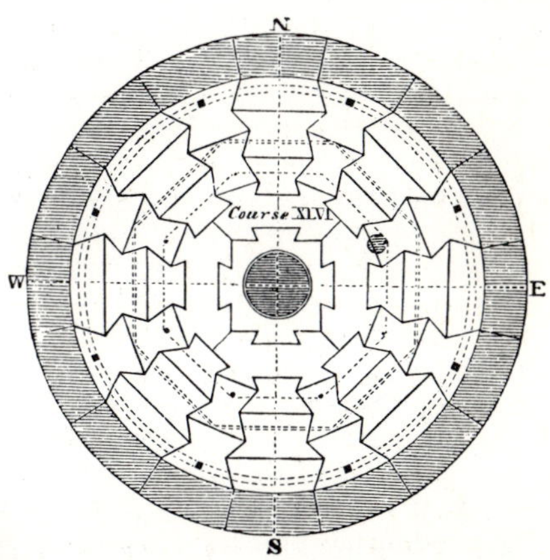

Two sections through the Eddystone Lighthouse, showing dovetailing of stonework and rock foundation

within months, be known by a limited circle throughout the country. As a result, Smeaton was constantly hearing of the problems and projects of the times, and giving his restless mind to the solutions. Instrument making brought him in touch with the leading astronomical workers so that, as his Royal Society papers show, he took a deep interest in the subject for the rest of his life. Apart from that, however, there was no subject of a mechanical or constructive nature which failed to catch his interest or upon which he would not argue, as Holmes says:

> sometimes breaking out harshly when anything was said that did not tally with his ideas, and he would not give up anything he argued for until his mind was convinced by sound reasoning.

He impressed every man of scientific understanding he met, and as a result when any major project was mooted, his name was usually put forward, as in this case by the President of the Royal Society. A great factor in his favour was his finanacial independence which made him free to give time to anything that interested him.

But Smeaton, thinking he would be asked to compete in offering a design, noted in his diary that he "received the call without joy". Still, he wrote back that if he were "absolutely chosen to this business" he would consider it an honour. Although a man of obsessional energy who never left anything he had undertaken until it was finished, he had no stomach for open competition and no need to engage in it.

The reply reassured him: the task was to be his alone; and Smeaton set out for Plymouth forthwith to start the work which was to make him internationally famous.

The design of the Eddystone lighthouse, the first of its kind, is too well known to need detailed description here. Smeaton realized that weight was the first requirement in a structure constantly subjected to the force of running seas, and therefore built a core of dovetailed Portland stone, surrounded by an outer crust of moorstone, which is impervious to the action of marine life. Each course was cemented in place and then doubly secured to the one below by hard oak trenails, hammered down through holes. Smeaton made his own cement, there being nothing at that time which would withstand constant damp. It was a mixture of lime and Italian volcanic ash and was probably the first quick-drying cement known, at least since Roman days.

So far so good; but the man who had the inventive faculty to produce such a design had also to have the hardihood of a working sailor plus high physical courage, or the fabric would never have been built. On his first two-months visit to Plymouth the weather allowed only ten trips to the reef during only four of which he could actually get on to the rock. The longest period of work in daylight was four hours and the average journeys out and back took six hours when the weather was fair. These were the working conditions in which cut stones, averaging $1\frac{1}{2}$ tons each, were ferried out to be hoisted on to the rock.

Smeaton himself superintended every part of the work; yet his only previous experience of the sea had been as a passenger on a Channel packet. He found an invaluable henchman in Josias Jessop, a shipwright, who had been in charge of maintenance of the earlier building.

It is a measure of Smeaton's character that in his diaries and 'narrative' of the work at Eddystone he merely records details of weather as simple facts and does not elaborate on the grinding hardships. From the first his mind is preoccupied with the design, then the proving of models and finally in the building itself. He made a careful study of the plans of the previous structures and noted, typically: "They afforded me just so much light as to discern the want of more information".

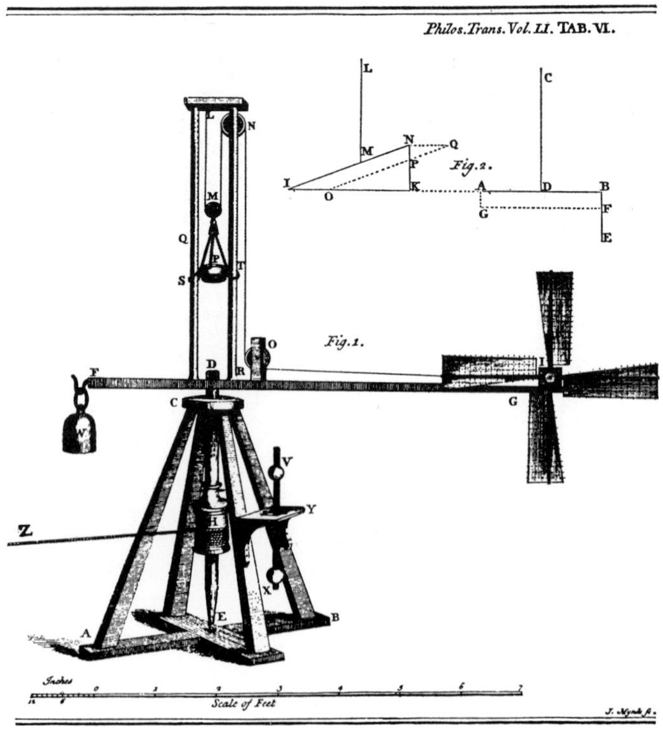

By pulling the cord of this windmill test rig, the jib arm was rotated to produce wind on the sails, whose performance was measured by the weight on the pulley

He made his own first direct experiment by taking a pick to the gneiss rock and noting that it took him a minute to dig 5 in. This was the sort of fact upon which he could base a wage and time schedule. Of the business of model-making he wrote:

> Those who are not in the practice of handling mechanical tools, themselves, but are under the necessity of applying to the manual operations of others will undoubtedly conclude that I would have saved much time by employing the hands of others in this matter . . . but . . . provided I could work with as much facility and despatch as those I might happen . . . to employ I should save all the time and difficulty, and often the vexation, mistakes and disappointments that arise from a communication of one's own ideas to others.

Smeaton completed the structure with his own hands in August 1759, three years after it was begun, when he fixed the gilt ball on the lantern apex, 91 ft above sea level. The great structure was still solid and undamaged more than 100 years later when it was decided to build anew on a different part of the reef because the foundation rock was becoming undermined.

Wind and water mills

The year 1759 was one of double triumph for him since it also marked the award to him of the Royal Society's Gold Medal for the paper *An Experimental Enquiry concerning the Natural Powers of Water and Wind to Turn Mills and other Machines Depending on a Circular Motion.*

Smeaton had built two accurate and ingenious rigs for testing water wheels and windmills, but he was among the very first to realise the dangers of the scale-effect. Therefore, although the testing was elaborate and the

Similar to the windmill rig, this watermill equipment tested the blade performance. A hand pump provided a constant head and the performance was tested by winding up the weight

systematizing of the resultant figures equally so, Smeaton decided to wait until he had built full-sized water wheels and mills before publishing the theories at which he had arrived.

In the event mills built according to his designs proved one third more efficient than any constructed before. In all he built four windmills and 43 water mills in the course of his life.

In both test-rigs his first object was to provide himself with a fluid force which could be accurately measured and sustained over a sufficient period. In the case of the water wheel, this was achieved cleverly enough but the test model for windmill sails was a piece of unique ingenuity. Since he could not hope to find a spot where the wind blew with complete regularity, he set about providing it inside his own workroom, long before anything like a wind-tunnel had been thought of.

The model, as illustrated by him, is shown in the illustration.

A pull on the cord *Z* will rotate the whole upper structure. The sails, as a result, revolve and wind up the cord passing through the pulley *M* which rises together with its attached weight-tray. Smeaton found that his assistant, after a little practice, could exert an even pull on the driving cord to obtain one revolution per second, as timed by the pendulum. This gave a velocity at the centre of the sails of 6 ft/s. The windmill sails were made adjustable in angle, and various shapes and constructions in wood and cloth could be tested.

Smeaton also used the figures from a long series of experiments as a basis for a reasoned mathematical theory on the effect of air as a mechanical power in the production of circular movement.

His model test rig for water-wheels, also illustrated here, was even more ingenious. The bearing-plates were adjustable vertically so that the wheel blades could be raised or lowered in relation to the 'mill race'. Water from an upright tower was released by a sluice so that it ran from an orifice beneath the wheel-blades into a reservoir from which it was recirculated. To ensure that the volume of the water in the tower was not materially affected by the shaft of the hand pump, the pump-shaft was made as big as the valve piston at the base. The water level was shown by means of a float and calibrated measure.

On the shaft of the water wheel was a drum for winding a cord which, again, served to raise a weight. By means of a dog-clutch, the winding action could be started and stopped at will. An upper trunk could be attached to supply water for testing overshot wheels.

The best design of overshot wheel proved to be almost exactly twice as efficient as the best undershot type.

In making his tests, Smeaton allowed for the friction in the mechanisms. He did this by timing the mill running free of any weight-load and comparing the result with the speed at which the blades or sails ran slightly faster than the fluid acting upon them.

Scientific work

Smeaton married in 1756 and went to live at the family home at Austhorpe which had been left him by his father. Here two daughters were born both of whom kept house for him after his wife died in 1784.

At Austhorpe he built a tower of four stories, at the top of which was his observatory. The tower, built separately from the house, was also his workshop. On the ground floor was his forge. On the second were arranged all the beautiful models of the lighthouses, bridges and mechanical inventions for which he was responsible. The third floor was his study, leading up to an attic storeroom which also served as an observatory.

There followed papers to the Royal Society on various astronomical subjects, astronomical instruments and also a new hygrometer. In his paper to the Royal Society of March 21, 1771, Smeaton refers to this hygrometer using a hempen cord. In view of his work on windmills he had wanted some means of measuring the "disposition of air to part with or precipitate water".

The hempen cord method failed; it would stretch unevenly, no two lengths acting similarly. He therefore set the idea aside until 1788 when he made two of what must be the earliest hygrometers—two because he needed to check one against the other. He used lengths of flax, first boiled in salt water for half an hour and then stretched for a week.

In each instrument a pound weight was hung on the end of the flax cord, the stretching or shrinking of which was recorded by a simple lever-pointer moving against a calibrated scale. In his paper Smeaton admitted that the method was inaccurate. However, the two instruments agreed on average to within a single scale division.

Civil engineering

The list of his civil engineering works is so long that only an outline can be given here. He planned improvements to the navigation of several rivers and was consulted about the draining of the Fens. He strengthened the piers of old London Bridge in 1777 and designed a new pumping engine,* placed in the fifth arch, to supply the city with water. His

* *cf p 20, Engineering Heritage, vol. 1.*

efforts prolonged the life of the bridge for sixty years more.

Smeaton built the bridges at Perth, Coldstream, Banff and Hexham. He redesigned the water supply for Edinburgh and, in partnership with Holmes, for Deptford and Greenwich. He began to build the Forth and Clyde Canal which gave a navigable passage from ocean to ocean. He built lighthouses and harbours at St Ives, Cornwall and Ramsgate and was consulted on a host of others.

A heavy blow struck him in 1782 when a storm flood swept over the Tyne and brought down his bridge at Hexham. Afterwards he discovered his mistake in setting the foundations on gravel. He wrote, typically, to Mr Pickernell, the resident engineer:

> All our honours are now in the dust; it cannot now be said that in the course of thirty years practice . . . not one of Smeaton's works has failed. Hexham bridge is a melancholy instance to the contrary.

It was, however, the only failure of his life. At this period he had to be much in London, appearing before Parliamentary Committees in connection with the projects his clients put forward. Holmes says that:

> . . . it was his constant custom when applied to do a plan or support any measure, to make himself fully acquainted with it and see its merits before he would be engaged in it. By this caution added to the clearness of his description and the integrity of his heart he seldom failed having the bill he supported carried into Act of Parliament . . . in the Courts of Law he had several compliments paid him from the Bench by Lord Mansfield and others for the new light he threw on difficult subjects.

At this period he was in the habit of meeting brother engineers at the Queens Head Tavern in Holborn in the evenings, after Parliamentary work. These meetings became at first informally known as the Smeatonian Club (the name of which survives to this day) and later a list of members and objects of association were drawn up. After Smeaton's death the members formed the nucleus of the Institution of Civil Engineers which was founded in January 1818.

Towards the end of his life Smeaton refused almost all offers of further business because he had decided to write an account of all his major works. His *Narrative of the Construction of Eddystone Lighthouse* . . . was, however, the only descriptive writing he achieved while he lived—there are, of course, his voluminous reports, done in the course of his work, and his papers published by the Royal Society. In the introduction to his *Narrative* he shows an (entirely unnecessary) sense of inadequacy about lucid and clear exposition.

> I am convinced that to write a book tolerably well is not a light or an easy matter: for as I have proceeded in this work I have been less and less satisfied with the execution. In truth I have found much more difficulty in writing than I did in building as well as a greater length of time . . . It is true that I have not been bred to literature but it is equally true that I was no more bred to mechanics.

On 16th October 1792 the state of his health is shown in a letter to Holmes in which he says:

> 'I conclude myself to be nine tenths dead and the greatest favour the Almighty could do me (as I think) will be to complete the other part.'

Two days later, while walking in the garden at Austhorpe, he undoubtedly suffered a stroke—the contemporary report says 'struck with palsy'—from which he never recovered. He died on the 28th of the month in his 68th year and was buried in Whitkirk Parish Church. It seems he had suffered from gastric ulcers since he wrote to a friend, during the last year of his life:

> '. . . . unless the dreadful disorder in my stomach can be effectually removed which consists in an over-quick digestion and is medically called Fames Canina I believe my mental faculties will be of little further use . . . you will scarcely be able to conceive what a task it has been to get this letter written.'

An appalling circumstance—appalling to a biographer—followed his death. The contents of his tower study were accounted as rubbish and a mass of documents, plans and letters, together with models and tools burnt in the yard.

Fortunately there were those in high places with a clearer appreciation of his worth. When his Eddystone lighthouse was dismantled after the erection of the new one in 1886, the stones of the upper section were carefully numbered and brought ashore. There they were re-erected so that the nation has a memorial of him on the greensward of Plymouth Hoe.

REFERENCES FOR FURTHER READING

1. SMEATON, JOHN, *Narrative of the Building and a description of the Construction of the Eddystone Lighthouse with stone.* 2nd Ed. 1793.
2. Smeaton's Reports. (1812) Society of Engineers.
3. *J. Smeaton. Diary of his journey to the Low Countries*, with introduction by Arthur Titley. 1938.
4. Holmes' *Short Narrative of the Genius, Life and Works of the late Mr John Smeaton.* 1793.

Smeaton's house at Austhorpe; workshops and observatory were installed in the tower

Of all subjects that concern engineers the story of materials provides perhaps the best guide to technical progress through the ages. For thousands of years engineers built machinery in a few basic materials. Today the material itself is often designed to fit the application and the designer can have considerable confidence in the theoretical knowledge gathered in the course of the last two centuries.

Materials Theory and Practice

by J. C. Levy, Ph.D. (Associate Member)

Table 1—The Time Scale

Approx. Date	*Materials*	*Production Methods*	*Typical New Applications*
B.C. 100 000	Wood, flint, emery, skins, hair, bone, gut. Clay, reeds, leather	Manual flaking, carving, hammering. Firing	Flint axes, knives, etc., skins for clothing. Tools and weapons of wood, bone stone. Pottery, ropes and fishing tackle.
8000	Glass Native copper Native gold Meteoric iron Flax Bronze	Weaving, basketry, reduction of copper ores, simple loom, kilns, pottery wheel.	Sandals, buckets, jewellery, wheels, wood and stone construction, metal tools and weapons, statues, textiles.
3000	Lead, silver, cotton	Casting, forging, soldering, wire drawing.	
2000	Tin, wool, glue, iron and steel	Iron smelting with charcoal. Cementation process.	Glass vessels, kiln-baked bricks. Wooden machines and wagons.
1000	Parchment, papyrus, Iron Age (Britain), brass, coal (China)	Welding, quenching of steel, glassblowing. Lathe for wood turning.	Horseshoes, ploughs, swords of iron, leather bellows and pump valves, iron nails, chains, hinges.
0	Paper, gunpowder (China) Roman concrete	Tempering of steel.	Pipes, water mills, window glass, windmill (Persia), bronze bells.
1000	Rubber (S. America) Cast iron (Europe), tin plate	Coal for smelting. Water-driven iron works and sawmills.	Spectacles, wrought iron and cast bronze cannon, printing, instrument making.
1500	Zinc (India) coke	Iron rolling.	Cast iron cannon, colliery rails.
1700	Cobalt, nickel manganese, molybdenum titanium, chromium	Iron smelting with coke, crucible steel, puddling, factory system.	Newcomen's engine, improved machine tools, Watt's engine, achromatic lenses, cast iron machine parts.
1800	Sodium potassium, aluminium magnesium, portland cement, cheap steel, alloy steels, dynamite, viscose, industrial diamonds	Galvanizing, electro-plating, rubber vulcanization, Bessemer and open hearth processes, deep drawing, electroytic Cu, paper from wood, steel rolling, electrolytic Al, Mond nickel.	Railways, wire ropes, diamond tools, cast steel guns, iron ships, armour plate, electric traction, reinforced concrete, steam turbine, petrol engine, pneumatic tyre, diesel engine, aircraft.
A.D. 1900			

Throughout history the evolution of technology has been controlled in part by the materials available. Stone Age materials were an obvious limitation on primitive man and, despite all their variety, modern materials still limit our own engineering efforts.

In retrospect, three ages of engineering materials can be identified. The first began about 3500 B.C. and continued until the Industrial Revolution. It was characterized by the use of wood and stone for machines and structures, plus the working of metals on a small scale. The second began in the 18th century with the large-scale production of coal and iron and the application of mathematics and mechanics to the engineering use of materials. We are now in the midst of the third age in which elaborate stress analysis and testing techniques match the variety of materials available and in which a scientific understanding of the structure of matter has permitted chemical 'tailoring' to produce materials for specific purposes.

In addition to strength and stiffness, engineers are nowadays concerned with the electrical, magnetic, nuclear, thermal, optical and acoustic properties of materials, besides cost, durability, weight and appearance. The story leading up to this complex state of affairs goes back nearly half a million years.

The first materials

The activities of Stone Age men were first directed towards hunting and later to agriculture and the domestication of animals. They could call only upon those materials which were easily obtainable from their surroundings; flint from the rocks or from the beaches, wood from the forests and animal derivatives like bone, hair, skin and gut.

Between 300 000 and 100 000 years ago Stone Age culture produced a range of flint knives, axes, scrapers, hammers, needles, harpoons and saws. By 20 000 B.C. the wooden bow and arrow had been invented; 12 000 B.C. saw the glazing of small pottery objects and 9000 B.C. the production of properly tanned leather for clothing and buckets. Ropes of hair or hide, made by plaiting or twisting, were used for haulage of sledges and the manufacture of fishing tackle. Pieces of ornamental glass found in Egypt date from 7000 B.C.[1]

By 3500 B.C. the Sumerians in the Middle East had developed simple wooden machines such as the wheel and axle, the wedge, the lever and the basic loom. They were the first to use metals systematically—especially copper, a metal easily extracted from its ore and readily cast, forged and welded. It is worth noting that examples of all the three main classes of materials were now available; metals (copper and gold), organics (wood, leather) and ceramics (stone, clay, glass). But these had been acquired over an immense period of time. Another 1500 years passed before the knowledge of bronze became widespread and a further 500 for iron to become well established, as can be seen from Table 1.

The use of stone for building started in N. Africa and the Middle East about 3500 B.C. It then spread to Western Europe which has about 2000 tombs and temples of roughly dressed stone erected before 1000 B.C., of which Stonehenge is an example.

The more advanced use of stone for tall structures was an innovation of the Egyptians about 2700 B.C. Their vast projects, requiring stones weighing up to 1000 tons each, inevitably made severe demands on mechanical engineering skills. Egyptian carpenters, who excelled in their craft, produced heavy haulage equipment, including rollers, levers, rafts, sleepers, sledges. They also manufactured plywood, built chariots with light spoked wheels (secured with animal glue) and constructed comparatively large ships which travelled the Eastern Mediterranean.

In a way, wood was the mild steel of the age—and indeed of every age up to 150 years ago. A general purpose material, it could withstand tension, compression, bending and twisting. Unlike stone, it could be deformed considerably without fracture and was light for its strength.

Besides wood and stone, the Egyptians had supplies of bone, horn and ivory for rings, needles and tool handles and they used papyrus for the manufacture of sails, cords and writing material. When working gold they exploited the processes of beating, wire drawing and casting, including the use of the lost-wax technique. The use of pure tin also dates from around this period.

The civilizations of Mesopotamia continued to flourish. By 1000 B.C. dried clay for buildings had been replaced in Assyria by kiln-baked bricks, set in bitumen obtained from natural seepages. Bitumen was also used for water-proofing canals and for the caulking of boats. Iron smelting, which requires a higher temperature than for the production of copper or bronze, was introduced by the Hittites about 1400 B.C. in primitive hearths such as that shown in Fig. 1. Steel was made by the laborious cementation process in which the carbon content was changed by repeated heating with charcoal and hammering, an air blast being provided by leather bellows.

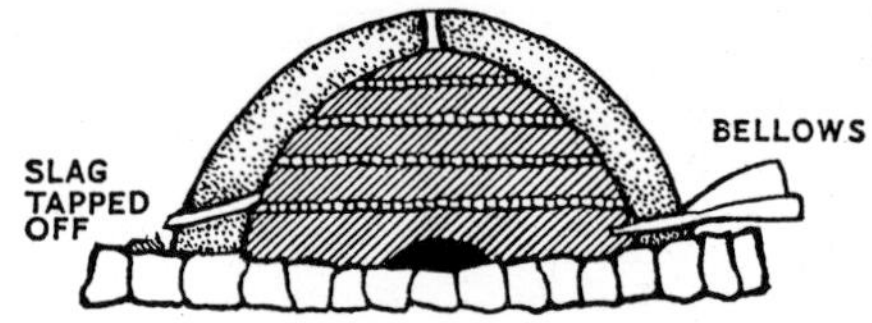

[Courtesy: Oxford University Press]

Fig. 1. A primitive blooming hearth for wrought iron smelting consisted of alternate layers of charcoal and ore on a stone platform

Although early iron had disadvantages compared with bronze (it rusted, for example, and was inclined to be brittle) it provided better tools such as hammers, saws, rasps and axes. By 1000 B.C. iron making had penetrated to Austria and by 500 B.C. there were horseshoes and ploughs of iron.

Greece and Rome

The movement of large armies now led to a more general pooling of knowledge. Technological fall-out is no new phenomenon. The Egyptian empire declined in 500 B.C. and was supplanted by the Persian, in turn defeated by the Greeks under Alexander the Great whose successors were subdued by Rome. The rise of Rome began in 250 B.C. and during the following centuries the Roman Empire was established from Britain to Egypt and from Spain to Turkey.

Advances in smelting spread throughout Europe, even ahead of the Roman armies. Both the Greeks and Romans manufactured improved iron, hardened by quenching, for agricultural implements such as spades, forks, sickles, shears and also tools, lifting tackle and clamps. Fig. 2 shows a Greek smithy and forge, with tools, including hinged tongs. Naturally, iron was also used for weapons like spears, swords, shields and tips of battering rams.

The Greeks had a practical understanding of the mechanics of simple machine elements, classified as the lever, the wheel, the screw, the wedge and the pulley. Their wagons were of superior design and wooden machinery included lathes for wood-turning and screw presses for grapes and olives, the parts sometimes being reinforced with iron to resist wear. There were also iron chains and hinges and even the occasional use of wrought iron beams in buildings. Cast iron, however, was unknown.

Glazing of windows was practised from the first century B.C. and for grinding cereals the Greeks developed the rotary stone quern and the donkey mill.

[Courtesy: Oxford University Press]

Fig. 2. Paintings on Attic vases of a smithy (left) and a forge, showing a number of typical iron implements

[Courtesy: Oxford University Press]

Fig. 3. Constructing a stone bridge and wooden walls under fire from the enemy during the siege of a city by Charlemagne

The writings of Philo of Byzantium of the second century B.C. reveal an early appreciation of the elasticity of metals. He described the testing of swords by bending them and also suggested the use of bronze springs for catapults.

The Romans carried out extensive mining for iron, copper, tin, mercury, gold, silver and lead. Metals and metallic ores were used in a variety of combinations—tin and lead for solder, copper and zinc for brass. Lead became relatively plentiful as a by-product of silver mining and was used for water pipes in Imperial Rome though pipes were also made from stone, earthenware, and copper. Bronze was employed for locks, pipe fittings, bearing bushes and the barrels of piston pumps. The tempering of steel was a Roman innovation.

The dark ages

The technical knowledge of the Romans was largely lost when their empire declined but the use of waterwheels, including many with crude wooden gearing, was gradually extended. An advance in the use of metals was that bronze bells were cast in 800 A.D. and at about the same date, paper came from China where it had been woven from vegetable fibres as far back as 100 A.D.

China in fact produced many inventions which reached Europe through the Middle East after a delay of some 9 or 10 centuries. Silk reeling machinery and the mastery of cast iron are examples in the materials field. The Chinese are also credited with the mining of coal as early as 1000 B.C., compared with about 1200 A.D. in the west.

The period from 1000 to 1400 A.D. was marked by advances in the design of machine elements such as shafts, gears, levers, brakes and lifting jacks. By now water power was being utilized in a number of industries so that tanning mills, paint mills, saw mills and textile mills shared in the gradual development of machinery. Wood and stone remained the predominant materials of construction and some typical activities employing both are shown in Fig. 3. However, iron became increasingly common for chains, bolts, gudgeons, pivots and other heavy duty parts.

Although production of cast iron became possible in Europe by 1400, due to the introduction of more powerful water-driven bellows for blast furnaces, the general replacement of wood by cast iron did not begin until the 18th century. The few tens of horsepower available from even the largest wind or water mills could be satisfactorily transmitted by timber shafts and gears so that there was little incentive to change to iron.

The ready availability of wood and the accumulated skills of the wheelwright, the cooper and the shipbuilder were additional factors in the persistence of wooden construction.

However, in two industries, clockmaking and armaments, wood was inadequate and the use of metals proceeded more rapidly. Clocks, and later watches, the first all-metal automatic machines, were brought to the point of practical utility in the 13th to 15th centuries. New armaments included improved chain mail in the 12th century, plate armour and the steel crossbow in the 14th. Techniques originated for the casting of bells were, by 1350, being employed for the casting of bronze cannon and, rather later, for iron cannon and shot.

Glass now found new applications in optics. The first real lenses had been made in the 11th century and were followed by spectacles in 1290. Improved glass making and blowing also had beneficial repercussions on chemistry and physics: transparent and chemically inert vessels conferred obvious benefits in the examination of various reactions and effects.

The Renaissance

The invention of printing with movable type was perhaps the outstanding technical breakthrough of the 15th century. Printing involved advances in the casting of metal, the construction of presses and the production of paper through which engineers contributed to the spread of knowledge which led to the Renaissance.

The scientific era in strength of materials was foreshadowed by Leonardo da Vinci (1452–1519) who described mechanical tests on wires, beams and columns. He invented the first tensile testing machine, depicted in Fig. 4: when the test wire AB broke, the supply of sand or shot was automatically cut off. From his writings it is evident that Leonardo foresaw the need to apply science to engineering in order to relate, by calculation, the loads carried by a structure to the loads carried by its individual members and to the strength of the material from which those members were made.

The first surviving record of a scientific attempt at such calculation is contained in Galileo's book *Two New Sciences* of 1638 in which he formally initiated the subject of stress analysis. Galileo carried out many experiments and developed a theoretical treatment of the bending of beams, only narrowly missing the correct solution. It may be said that the work of Leonardo and Galileo on these strength of materials problems pioneered the whole scientific method of experiment and the formulation of empirical laws in mathematical terms.

A few decades later, Hooke in England and Mariotte in France established the idea of linear elasticity for wires, beams and springs. Mariotte also began the study of pipes under internal pressure and of plates carrying lateral loads. His attempt to improve Galileo's theory of bending was unsuccessful but the correct solution was achieved by another Frenchman, Parent, in 1713, unfortunately without receiving wide recognition.

In the late 17th and early 18th centuries the Bernoulli brothers, Jacob and John, analysed the shapes of bent elastic beams. Jacob derived a deflection equation which was correct in its general form and Daniel, the son of John, also added to the theory of elastic curves. Their work was carried much further by a Swiss, L. Euler, a pupil of John Bernoulli, who developed the basic ideas on the buckling of columns. To all intents and purposes the modern theory of simple bending was now complete, though much work remained to be done on interpretation and application.[2]

As regards materials used, there were few new developments. Fig. 5 shows that wood was still the principal material for machine parts other than fasteners and cutting edges.

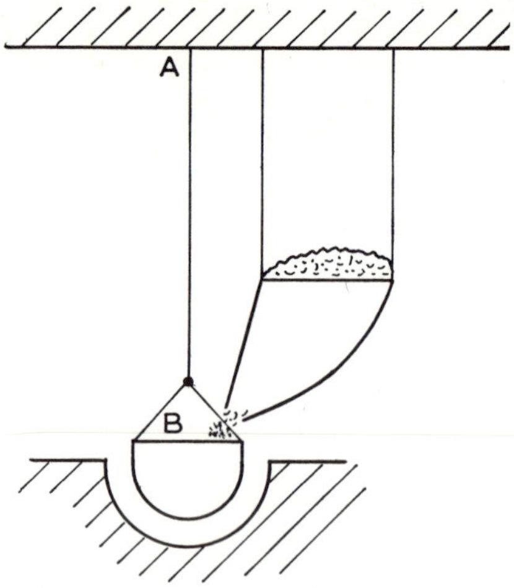

Fig. 4. Probably the first materials testing machine: Leonardo's sketch shows a hopper feeding sand into a vessel suspended by the wire under test; the amount of sand run out indicated the load to fracture

The Industrial Revolution

In 1700 the Industrial Revolution was in the making. The old technology based on water, wind and wood began to merge with the technology based on coal and iron.[3] The social, political and economic pressures brought about by increased trade, the new industrial towns and, as ever, military requirements, generated a greater demand for metals, particularly iron. The production of iron was, however, largely dependent upon charcoal, now in short supply due to the depletion of forests.

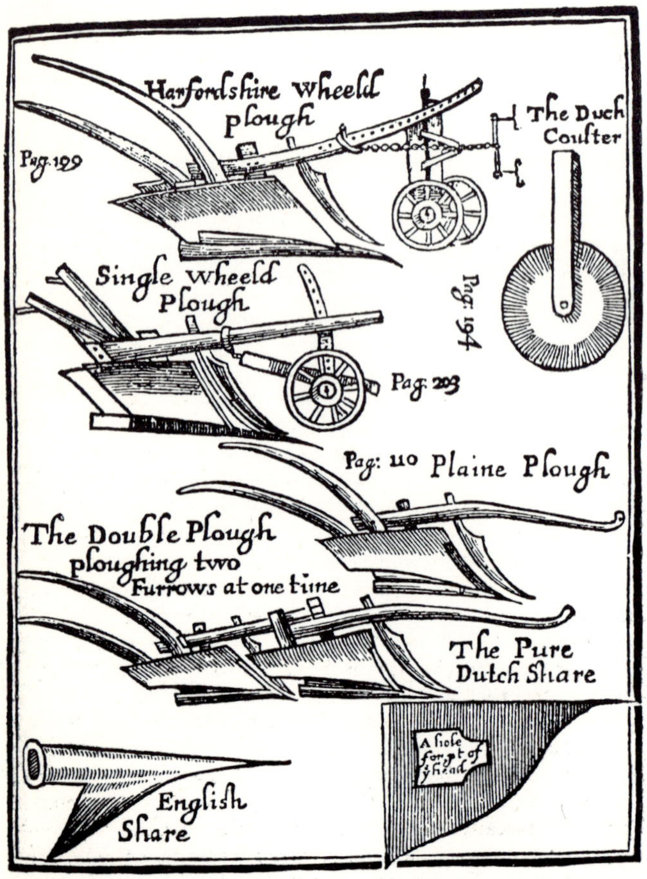

[Courtesy: Andrew Melrose]

Fig. 5. Seventeenth century ploughs show the extensive use of wood for anything but fastenings and cutting edges

Coal had been tried intermittently for smelting since at least the 11th century but was not altogether satisfactory due to sulphur content and its adverse chemical reaction. Consequently, when the Darbys of Shropshire succeeded in using coke for smelting commercially, Britain gained a decisive technological lead and rapidly overtook Sweden, France, Russia and Germany in iron production. Another significant advance was the introduction by Cort of the puddling process in 1784 which produced a superior wrought iron very economically.

Concurrent with improvements in the quality and quantity of iron were steady advances in machine tools to shape it and these, in turn, aided the development of the steam engine as a source of power. The first practical engine built by Newcomen and the very much improved designs of James Watt helped to secure British industrial superiority from 1770 until 1850. The progress of steam power on the one hand and iron making on the other set up a series of interlocking developments. Better metals led to better engines, and vice versa.

In the latter part of the 18th century, iron was used increasingly for items like pipes, vats, forge hammers, guns, ploughs, and colliery rails. The year 1769 saw Smeaton's first cast iron mill-shaft and by 1788 Boulton and Watt were producing cast iron wheels, gears and shafts for the more efficient transmission of steam and water power. The Albion flour mill, designed by Rennie in 1786, was said to be the first in which all the machinery was made of metal. The first cast iron bridge, which weighed 400 tons, was erected by Abraham Darby III and John Wilkinson at Coalbrookdale in 1779.

The search for other materials and new applications now began in earnest. Zinc was smelted in Wales in 1730, having previously been separated in India in the 16th century. Copper coinage in England dates from 1797 and silver-plated copper from 1742. Cobalt was being produced in Sweden, nickel was becoming well-known and the lead chamber process for sulphuric acid, established in 1746, helped lay the foundations of the chemical industry. Platinum was mined in South America by 1736 and, soon afterwards, in Russia. About 1800 Davy separated potassium, barium, sodium, strontium and magnesium by electrolysis.

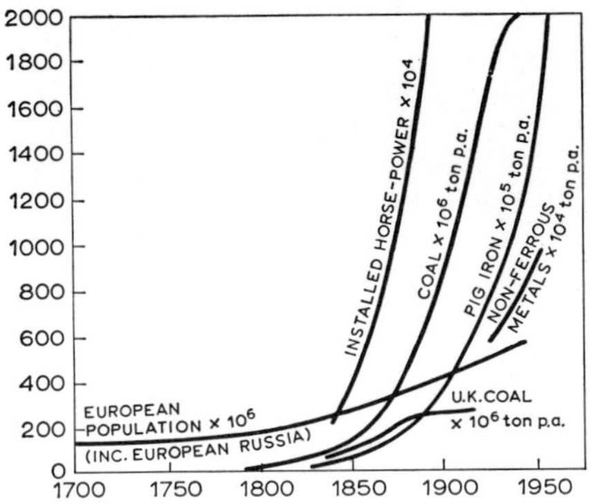

Fig. 6. Material and power consumption compared with population growth

Meanwhile steel production languished. In 1737 total British output was only 1000 tons, made by the ancient cementation process. The year 1744 saw the introduction of a rather better product, Huntsman's crucible steel, in which iron was smelted and refined in clay crucibles, with the aid of intense heat. But this process was very expensive: cheap steel was not to arrive for another hundred years.

Fig. 6 shows how production of coal, metals and steam engines soared with the impetus given to engineering by the Industrial Revolution.

To aid the design of machines and structures it was essential to obtain more information about the safe loads that could be carried by metals. Consequently the 18th century saw the development of properly designed test equipment for the application of tension, compression or bending to prepared samples of material. An outstanding theoretical and practical worker in this field was Coulomb (1736–1806) who had a clear and correct idea of bending theory and was one of the first to study scientifically the effect of heat treatment on the strength of iron.

By the end of the century the subject Strength of Materials had become sufficiently formalized for Girard to write the first book devoted to it. However, he perpetuated the erroneous bending theory due to Mariotte.

The 19th century theoreticians

Symptomatic of the need to apply scientific method to engineering problems was the founding of the *École Polytechnique* in Paris in 1795, which marked the beginning of engineering education as we know it. The *École* produced many notable engineers who concerned themselves with materials and structures.[2, 4] Among them were Poisson, Navier, Poncelet, Cauchy, Lamé, Clapeyron and Saint-Venant whose lives spanned the years 1781–1886 and who laid the foundations of much modern theory relating to beams, shafts, trusses, arches, shells and plates. They directed attention to the equilibrium of each small element of a piece of material and to the compatibility of its deformation with that of its neighbours, founding that branch of the subject known as the Theory of Elasticity. These workers also studied the various theories of strength under combined stresses that were in the process of development and the effect of impact, vibration and dynamic loads on structures.

Although French theoreticians were pre-eminent, British engineers undertook a great deal of experimental work which more than compensated for any analytical shortcomings—and indeed enabled theories to be developed later. Many of their experiments were related to the needs of the railways which, by 1850, had spread over much of Britain, the continent of Europe and the U.S.A. The design of bridges to carry heavy, moving loads necessarily received close study. For example, pioneering work in stressed-skin construction was undertaken in the design of the famous Britannia tubular bridge (1849) at the suggestion of Robert Stephenson. The experimental work on models was carried out by Fairbairn and Hodgkinson who thus were the first to encounter the phenomenon of buckling in such structures.

The great number of stress cycles that railway machinery and structures had to endure brought fatigue into prominence. Experiments in England (Fairbairn) and Germany (Wöhler) showed that loads carried safely under static conditions could cause fracture if repeated enough times, especially in the presence of sharp corners or sudden changes in section. The same problem of fatigue had already been met in mine hoist machinery when wire ropes inexplicably broke in service though the proof load had been easily carried. Such service failures, and the growing necessity to save weight for the economical utilization of power, were the driving forces behind the interest in strength theory which was now in evidence throughout Europe.

In Germany, the polytechnics developed close ties with industry. At Königsberg, F. Neumann (1798–1895) studied the effect of non-uniform temperature distribution in solids and also discovered the phenomenon of double refraction in transparent materials—later to be the basis of an important experimental method, photoelasticity.

In England, Thomas Young (1773–1829) founded the Cambridge School of Materials and Structures. He studied elasticity and impact problems, the buckling of columns and the twisting of shafts. J. Clerk Maxwell (1831–1879) worked both at Cambridge and at King's College, London. Among his many achievements were the analysis of photoelasticity, contributions to the theory of plates, rotating discs, and pressure vessels and graphical methods for the analysis of structures.

In Russia, an engineering college was established in St Petersburg in 1809 through the co-operation of French engineers. Both Lamé and Clapeyron taught there but the tuition was soon taken over by the Russians themselves, notably M. Ostrogradsky, whose pupil, D. Jourawski, was responsible for a remarkable analysis, now in common use, of the shear stresses in beams.

During the first half of the century the quality of iron was steadily improved by progress in puddling and blasting techniques and in the design of furnaces. To handle and form the larger ingots that could now be produced, the processes of rolling, drawing and forging were brought under better control and a range of metal-framed machine tools was developed that could form and cut metals with speed and precision. Many prominent engineers figured in this work, including Maudslay, Whitworth and Nasmyth.

After 1850, American advances in machine tools led the rise in transatlantic technology. The universal milling machine, capstan and automatic lathes, twist drills and grinding with abrasives like silicon carbide and aluminium oxide, promoted the production of accurate, interchangeable metal parts; initially for pumps, guns and clocks, later for most other machinery. The forming of metals was advanced by the deep drawing process, originally introduced for the manufacture of cartridge cases.

Mass-produced steel

A decisive technological change came about with the conversion of iron into steel by the Bessemer process in 1856. In this process, invented for the manufacture of heavy guns, steel could be produced cheaply by blowing air through melted pig iron. It was soon followed by the Siemens-Martin open hearth furnace which allowed scrap and low-grade iron to be used for steel making. In the form of wire, plate and bar, steel was soon adopted for a wide range of manufactured articles.

However, an initial hesitancy had to be overcome because the mechanical properties of steel were relatively unknown. Its use for bridges was not authorized in England until 1877 although Bessemer steel rails had been laid at Crewe in 1862 and had proved to last many times longer than iron. However, by 1882 the Forth railway bridge was begun and this alone used 50 000 tons of steel. In 1890 rolled steel joists and seamless steel tubes were produced and by 1900 ingots weighing 130 tons could be rolled. That year Britain produced 4 m. tons of steel out of the world's total production of 28 m. tons.

In shipbuilding, iron-framed construction with riveted plates originated about 1850. I. K. Brunel's huge iron ship, the *Great Eastern*, laid down in 1854, was double skinned and displaced 19 000 tons. The first all-iron warship, launched in 1861 had 4½ in thick plates, and in 1868 the Royal Navy had an iron dry dock.

Alloy steels containing chromium, tungsten, molybdenum, manganese and nickel were available from 1868 and tinplate, first made in 14th century Bavaria became almost a British monopoly until 1900. Another protective method, galvanizing by the hot dip method, dates from 1838.

New materials

Significant advances were made in the technology of non-ferrous metals. Bronze and brass were used for steam condenser tubes and for the blading of water turbines and pumps. In 1865 the electrolytic refining of copper improved the electrical properties. The first Atlantic cable of 1858 had had only half the possible conductivity although the copper used in it was 98·5 per cent pure.

The latter part of the 19th century saw the discovery or first application of a large number of new engineering materials. Aluminium, known since 1827, came into greater use when it too was produced electrolytically. The production of nickel was revolutionized by the Mond process, introduced in 1890.

Rarer metals with high melting points, osmium, tantalum and tungsten, were employed at the end of the century for the filaments of electric lamps, representing perhaps the first use of metals for service at very high temperatures. They replaced the old filaments which had been made in 1880 by carbonizing threads of cellulose; a process which, as it happened, led directly to the production of viscose, the first artificial fibre, from wood pulp. So the rayon or artificial silk industry was born.

The production and application of coal, rubber, glass, ceramics, brick, stone, concrete, bitumen and paper all underwent considerable development. Although some of these are not used structually, their strength has always been important to mechanical engineers. The strength of coal and rock, for example, is a vital factor in the design of coal cutting machines, as is the strength of paper in the design of printing presses, and the strength of threads in the successful operation of textile machinery.

Rubber was known in South America in the 13th century and in Europe by 1750, but early products, such as rubberized fabric and springs, were unsatisfactory because the crude rubber melted and dissolved easily. Goodyear overcame this difficulty in 1841 with the process of vulcanization in which crude rubber is heated with sulphur. By 1856, rubber belting, hoses and washers and solid rubber tyres were produced. Dunlop's pneumatic tyre appeared in 1888 and at about the same time rubber began to be used on a large scale for electrical insulation.

Availability and fitness for purpose were always the ruling factors in the selection of materials but prior to the steam age the choice of wood, stone and a small amount of metal offered limited scope for choice. In the 19th century relatively large quantities of materials with different properties became available for the first time. Moreover, as machines increased in power and complexity, so material selection became more critical because the consequences of failure were progressively more serious.

Victorian engineering appears at its most characteristic in the steam locomotive. Fig. 7 shows a model of the 1880's, designed to negotiate a gradient of 1 in 45. The caption reproduces an extract from the schedule of materials, somewhat vague to modern eyes but of course terms like 'best iron' or 'Farnley iron' carried a definite connotation, just as the trade names of manufacturers do nowadays.

All sections of industry benefited from new materials. In boiler making, better iron and steel, together with improved stress analysis and inspection, helped reduce the risk of explosion. In mining, wrought iron bits in drills were supplanted by steel tips which, in turn, were replaced by diamond edged tools in the 1860's. A landmark affecting the whole field of machine design was Isaac Babbitts' patent of 1839 for a tin-antimony-copper bearing which combined score and corrosion resistance with good deformability and reasonable fatigue strength.

Theory and testing

The continuing development of railways, marine engineering, mining, armaments and mass-production machinery encouraged theoretical and experimental work on the strength of materials in all the industrial countries. In Germany, O. Mohr improved the graphical analysis of beam and bridge structures and also devised an elegant graphical representation of the state of stress at any point. A great deal of work was carried out by Föppl, Grashof and Winkler on the stress analysis of machine parts such as hooks, springs, plates, tubes and rotating discs. H. R. Hertz (1857–1894) analysed, with the aid of the theory of elasticity, conditions in the vicinity of a concentrated load, a step towards understanding

[Courtesy: Engineering]

Fig. 7. A passenger and goods locomotive designed by William Thow of Glasgow, c. 1887. An extract from the materials list is given below

Boiler barrel	'Best iron'	Firebox flanged plates	Farnley iron
Outside firebox	'Best iron'	Inside firebox	Copper
Main frames	'Best iron'	Tubes	Brass, solid drawn
Axles	'Best iron'	Axle boxes	Wrought iron
Tyres	Crucible cast steel	Wheels	Hammered scrap
		Rails	61 lb steel Vignolles

the action of a wheel on a rail and of ball and roller bearings.

At Cambridge Lord Rayleigh analysed the vibration of bars, plates and shells. One of his methods was later extended by W. Ritz to form a powerful procedure for structural analysis and another method was made possible by an Italian, A. Castigliano, who stated an important theorem connecting the deformation of a structure with its elastic energy.

The question of stored energy had previously been discussed by Clapeyron and Kelvin but Castigliano brought it into the realm of usefulness in structural analysis. Shortly afterwards, Engesser generalized the theorem for non-linear behaviour.

Mechanical testing of materials made considerable progress during the decades before 1900. Attention was paid to tension, compression, bending, shear, torsion, hardness, impact and fatigue; also to the effects of various heat treatments and production processes on mechanical properties. A leading figure in this work was J. Bauschinger of Munich who invented sensitive instruments for the measurement of very small deformations and carried out basic fatigue studies.

A Swede, J. A. Brinell, invented an indentation hardness test which quickly became standard; and, in 1883, Tetmajer devised a pendulum impact machine, versions of which are now in common use to detect the dangerous condition of brittleness in metals.

The high-powered metallurgical microscope made it possible to study the micro-structure and deformation process of metals.

The modern era

The turn of the century saw the era of coal and iron give way to one based upon steel, light alloys and electricity. New methods of power generation, including the internal combustion engine, the steam turbine and hydro-electric plant, stimulated research into materials able to resist temperature, corrosion and wear.

The enormous development of semi-automatic machine tools and other manufacturing techniques also led to demands for new alloys, both for tool bits and workpieces. In constructional work, steel and ferro-concrete found wider use at the expense of wood, brick and stone though lightweight wood construction did take on a new lease of life, together with cotton fabric, for aircraft structures.

Various new government establishments were concerned with the testing and development of materials, typical of which were the National Physical Laboratory in England and the National Bureau of Standards in the U.S.A. Learned institutions the world over, including the Institution of Mechanical Engineers, took much interest in the study of materials.

The seeds of the next materials revolution had been sown when Mendeleyev, a Russian, devised the periodic classification of the elements and predicted the existence of germanium, scandium and gallium. In 1912 Von Laue showed how X-ray diffraction could be employed to investigate the arrangement of atoms in materials and in 1911–12 the work of Rutherford and Bohr led to the planetary model of the atom itself. This provided the key to understanding the structure and properties of materials and also to the synthesizing of many new ones. Already there are more than 2000 specifications of steel alone and an American glass company boasts a file of 100 000 different formulae for glass and ceramics.

Nevertheless, it is still the materials which impose the greatest limitations on mechanical engineering advances. To draw examples from different fields one only has to think of the temperature limits imposed on heat engines, including rockets and M.H.D. plant, weight and fatigue problems in vehicles and rigidity requirements in accurate production methods.

Technological progress is clearly linked to the development of new materials, but of equal importance is the efficient use of existing ones. This is a task which, directly or indirectly, concerns almost every mechanical engineer. As in the past, a successful approach depends on experience, judgement, and a mature outlook on questions of economy and safety; though today there is no doubt that the designer can have much greater confidence in the results of calculations as theory becomes more complete and test facilities more comprehensive.

REFERENCES FOR FURTHER READING

1. Derry, T. K. and Williams, T. I., *A Short History of Technology*. 1960 Oxford Clarendon Press.
2. Timoshenko, S. P., *History of Strength of Materials*. 1953 McGraw Hill.
3. Mumford, L., *Technics and Civilization*. 1934 Routledge and Kegan Paul.
4. Todhunter, I. and Pearson, K., *A History of the Theory of Elasticity and of the Strength of Materials*. 1886 Cambridge University Press (2 vols).
5. Burstall, A. F., *A History of Mechanical Engineering*. 1963 Faber and Faber.
6. Armytage, W. H. G., *A Social History of Engineering*. 1961 Faber and Faber.

THE GREAT MASTERS

In the days when engineering laboratories were practically unknown, Fairbairn was among the first to carry out extensive and systematic experiments on the strength of materials and structures. During his long life he became not only a great manufacturer and designer but fulfilled his greatest ambition—to become a famous man of science. His range of interests and his versatility extended throughout and beyond the whole field of mechanical engineering.

William Fairbairn—Experimental Engineer

by A. I. Smith, B.Sc. (Associate Member)

The third President of the Institution was William Fairbairn, described by Rankine, as 'one of the most eminent engineers and cultivators of mechanical science'—an appropriate description, since Fairbairn had himself been associated with the 'friends of science', who helped to found the British Association in 1831 with the object of cultivating science.

It is to these two engineers, Fairbairn and Rankine,[1] born within fifty miles of each other—that we owe, perhaps more than to any others, the establishment of modern mechanical engineering science. Rankine made unique contributions on the theoretical side, and Fairbairn on the practical and experimental.

At the beginning of the nineteenth century the counterpart of the modern engineer was the millwright. The word 'engineer' was used almost exclusively for military personnel so that the term 'civil engineer' was introduced to cover all non-military work. Later, mechanical engineering became recognized as a branch in its own right, and Fairbairn played a leading part in this development. As *The Engineer* put it—'Fairbairn found millwrighting a second-rate trade. He abolished the millwright and introduced the mechanical engineer'.

He became famous as a manufacturing engineer but his unique place in history is due to his long series of experimental researches. He was one of those giants of engineering who, in their generation, were able to dominate almost the whole field.

The results of his work were made available in some eighty papers and half-a-dozen books; about three-quarters of his papers were published after his sixtieth birthday, and about a third after his seventieth. Fairbairn himself was not a theorist, being largely self-taught and without academic qualification, but he provided much of the experimental information required to unify theory and practice—an objective which he regarded as one of his greater missions in life.

Apprenticeship

William Fairbairn was born on 19th February 1789, at Kelso, Roxburghshire, and lived for eighty-five years. Fortunately he left some autobiographical notes,[2] written in 1851 during a long sea voyage, and these help to reveal something of his early struggles and his character. From the age of four, he attended a small day-school and then the parish school. Until he was fourteen, all his clothes were spun and woven by his mother.

In 1799 his father was offered joint ownership of a farm about five miles from Dingwall in the north of Scotland, and in October the family, including five children—the youngest only six weeks old—left Kelso for the 200-mile journey in an old cart covered with canvas. While in the north, William attended the local school and learned the rudiments of arithmetic and English.

It was at this time that he first gave promise of mechanical aptitude. It was William's duty to look after his fifteen-month old brother. To save himself the trouble of continually carrying his brother on his back, he built 'a respectable wagon' with the meanest of tools.

He recalled that 'the greatest difficulty was the wheels, which I surmounted by cutting sections from the stem of a small elder tree, and with a red-hot iron burnt holes in the centre to receive the axle'.

The crops failed in 1800 and 1801, and his father moved to Yorkshire in 1803. On the way he left his family at his former home town of Kelso. About this time, William was selected through the influence of his uncle's employer, Lord Seaforth, to go to India as a cadet; fortunately for the future of engineering this was prevented through unforeseen difficulties, and at the age of fourteen he got employment at three shillings per week on the new bridge then being built at Kelso by Rennie.

In 1804 his father was appointed steward of a farm belonging to the owners of Percy Main Colliery, near North Shields, and this led to William being apprenticed to the millwright in charge of colliery engines and machinery.

During his seven-year apprenticeship his weekly wage rose from five to twelve shillings. Conscious of his very limited education, he set himself a course of evening study, the syllabus of which shows a desire to develop literary appreciation as well as mathematical studies.

Although mechanical things fascinated him, he also seems to have had a keen interest in music, and bought an old Hamburg fiddle for two shillings and sixpence.

'After tormenting the rest of the house with discordant notes for two months, the very author of the mischief tumbled to pieces in my hands, to the great relief of every member of the family.'

However, Fairbairn made himself a new fiddle ('this operation cost me five weeks' hard labour chiefly at night') and persevered with his attempts to become a distinguished musician. He took lessons from an old man who played at social functions, but alas—'After two or three lessons he offered me twenty shillings for my violin, and advised me to abandon all thoughts of ever rising in the profession'.

Fairbairn was transferred from the workshop to take charge of the steam engine and pumps at Percy Main Colliery, where he met George Stephenson, who was in charge of an engine at Willington Ballast Hill, only a mile or two away. He often visited Stephenson's cottage and the two men—who were of an age—formed a life-long friendship.

His apprenticeship was completed in 1811 and he obtained employment, first at Newcastle and later at Bedlington. In the latter place he met Dorothy Mar whom he married five years later. She was the daughter of a farmer at Mar's Banks, and Fairbairn later recalled:

'To these banks I used to resort as a sort of pilgrimage to contemplate the spot which had formerly been the residence and playground of the object of my affections. In this way, and in many other ways, I entertained wild and romantic notions.'

Work in many places

When his work at Bedlington was finished, Fairbairn

'... took leave of Miss Mar, after an interchange of unalterable affection and, with half-a-dozen new shirts, a new suit of clothes, a watch, and four pounds in my pocket, I embarked on a collier on December 11th, 1811, at North Shields for the metropolis.'

He was accompanied on the voyage—which took twelve days—by a friend called Hogg. They arrived in London late in the evening and his first impressions of the city were far from encouraging. Awakened by noise he found that a whole family next door to his lodging had been murdered!

The two friends tried to get employment with Rennie who had just begun the construction of the Waterloo Bridge. The great man accepted them, subject to the approval of the Millwrights' Society. Unfortunately, the next meeting of the Society was not for about a month, and Fairbairn and Hogg had to live in penury in a garret for the next three or four weeks. They even burned the frame of one of their landlord's beds as fuel. At the end of the month they were rejected by the Society. They obtained temporary employment at Cheshunt and then returned to London and had the good fortune this time to meet a friend of Hogg, who was secretary of the Independent Millwright's Society, reputed to be 'a society founded on liberal principles, and greatly superior to the vagabonds of East Cheap'. They were accepted by this Society, and worked in Grundy's Patent Ropery at Shadwell, and Penn's of Greenwich,[3] for about eighteen months, for wages of two to three pounds per week.

Fairbairn met a clergyman who had a special interest in agriculture, and they jointly designed a steam digging machine, but this was not a success. Later he constructed a sausage-making machine for a butcher in the Tottenham Court Road; this device worked well and he was paid thirty-three pounds for it—his first commercial success.

In 1813 Fairbairn left London to seek employment elsewhere. He visited Bath, Bristol, South Wales and Dublin, where he worked for a few months in the Phoenix Foundry.

However, he returned to England and settled in Manchester in 1814. During the next two years he worked for a Mr Parkinson and resumed private study in his spare time. He had saved thirty pounds and married Miss Mar by the time he was twenty-six. The following year his wife almost died in childbirth, and shortly afterwards the nurse accidently set fire to the bed curtains and only prompt action by Fairbairn prevented the tragic end of wife and child. It was his marriage which stimulated him to set up his own business.

Fairbairn and Lillie

Fairbairn resigned as a draughtsman and persuaded an old workmate, James Lillie, to set up in partnership with him as millwrights. Their first workshop was 'a miserable shed' in High Street. They undertook a few small jobs, built a lathe, and engaged an Irish labourer as its motive power. Their first chance came with an order to renew millwork and shafting in a large cotton mill. Fairbairn later recognized this as the turning point of his career.

'I may safely ascribe my own success in life ... to the saving in power effected by increasing threefold the velocity of the shafting in mills, more than forty years ago. The introduction of light iron shafting not only enabled the manufacturer to effect a considerable saving in cost, but a still greater saving was effected in power, whilst it relieved the mills from the ponderous wooden drums and heavy shafting then in use.'

These improvements led to many orders and a transfer of the business to better premises in Mather Street. In 1822 they bought a second-hand Boulton and Watt steam engine, and in 1824 started the Canal Street Works at Ancoats which were to be famous for half-a-century as one of the foremost engineering works in Europe.

The firm obtained a most important contract in 1824—to build new waterwheels for the Catrine cotton mills of

James Finlay*. This was very successful, largely because of their improvements in wheel design, and considerably enhanced their reputation. In the same year Fairbairn visited Escher at Zurich to erect new waterwheels. This led to numerous other Continental orders, and by 1830 the firm had three hundred employees and a bank balance of forty thousand pounds.

In 1831 Fairbairn undertook what were probably his first important commissioned experiments. He was asked by the Forth and Clyde Canal Co. to investigate the application of steam power to canal barge propulsion. The canal authorities were concerned about the competition of the railways and it was thought that a substantial increase in speed of canal boats would be the answer. Fairbairn undertook a series of experiments and built a 68-ft long iron boat, the *Lord Dundas*, which was steam-powered. He was not satisfied with the results, but the details of these trials formed the subject of his first published paper in 1831.

Failure at Millwall

In 1832 the partnership was dissolved, partly due to disagreement about an unsuccessful business venture in converting a dyeworks into a cotton mill.

The new position of sole authority offered Fairbairn greater scope. Ten iron steamers had been built at the Manchester works, but the great lack of data for design purposes made Fairbairn resolve to undertake experiments on the strength of malleable iron.

In 1835, when it became obvious that the building of ships in sections at Manchester and their transport overland was uneconomic, Fairbairn established a shipyard at Millwall on the Thames. During the next thirteen years more than 100 vessels were built there, and it was reputed to be the first major iron shipbuilding yard in Britain.[3] Nevertheless, the Millwall works were never a financial success, due to increasing competition on the Thames and the great difficulty of managing two works nearly 200 miles apart.

The Millwall yard was eventually disposed of in 1848 and became part of the yard in which Scott Russell built Brunel's *Great Eastern.* The £100,000 loss at Millwall was made good, however, by the great prosperity of the Manchester works. Construction of steam engines and boilers had begun there in 1832, and of locomotives in 1837. More than six hundred of the latter were built, and Fairbairn is credited with the introduction of the tank loco.

Strength of materials

Amid all his responsibilities as a leading manufacturer, Fairbairn was able to undertake a long series of experimental investigations. In 1835 the British Association commissioned him and Eaton Hodgkinson (a man of greater theoretical knowledge, who later became a professor of engineering) to investigate the effect of the recently introduced hot-blast process on the properties of cast iron. Hodgkinson undertook to test in tension and compression, and Fairbairn in bending, a large number of samples from most British iron works. Typically, Fairbairn did additional tests on his own initiative—in this case on the effects of high and low temperatures and prolonged loading on strength. The former were the first systematic tests on the subject; the latter, in effect creep tests, lasted seven years and were to remain without parallel in this country for about a century. Another important investigation he undertook about this time was on

* *cf. page 27, Engineering Heritage, vol. 1,* pp. 30/31

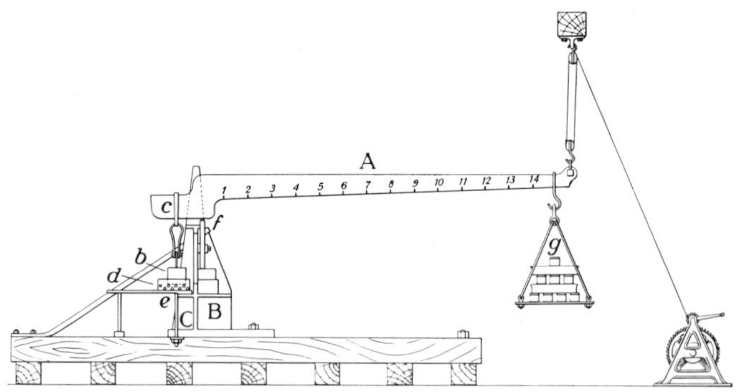

A—lever B—frame *b*—oil bath *c*—shackle *d*—fire grate e—screwed rod *f*—fulcrum *g*—weights

Fairbairn's famous lever testing machine was used to apply loads in many kinds of experiments: here it is equipped for tensile testing of hot wrought iron

riveted joints for iron ships and boilers. He also invented a steam riveting machine—stimulated by a strike of the riveters at his works.

He visited Turkey at the request of the Sultan to inspect and advise on the modernization of all government factories and installations in that country; this visit led to many orders for machinery, etc. The following year he was consulted by the Government on the safety of workpeople in factories because accidents were increasing. He submitted a long report and recommendations to the Board of Trade. In 1842 he investigated the prevention of smoke in steam boilers and in 1844 he introduced the Lancashire, or double-flue, boiler.

The large number of disastrous fires in wooden factory buildings led to the adoption of fire-proof construction with cast-iron beams and columns. But numerous fatal accidents occurred due to sudden breakage of cast-iron beams. Fairbairn published several papers and a book in which he explained the principles of the safe use of iron in buildings.

Tubular bridges

At the Millwall works between 1845 and 1847, Fairbairn carried out the most celebrated of all engineering experiments on the grand scale.

Robert Stephenson had the difficult task of erecting railway bridges across the Menai Straits and at Conway in North Wales. As large suspension bridges were considered unsuitable for heavy railway traffic, he conceived the idea of a huge fabricated iron tube through which the trains would pass. The idea was proved practicable by Fairbairn. Two dozen preliminary tests settled the basic form of the tube; these were followed by others on a model of one-sixth full-size (78 ft long, 4 ft 6 in. deep, and 2 ft 8 in. wide), supported at its ends, and incrementally loaded at mid-length; the deflection was measured at each load. These experiments cost several thousand pounds and led to the rectangular cellular construction used in the Conway and Britannia bridges and about 1000 others built by his firm.

By development of the basic design he was able to increase the load at failure by a factor of about 2·4. These tests also provided the first demonstration of the weakness of thin sections due to buckling in compression. Fairbairn also investigated the effect of long-continued loading on the riveted structure.

The final experiment was—

'. . . of a gigantic character, namely upon the tube itself. It was resolved, in order to make security double sure, that an experiment should be made upon the first of the Conway tubes; that temporary piers should be built under it at each end (corresponding with the span of 400 ft); and that deflections should be taken with variable and increasing loads in the usual way.'

This impressive test was undertaken on site at Conway. The riveted wrought-iron tube weighed 1300 tons and deflected 8 in.; loads were increased by 300 tons and the deflections measured. Lateral deflections due to oscillations produced by a severe gale and the distortion of the tube due to the sun shining on the upper surface and on one side were also measured.

Unfortunately Fairbairn's enthusiasm led him to release—unknown to Stephenson—the results of the test of the Conway tube to several of the leading scientists in the land, and this caused one of several misunderstandings between

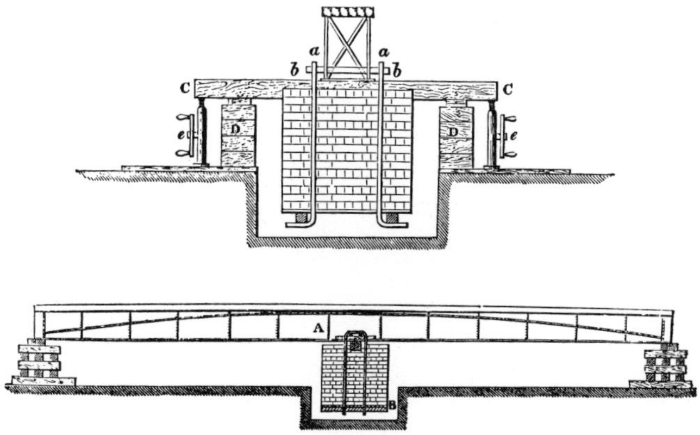

How Fairbairn tested the model tubes for the Britannia Bridge

A portion of the actual bridge structure resting on the central pier in the middle of the Menai Strait: note the cellular construction of the tube

the two great engineers which culminated in Fairbairn' resigning his position of engineer-in-charge of the bridges, after the first Conway tube had been safely lifted into position. Fairbairn himself commented:

'Labouring under the prognostications of failure it is perhaps not extraordinary that in the moment of success I should have betrayed some excitement and have made use of certain hasty expressions, which in calmer moments would not have escaped me.'

These two great bridges of advanced design created very great interest.

Fairbairn was elected a Fellow of the Royal Society in 1850, in recognition of his work on strength of materials and other aspects of practical science. He visited Sweden and Russia that year and inspected many engineering works.

Between 1851 and 1857 he collaborated with Hopkins of Cambridge in experiments on the effect of high pressure on the melting of various substances. Hopkins was interested in theories of the formation of the earth and considered that these depended on knowledge of the influence of high pressure on the fusion of the materials in the crust. Fairbairn developed equipment to measure the effect, using his famous lever testing machine to produce the loads.

In 1852 Fairbairn was largely responsible for the layout and machinery for the Small Arms Factory at Enfield and he also completed the installation of the then most up-to-date mill in the world at Saltaire, near Bradford.

The following year he partly retired and handed over the principal reins of management to his sons. That year he was President of the Mechanical Section of the British Association and in 1854–1855 he succeeded the Stephensons as President of this Institution.

Steam and high temperature work

Many serious boiler explosions involving loss of life occurred about the mid-century, and Fairbairn investigated several of them. His findings led him in 1855 to found the Association for Prevention of Steam Boiler Explosions which first sponsored regular inspection of boilers. He tried unsuccessfully to persuade the Association to undertake tests of boiler strengths; later he made these himself and published the results in a classic paper in 1858. He demonstrated, for the first time, that the resistance of a tube to external pressure decreases as its length increases, thus explaining the cause of many explosions, and pointing the way to their prevention.

Feeling the need for a personal assistant in his experimental work, he engaged W. C. Unwin, then a lad of eighteen, who later became manager of his engine shop, and eventually a notable Professor of Engineering at London, and President of the Institution in 1915–16. Another famous pupil was Lord Kelvin's brother, James Thomson, who followed Rankine at Glasgow and is noted as the inventor of the vortex turbine.

In 1856 Fairbairn published the results of tests, in his lever testing machine, of the effects of high temperature on the tensile strength of wrought iron. These were the first systematic tests on this topic and demonstrated the now well-known rise in strength of certain ferrous materials at slightly elevated temperatures. He appears to have revelled in the challenge of unusual assignments. An interesting paper, presented to the Institution in 1858, described the conversion of two steamers to floating flour mills and bakeries for supplying the troops with bread during the siege of Sebastopol. The whole undertaking had been completed within three months.

Experiments with the canal boat 'Swift': a mercury dynamometer and stop watch were used to determine power required at various speeds

He published several important papers on the properties of steam as a result of collaboration with Thomas Tate, Mathematics lecturer at Battersea Training College. These papers provided engineers for the first time with reliable data on the specific volume of steam over an extensive range of pressures and temperatures.

A wide range of activities

Between 1859 and 1865 Fairbairn was interested in the Atlantic cable and served on a Board of Trade Commission enquiring into the best type of covering for submarine cables; again, characteristically, he made his own experiments and we find papers on the permeability, under high hydrostatic pressure, of various kinds of insulators, as well as on the mechanical strength of cables.

In a two-year series of pioneer experiments he investigated the effects of repeated loading on a riveted model girder, using a waterwheel as a source of continuous motive power. The resulting papers are the first in the Proceedings of a British scientific society on fatigue experiments, although Fairbairn discusses some tests undertaken a few years previously by James and Galton.

At the age of seventy-two, he was the first engineer to be elected President of the British Association. He was also offered, but declined, a knighthood. He published that year his book on the history and properties of iron and also the first part of a work which he had long intended to write: *Mills and Millwork*, in effect a manual of mechanical engineering practice; its publication is a landmark in the history of our profession.

Between 1861 and 1867 he served on the Special Committee on Iron which had been appointed to advise on the armouring of ships; the Committee investigated the types of armour, their backing and mounting, the laws of penetration by projectiles, and undertook many firing trials. As usual, Fairbairn not only participated in these tests but undertook additional experiments of his own and published several papers on armour and its penetration. He was a great protagonist of the fast, locally armoured ship of moderate size.

His *Treatise on Iron Shipbuilding* was published in 1865 and he made no small contribution to naval design. He was the first to demonstrate that ships must be designed to withstand repeated longitudinal bending due to the varying buoyancy conditions as they pass through the waves.

In his eightieth year Fairbairn was created a Baronet. Even at this advanced age he remained mentally active: he was keenly interested in the use of high pressure in steam boilers and took out several patents. About this time he also wrote a large section of a book on the industries of Lancashire and Cheshire.

In 1873 Fairbairn attended the opening of new buildings at Owens College when he caught a bronchial chill from which he never fully recovered. In that year his last contribution to scientific literature was presented to the Royal Society—a paper dealing with the durability of iron ships with riveted joints.

He died on 18th August 1874, at Farnham, Surrey, and was buried at Prestwich Parish Church. It was estimated that between 30 000 and 50 000 attended the civic funeral in Manchester.

Character

In daily life Fairbairn was a man of regular habits; a friend recalled that in the walk from his home—*The Polygon*—to his works he had a favourite route and definite places for crossing the roadway, which he adhered to, even to the point of pushing his companion into the mud. He was punctual and, according to Unwin, commanded respect without domineering.

A man of simple tastes, he took little exercise other than walking, and participated in no sports. His holidays he spent in the English countryside, or revisiting the haunts of his childhood in Scotland. His literary preference was for history and biography; he had little interest in fiction. He was religious to a degree, worshipped in the same Congregational Chapel in Cross Street for fifty years, and often referred to Divine Providence. He appears to have been generally popular and was constantly in demand as a lecturer.

The record suggests, however, that he may not have been the easiest of men to get on with. He had disagreements with his partner Lillie, with Hodgkinson his collaborator in his early research on materials, and with Robert Stephenson.

It may well be that, persisting in the course which he was convinced was right, he expected others to be as absolutely straight as he himself was. The only criticism which his biographer[1] allowed was that he had

> '. . . perhaps an excessive ambition for popularity and fame, but this foible had one redeeming feature, namely, that he aimed not so much at obtaining the applause of the million as at standing well with the good and the wise.'

Except when he was on one of his numerous travels, he was at work every day from 9 to 6. His many papers and reports were mostly written in the evening at *The Polygon*. Even in retirement he retained an office at the works—at least past his seventy-fifth year.

He was said to have had 'a wonderful eye for proportion

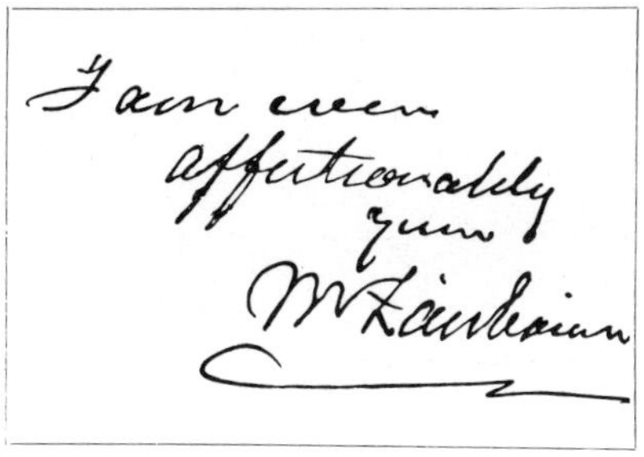

Facsimile of the inscription on the portrait

and the mechanical fitness of things'. His hand sketches were admirably clear and he had a team of eight or ten draughtsmen working directly under him when he was at the height of his powers.

Fairbairn appears to have remained young in spirit into his old age—he described himself, at the age of seventy, as 'a happy and contented old codger'. He was very fond of children and even in his later years delighted to romp with his grandchildren.

Having achieved success as a leading manufacturer, he developed the ambition to be 'a famous man of science'. What were the attributes which contributed to his success? A strong physical frame supplied him with indefatigable energy; this was combined with great personal ambition, perseverance and determination to excel. It was not that he had an easy life;

> In the fits of melancholy which sometimes overtook me, I imagined every contingency . . . I resolved to stem with an energy that nothing could crush, the tide which set in against me, and I set to work with redoubled energy . . . It was this determined perseverance that enabled me to keep the wolf from the door.

Fairbairn acknowledges the inspiration he received from the achievements of the great scientists and engineers of the past: 'the examples of Franklin, Ferguson*, and Watt are always before me'.

Literary talents

His writings are remarkable for a man who was largely self-educated, and many passages in his papers possess considerable literary merit.

During his apprenticeship, Fairbairn had fallen in love with a girl he met at church, and although this romance came to nothing he ascribed his literary ability to this attachment which first led him to read, and base his own efforts on, a correspondence between two lovers, Frederick and Felicia, published number by number in the *Town and Country Magazine* for 1782.

> 'I commenced the correspondence by first reading the letter [from Felicia] in the magazine and then shut the book for the reply, and wrote the letter that Frederick was supposed to have written. I then referred to the book, and how bitter was my disappointment at finding my expressions unconnected and immeasurably inferior to those of the writer. Sometimes I could trace a few stray expressions, which I thought superior to his but, as a whole, I was miserably deficient. In this way did I make love, and in this way I inadvertently rendered one of the strongest passions of our nature subservient to the means of improvement.'

Later, Fairbairn applied the same technique to more serious authors—Gibbon, Hume, Milton, Shakespeare, Burns, whose works he was able to borrow from the library at Shields.

The following passage on the use of iron in ship construction was written when he was about seventy-five.[4] The theme is perhaps not inappropriate to our present age.

> 'It is difficult for mankind after a certain age to divest themselves of the opinions which in the early stages of their professional career they were taught to believe were the only true principles on which they could practice and exercise their skill. Doubtless an adhesion to these principles, as far as they go, is commendable, but the progress of science and the adoption of an entirely new material of construction introduces new laws and new principles, widely different from those previously in use. It is not therefore surprising that new constructions and new material should require new heads and hands to carry out measures calculated to set aside old customs, and supplant them with entirely new and more important structures. Younger men enter upon the stage of life with new ideas and readily adopt the practice of new principles.'

And aware of his own declining physical powers in his old age, he wrote: 'Men only live while they are useful'.

Final assessment

When Fairbairn died in 1874, full of years, full of great works and honours, he had more than fulfilled his early ambition to be his own master: he had surpassed also his later ambition to be 'a famous man of science'.

In his engineering creations he was advanced in outlook, bold in conception, and courageous in design, but this courage was not foolhardy: it was born of experiment and factual knowledge of the properties of materials. He was, perhaps, the great master of engineering experiment in an age when the engineering laboratory was still unknown.

His influence over almost the whole field of mechanical engineering is reflected in the tribute paid to him in *The Engineer*, shortly after his death, in words which might well serve as his epitaph:

> . . . it is difficult to discover a branch of the art of mechanical engineering to which Fairbairn has not contributed something. His footprints may be found on every path that the engineer can tread, and the sands of time will never efface them.'

—a tribute which could be paid only to a very select company of engineers; a tribute which even Fairbairn himself would surely have considered adequate.

Four Presidents of the Institution: left to right, John Penn, Joseph Whitworth, Robert Napier and William Fairbairn

REFERENCES FOR FURTHER READING

1. RANKINE, W. J. M., 'Sir William Fairbairn'. *The Imperial Dictionary of Universal Biography*, vol 2. c. 1870. William Mackenzie, London.
2. POLE, W., *The Life of Sir Wm Fairbairn.* 1877 Longmans, Green and Co.
3. SMILES, SAMUEL, *Industrial Biography: Iron Workers and Tool Makers.* 1876 John Murray, London.
4. FAIRBAIRN, WILLIAM, *Treatise of Iron Shipbuilding, its History and Progress.* 1865 London.

* One of the foremost millwrights of the eighteenth century.

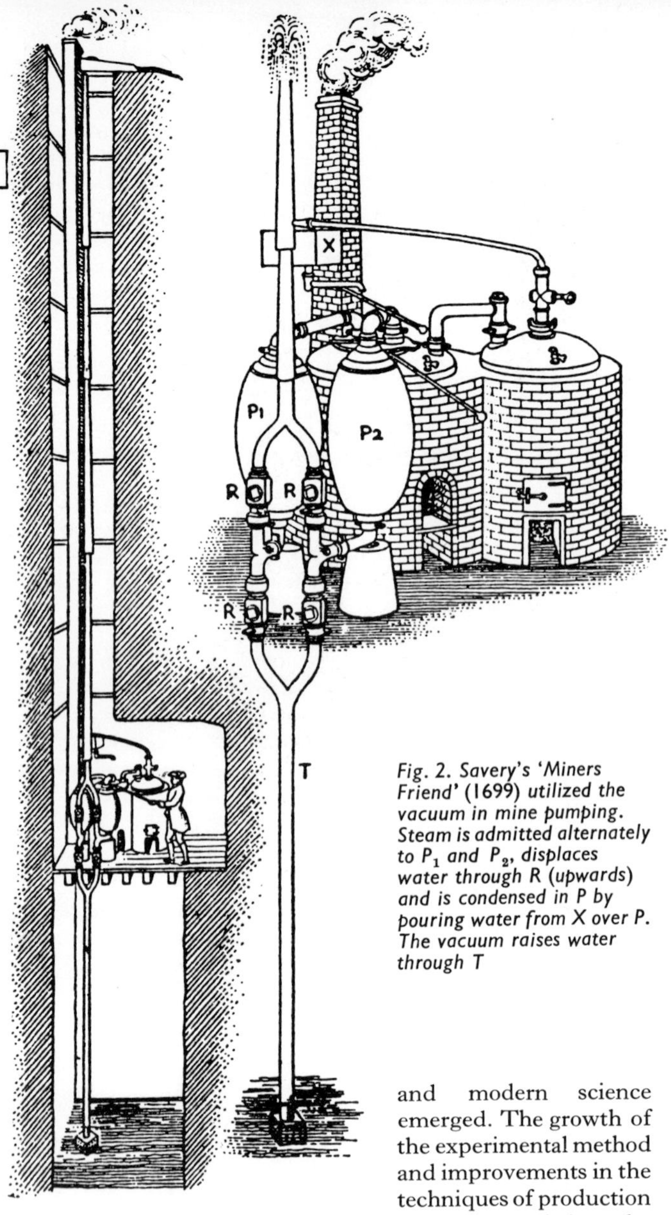

Fig. 2. Savery's 'Miners Friend' (1699) utilized the vacuum in mine pumping. Steam is admitted alternately to P_1 and P_2, displaces water through R (upwards) and is condensed in P by pouring water from X over P. The vacuum raises water through T

Some 200 years ago, James Watt started to repair a model steam engine. This article traces the development of ideas concerning the nature of heat and its conversion to work in an engine. This conversion became important only as a result of Watt's work and the Industrial Revolution, although fires had been kindled by expending work since the dawn of civilization. The economics of the steam age demanded a quantitative treatment and engineers played a key part in creating the new science.

Thermodynamics—a Daughter of Steam

by Derek Bradley, B.Sc., Ph.D. (Associate Member)

The importance of fire to ancient civilization is reflected in its being one of the four material elements of the Greeks (the others being earth, air, and water) of which all bodies were believed to be composed. The scheme shown in Fig. 1 persisted in the philosophy of the medieval alchemists. During the Renaissance the patterns of trade changed, new techniques developed, new production centres were created, and modern science emerged. The growth of the experimental method and improvements in the techniques of production meant that ultimately the nature of heat was settled on the basis of experiments, rather than by philosophic speculation.

Fig. 1. To the Greeks, fire was one of the four elements of matter. Each was formed by a passive material containing a pair of fundamental qualities; thus fire is hot and dry. The caloric theory descends from this concept

The need for metals led to the opening of new mines, the miners combining together in companies and conducting operations on a large scale. As the mines became deeper, pumping and hauling gear were improved with sometimes as many as five hundred horses in use at a single mine as sources of power. It was well known to miners that suction pumps could not lift water more than about 32 feet.

Galileo and Torricelli investigated this and ultimately the nature of a vacuum was impressively demonstrated in von Guericke's famous experiment with the hemispheres at Magdeburg.

Phlogiston and caloric

Using an evacuating pump similar to that used at Magdeburg, Robert Boyle (1627–91) showed that combustion was impossible in a vacuum. He also showed that metals increased in weight on calcination and attributed this purely to the action of fire upon the metals. Although the conclusions he drew from this work were rather indefinite, when taken along with some of the results of his other work, they indicated to Boyle a more fundamental structure of matter than that given by the Aristotelian elements.

The discovery of oxygen by Priestley in 1774 was the final refutation of the idea that air was an element. But the nature of fire and heat and the distinction between them was still not clear. Priestley subscribed to the phlogiston theory, according to which all combustibles contain this invisible substance which is released on burning. A combustible may be extinguished in air if the air becomes saturated with phlogiston. Because combustibles burn more readily in oxygen Priestley concluded that this gas must contain little or no phlogiston, and for this reason he called it 'dephlogisticated air.'

The French chemist Lavoisier (1743–94) recognized oxygen as an element and showed experimentally that combustion and calcination were not due to the escape of phlogiston, but to the chemical combination of the substance with oxygen. The use of accurate balances had shown that heat was weightless and Lavoisier classed it as an imponderable element, which he called 'caloric'. Prior to the work of Joule on the conservation of energy, light, magnetism and electricity were also classed as imponderable elements. Lavoisier regarded the evolution of heat during combustion as due to the displacement of caloric from the interstices between oxygen particles.

In Britain Boyle was the first to use sealed alcohol thermometers and at the beginning of the eighteenth century Fahrenheit developed the accurate mercury thermometer. The ground was thus prepared for the experiments of Joseph Black. His work showed the distinction between temperature and heat and gave rise to the quantitative science of calorimetry, based upon the caloric theory.

Although there was a lack of clarity as to the nature of caloric, which was essentially descended from the Greek element fire, this concept enabled purely thermal effects to be investigated mathematically. How successful this approach could be when used along with the idea of conductivity is demonstrated by Fourier's *Analytical Theory of Heat* (1822). The crisis of the caloric theory (which, indeed, destroyed it) emerged in the attempts to explain quantitatively the production of mechanical power by heat in the steam engine. A dynamic theory of heat was required and this, in turn, required a dynamic theory of atomic structure.

The steam engine

At the beginning of the eighteenth century the problems of mine pumping had led Savery (whose ideas are illustrated in Fig. 2) and Newcomen to develop the steam engine. 'Raising water by fire' became technically possible and also economic in that an engine-and-coal power source was cheaper than a horse-and-fodder power source.

In the winter of 1763–64, when James Watt was called in to repair a model Newcomen engine at Glasgow University, he became aware of the large wastage of steam due to the heat loss to the cylinder walls.* He appreciated that a larger cylinder would reduce the relative loss and he sought a material which would absorb a minimum of heat. He conducted experiments on different wall materials and also on the properties of steam. This work was carried out in collaboration with Joseph Black, then Professor of Anatomy and Chemistry at the University, and as a result of it Watt hit upon the idea of a separate condenser to reduce the condensation in the cylinder.

Watt needed to know steam densities in order to decide upon cylinder volumes and latent heat in order to design condensers. Out of this work on steam properties arose 'Watt's Law' which, in modern terms, says that the enthalpy of dry saturated steam is the same at different pressures. According to Robison, Black's close associate, it was Watt's engines

> more than all the love of knowledge, so boldly claimed by the eighteenth century, that spread the knowledge of the doctrine of latent heat, and the name of Dr Black.

But throughout his life Watt, who died in 1819, worked within the framework of the caloric theory. Caloric was added to the steam in the boiler and the same amount of caloric was removed from the steam in the condenser. It may be thought that measurements of engine performance would have shown this view to be erroneous. However, a Watt engine would convert only about two per cent of the energy released in the boiler into useful mechanical work and the small difference between the two energy transfers in boiler and condenser would be difficult to detect.

The equivalence between heat and work was recognized in economics, through the concept of engine 'duty', years before it was accepted in science. The duty was the amount of water in millions of pounds which could be raised one foot by burning one bushel of coal. Boulton and Watt supplied mine pumping engines to the Cornish mines and the mineowners agreed to pay the firm annually one third of the total savings as compared with the use of an equivalent Newcomen engine. Watt's improvements, including the separate condenser, had led to his engines having a duty about five times that of a Newcomen engine.

In his patent of 1782 Watt showed the theoretically possible improvements in economy by the employment of expansive working of steam. In this theoretical calculation Watt had to assume a law of expansion for the steam after cut-off, but after the development of the engine indicator by Watt and his draughtsman John Southern in 1803, it was possible to find the mean effective pressure experimentally.

It will be seen that steam engine practice of the period posed four fundamental questions:

1. Can the duty be improved and is there an upper limit to its value?
2. What are the properties of steam?
3. What is the law of expansion of the steam in the cylinder?
4. Would other working fluids be superior to steam?

Lazare and Sadi Carnot

A brilliant treatment of these problems was published by the young French scientist and engineer, a former student of the great École Polytechnique, Sadi Carnot, in a small book, *Reflections on the Motive Power of Fire.*[1] Several of Watt's engines had been exported to France and Carnot was struck by the potentialities of steam power. He wrote:

> To take away today from England her steam-engines would be to take away at the same time her coal and iron. It would be to dry up all her sources of wealth, to ruin all on which her prosperity depends, in short, to annihilate that colossal power. The destruction of her navy, which she considers her strongest defence, would perhaps be less fatal.

Although some reservations arise in the course of his treatment, Carnot's approach is also based upon the caloric theory. Sadi was a son of General Lazare Carnot, Napoleon's 'organizer of victory'. Lazare, a mathematician, published in 1803 an important work, *Fundamental Principles of Equilibrium and Movement.* In this, he discusses the efficiency of non-thermal machinery, showing that, because of energy loss in acceleration and friction, perpetual motion is im-

* *Article on Watt, p. 57, Engineering Heritage, vol. 1.*

possible. No doubt Sadi had absorbed some of his father's teachings, for there is a striking similarity between his treatment of the heat engine and Lazare's treatment of the waterwheel. Sadi writes:

> The motive power of a waterfall depends on its height and on the quantity of the liquid; the motive power of heat depends also on the quantity of caloric used, and what may be termed, . . . the height of its fall, that is to say, the difference of temperature of the bodies between which the exchange of caloric is made.

Lazare shows for hydraulic engines and Sadi for heat engines that, provided certain fundamentals are observed, the greatest mechanical effect will be produced regardless of design details. Sadi also showed that the mechanical effect produced in the perfect, reversible engine was independent of the working fluid. He demonstrated that the maximum power was obtained from a reversible engine. For, if it were not so, "this would be not only perpetual motion, but an unlimited creation of motive power without consumption either of caloric or of any other agent whatever." This is, essentially, the Second Law of Thermodynamics.

His book abounds in references to current engine practice. Carnot commends the expansive working of high pressure steam in the condenserless engines of Trevithick. This made possible an increase in power/weight ratio, and the development of the steam locomotive. He shows that the motive power "is fixed solely by the temperatures of the bodies between which is effected, finally, the transfer of caloric." This power is calculated for air, steam and alcohol and is found to be almost the same for the different fluids. Using a particular calorific value for carbon, he compares the duty of a good Cornish pumping engine with his theoretical value and finds that "1/20 only of the motive power of the combustible has been utilized."

Carnot died of cholera in 1832, aged only 36. It is clear from notes found after his death, and published in 1878, that he had later rejected the caloric theory in favour of a dynamic theory.

But Carnot's writings had no important impact for several years. There are no contemporary references to his work. He proposed experiments to find the relationship between heat and work, but in a period of French politics which presented difficulties for him, he was unable to carry them out.

James Prescott Joule

Some years later the brilliant experimenter J. P. Joule (1818–1889), conducted such experiments in Manchester, without knowing that they had been contemplated by Carnot.

As the wealthy son of a brewer he had both the money and the time to experiment in his own workshop. In 1834 his father sent him, along with his brother, to study under John Dalton. Their instruction finished three years later, owing to Dalton's suffering an attack of paralysis.

As Carnot appreciated the role of the steam engine in the Industrial Revolution, so Joule was excited by the potentialities of the new electro-magnetic engine. Faraday had discovered induction in 1831 and Sturgeon had developed the commutator in 1836. Joule began investigations on the electro-magnetic engine in 1838.[2]

At first he believed that electro-magnetic power would ultimately supersede steam, even envisaging perpetual motion based upon electricity. But this early optimism was soon tempered by a better understanding of the principles of the electro-magnetic engine. Three years later, as a result of experiments with a Grove's battery as a power source, he found that the duty of a steam engine was nearly five times that of the battery and motor, in terms of consumption of zinc.

However, he continued his investigations and, in some of the most exacting experiments he ever conducted, measured the heat produced in an electric generator, and derived his first value of the mechanical equivalent of heat of 838 ft.lb Btu. This result, of course, contradicted the reigning caloric theory, and when the work was presented at the British Association Meeting at Cork in 1843, it excited little interest.

Several years previously Dalton had measured the temperature changes of air due to compression and rarefaction. Fig. 3 shows how he interpreted his results according to the caloric theory.

Now his pupil set out to extend these investigations and to find the relationship between heat and work in an almost isothermal compression. Using a reciprocating pump, Joule compressed air into a receiver as shown in Fig. 4. The pump and receiver were surrounded by water, the temperature change of the air was negligible, he calculated the work done on the air, assuming isothermal compression, and derived values of J of 823 and 795. This assumed that all the work done on the air became sensible heat and none of it was latent. Or, as we now say, the internal energy of the air was

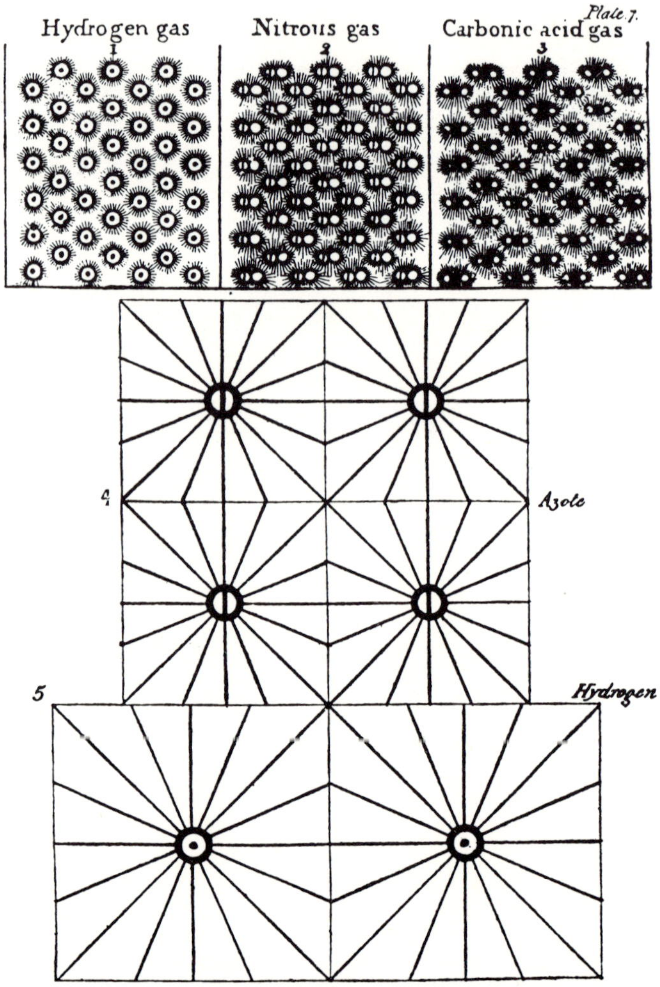

Fig. 3. Dalton's atomic structure (1810). Atmospheres of caloric are shown radiating from solid central atoms. The rays of caloric from the four atoms of azote (nitrogen) balance. The azote and hydrogen rays cannot readily match and this causes internal motion on mixing

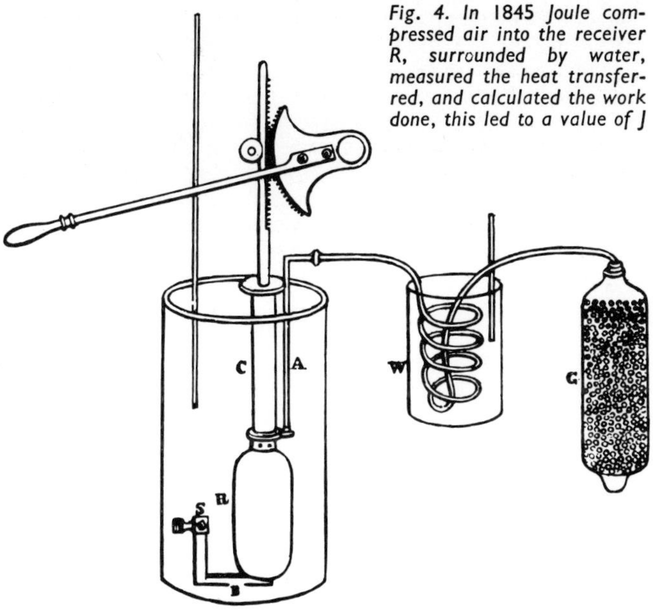

Fig. 4. In 1845 Joule compressed air into the receiver R, surrounded by water, measured the heat transferred, and calculated the work done, this led to a value of J

unchanged by the compression. The same assumption had also been made earlier by Mayer and was known in the literature as Mayer's hypothesis.

The grounds for Joule's assumption were the closeness of the values of *J* to the value obtained in the electric generator experiments, the supporting evidence of the well-known subsidiary experiments, in which air was expanded between two vessels, and his theory of atomic structure.

This theory was a radical departure from that which had been propounded by Dalton[3] who had subscribed to the caloric theory and had sought to combine the views of Newton and Lavoisier in an explanation of his Law of Partial Pressures. Newton had regarded air as an element with a force of repulsion between neighbouring particles. Dalton proposed that each particle, or atom, consisted of a small solid central atom surrounded by an atmosphere of caloric, the density of which decreased with increasing distance as shown in Fig. 3. He also proposed that unlike atoms did not repel each other.

Heating a gas would increase the repulsive forces and hence the volume or pressure. There could, however, be no satisfactory explanation of the behaviour of gases until the static theory of repulsions was abandoned and replaced by a dynamic theory. For example, how could Dalton's theory explain Joule's observation that there was no change in temperature when air expanded without developing mechanical power?

In the tradition of Descartes and Boyle, Joule sought explanations of his experimental results in terms of a dynamic theory of the atom. He developed "the beautiful idea of Davy" that the heat of elastic fluids depended partly upon a motion of particles round their axes, also taking into account Faraday's electro-chemical discoveries. He assumed that atmospheres of electricity revolved at high speed around their respective atoms; that the momentum of these atmospheres constituted 'caloric' and their outer velocities determined temperatures. If there were no cohesive forces between atoms and atmospheres, and between the atoms themselves then this model explained the perfect gas laws.

The paper describing this work of Joule was rejected by the Royal Society, but published in the Philosophical Magazine in 1845. At the end of the paper he concludes: "The principles I have adopted lead to a theory of the steam-engine very different from the one generally received, but at the same time much more accordant with the facts." In the same year at the British Association Meeting he presented for the first time results from his well-known paddle wheel experiments. This work was largely designed to convince the scientific establishment of the equivalence of heat and work. In 1847 he again presented his results to the B.A., but as his previous paper had aroused little interest, the Chairman requested him to confine himself to a short verbal description! Joule later wrote:

> This I endeavoured to do and, discussion not being invited, the communication would have passed without comment if a young man had not risen in the section, and by his intelligent observations created a lively interest in the new theory. The young man was William Thomson.[4]

Kelvin and Rankine

Thomson, the son of a mathematician who became a Professor at Glasgow University, was a brilliant student who, after graduating at Cambridge in 1845, went to work under Regnault in Paris for about four months. Regnault, under the auspices of the French Government, was investigating the properties of steam and other possible fluids for use in heat engines. This work provided data used by engineers for the remainder of the century. Under Regnault's influence, Thomson's earlier interest in thermal problems revived and he came across a paper by Clapeyron describing Carnot's work. In 1846 he returned to Cambridge and was appointed Professor of Natural Philosophy at Glasgow, at the age of 22.

After he had obtained a copy of Carnot's book, he published, in 1849, *An Account of Carnot's Theory of the Motive Power of Heat, with Numerical Results deduced from Regnault's Experiments on Steam.*[5] In this he expounds, in a mathematically more rigorous style than Carnot, the essentials of Carnot's book. He discusses Joule's experiments on the equivalence of heat and work and comes down on the side of Carnot and the caloric theory. He compiles tables from Regnault's data for steam, giving the 'mechanical effect' in ft.lbs. due to a 'Thermic Unit Centigrade' passing from a body, at various temperatures to a body at 0 deg. C. He concludes, as Carnot had done, by comparing the actual duty of steam engines with the theoretical duty.

The reluctance amongst other British scientists to accept Joule's findings and to reject the caloric theory may seem surprising today. This situation is discussed by J. G. Crowther in his stimulating articles on Joule and Thomson.[6] University science was not orientated towards industrial problems. There was not, in Britain, an equivalent of the École Polytechnique; scientists were more at home in the traditional fields such as astronomy and Newtonian mechanics. The caloric theory could solve the problems of simple heat exchange. But the heat engine inconveniently required a synthesis of both thermal and dynamic theories. And the locomotive age urged a solution. In the words of Osborne Reynolds:

> The condensing engine had not been obtrusive—it was not generally seen unless looked for. The locomotive is obtrusive; it will be seen: and by 1842 locomotives had obtruded themselves pretty well all over Europe. They immediately took their places as objects of as much wonder and interest to the grown people who saw them for the first time as they are still to the young; demanding the attention even of philosophers who had previously studied nothing lower than the planets.

Not only did engine practice pose the problem, it demonstrated the solution. As Mayer put it:

> A locomotive engine with its train may be compared to a distilling apparatus; the heat applied under the boiler passes off as motion, and this is deposited again as heat at the axles of the wheels.
> If falling force and motion are equivalent to heat, heat must also naturally be equivalent to motion and falling force.

It was indeed a railway engineer, later a Regius Professor of Civil Engineering and Mechanics at Glasgow University, who developed a single thermodynamic theory. William John Macquorn Rankine* analysed the nature of elasticity, both in the stresses and strains of solids and the pressure of steam on an engine piston. In 1850 he presented his great paper *On the Mechanical Action of Heat* to the Royal Society of Edinburgh.[7] In this he expounds his hypothesis of Molecular vortices, namely:

> that each atom of matter consists of a nucleus or central point enveloped by an elastic atmosphere, which is retained in its position by attractive forces; and that the elasticity due to heat arises from the centrifugal force of these atmospheres revolving or oscillating about their nuclei or central points.

He developed this theory mathematically and propounded a theory of heat in terms of mechanics, "the only complete physical science." In the case of a perfect solid, the elasticity of the atmosphere is negligible and the forces are between the atomic nuclei. The theory was later superseded by the kinetic theory of gases but his work at the time was a great advance. Rankine's model of the atom was similar to that of Joule but the depth of his quantitative treatment was new. He showed how Joule's results could be explained in terms of Newtonian mechanics.

Added to Joule's experimental evidence, and also the work of Clausius in Germany, Rankine's theory was an important factor in finally leading Thomson to a rejection of the caloric theory. Thomson developed his new synthesis in his celebrated paper *On the Dynamical Theory of Heat* which appeared in 1851, and in which the laws of thermodynamics are given in the form familiar to modern readers.

Carnot's function

The reciprocal of a function first introduced by Clapeyron became known in the literature as Carnot's function. The caloric theory defined it as the mechanical effect produced per unit of caloric transferred for each unit temperature difference between source and sink. From Carnot's reasoning it was a function of temperature only. The theoretical evaluation of this function which was the counterpart of the practical concept of engine duty, became one of the quests of the early thermodynamicists.

Joule suggested the function was equal to $\mathcal{J}/T$, where T is the perfect gas absolute temperature. This rested on the dynamic theory of heat and the validity of Mayer's hypothesis. Mayer, a German doctor, had concluded as a result of his medical experience and of Lavoisier's work on 'animal heat', that heat and work were equivalent. On the hypothesis that the internal energy of air was a function only of temperature, he derived the well-known relationship between specific heats and the mechanical equivalent of heat in 1842. In terms of the kinetic theory, his hypothesis was that there were no cohesive forces between atomic or molecular centres.

Even after he had rejected the caloric theory in 1851, Thomson was not convinced about Joule's value of the Carnot function. He knew from Regnault's data that the internal energy of steam was not in general a function only of temperature. It was decided to check Mayer's hypothesis and this led to the joint investigations by Joule and Thomson of the thermal effects of gases passing through small apertures.

If the hypothesis were correct and no energy were required to separate the molecules against forces of intermolecular attraction, then the temperature of the gas would be unchanged after the passage. If, however, there were a cohesive force, then expansion of gas would be accompanied by a drop in temperature. One of the principal objects of the research was to evaluate Carnot's function and the Royal Society made a grant towards the cost.

The experiments showed a slight cooling of air, and hence an attractive force between molecules. This Joule-Thomson Effect was to become the basis for the production of liquid air. According to Joule's initial assumption, the value of $\mathcal{J}/\mu$, (where μ = Carnot's function) should have been 288.85 deg. on the absolute air thermometer scale. The mean value obtained experimentally from these porous plug experiments was 288.82 deg.

Since the agreement was so close, Joule and Thomson resolved the problem by proposing a new absolute scale the temperature on which is simply $\mathcal{J}/\mu$. As the function is independent of the working fluid, this scale must also be independent of the properties of particular substances and hence more fundamental than scales based upon near-perfect gases. The possibility of such a basic scale had been foreseen by Thomson in 1848.

The formal definition of the new scale is, essentially, that the efficiency of a reversible engine working between a source at $T_1°$ and a sink at $T_2°$ is $(T_1 - T_2)/T_1$.

The concept of entropy

The first textbook on the new thermodynamics was Rankine's *A Manual of the Steam Engine and other Prime Movers*, published in 1859. In this he collects the results from much of his previously published work and arranges it in convenient form for engineers, alongside much practical information on prime movers. Sometimes Rankine's dual viewpoint of phenomena (on both the macro and molecular scales) was confusing, although it was this approach that had enabled him to make his great contributions. He defined a 'thermodynamic function', ϕ, which arose from his molecular theories and which had a constant value during an adiabatic change.

Its value could be calculated from the equation of state and thermal data. Equating values of ϕ and using the Clapeyron Relation, he was able to find the pressure/volume relationship during adiabatic expansion of steam. The ϕ function is, essentially, entropy. The modern term originates with Clausius who, in 1854, introduced a function called *Aequivalenzwerth* almost identical with Rankine's function. Clausius later called his function *Entropie*. Today engineers become familiar with entropy as a thermodynamic property arising from the Second Law and giving a criterion of reversibility. In the last century there were other interpretations.

Thomson used the function as an indication of the wasted work. It was clear that, although the conversion of work into heat can proceed to completion, the reverse process cannot. Thomson showed the magnitude of the amount of work "absolutely and irrecoverably wasted" in the heat-to-work conversion. If the constant temperature of the sink is T_2 and if the heat transfer to the fluid, dQ, from the source takes place at a temperature T then, for a varying temperature of heat reception, the amount of wasted work is $T_2\int\frac{dQ}{T}$. Thus

* *cf. Article on Rankine p. 32*

$\int \frac{dQ}{T}$, the entropy change during heating, is a measure of this work. In this manner Thomson, and also Clausius, gave the engineer a practical interpretation of the entropy function, albeit a somewhat pessimistic one.

The growing number of steam engine trials, carefully conducted by Willans and others, led the practical engineer to thermodynamics in order to explain his results The temperature/entropy diagram convincingly demonstrated the value of entropy in the interpretation of the trials. It was developed in the U.S.A. by Professor J. Willard Gibbs in 1873 and also, independently, by J. MacFarlane Gray in Britain; the latter brought it to the attention of engineers in discussions on steam engine trials in 1888 and 1889. Gray points out that one form of the Second Law is that each point on a temperature/entropy diagram is a characteristic state point. The diagram shows both the heat received and that rejected and also the work done, thus providing a valuable supplement to the pressure/volume diagram.

That the understanding of entropy has not always been as clear as it is today may have been due to a lack of precision about the terms in the First Law equation. The molecular approach led to some confusion, for sometimes heat was regarded as energy of internal motion, not different from internal energy. Work could also be both internal and external. In this century, consistent definitions have been given of heat and work with internal energy arising as a property from the First Law. With this greater clarity, a better understanding has been gained also of the Second Law and of the way in which the entropy function arises from it.

Conclusion

It has been shown how thermodynamics arose out of reciprocating steam engine practice during the last century. The analytical power of thermodynamics soon transcended the limited needs of mechanical engineers and was taken over into the pure sciences, but its origins were in technology.

In this century increases in steam pressures and temperatures have necessitated a knowledge of steam properties over an increasing range. Similar advances have occurred in our knowledge of the properties of the working fluids in internal combustion engines. There has been the development of internal combustion machines with an emphasis on high power/weight ratios. This in turn has underlined the importance of irreversible dynamic processes and brought thermodynamics into closer relationship with fluid mechanics and the chemistry of fuel reactions.

These new techniques have set new requirements for thermodynamics, just as the steam engine provided the need for the science itself.

REFERENCES FOR FURTHER READING

1. Carnot, S., *Reflections on the Motive Power of Fire*, editor E. Mendoza Dover Publications Inc., New York, 1960.
2. Joule, J. P. *Scientific Papers*. Taylor and Francis, London, 1884.
3. Partington, J. R., *A History of Chemistry*, vol. 3. Macmillan, London, 1962.
4. Joule, J. P. *Joint Scientific Papers*, p. 215. Taylor and Francis, London, 1887.
5. Thomson, W. *Mathematical and Physical Papers*, vol. 1. Cambridge University Press, Cambridge, 1882.
6. Crowther, J. G. *British Scientists of the Nineteenth Century*. Routledge and Kegan Paul, London, 1935; paperback, 1962.
7. Rankine, W. J. M. *Miscellaneous Scientific Papers*. Griffin and Co.. London, 1881.

THE GREAT MASTERS

Macquorn Rankine—A Founder of Thermodynamics

by E. A. Allcut, M.Sc., M.E., F.R.Ae.S. (Member)

Considered by Maxwell to be one of the three founders of thermodynamics, Rankine was also one of the pioneers of academic engineering education. Unlike many engineers today he was extremely versatile in his interests, being a competent poet, classicist, linguist and musician. If the portrait drawn in this article seems too good to be true it is because no one appears to have left an unkind word about him on record.

During the nineteenth century and in the early years of the twentieth, there was little specialization among scientists and scholars, so that most of the professions were dominated by men who were essentially versatile and imaginative individualists.

They might be described as 'personalities' or 'characters'.

In marked contrast personality today has become submerged in collectivism and individuals have been replaced by committees or boards. Conformity has become more important than ability and the probable ultimate result will be the control of society by electronic robots, fed with data by quasi-human beings of similar characteristics.

The early engineers, such as Watt, the Stephensons, Nasmyth, Ericsson and Perkins, were interested in many things that were outside the range of their normal professional duties and it seems probable that the mental flexibility thus engendered contributed materially to their professional development and prestige. On the other hand, the modern cult of specialization may be expected to produce mental stagnation and narrowness of outlook.

It is appropriate, therefore, that engineers should pay homage to the memory of W. J. M. Rankine, who was a classicist, a linguist, a musician, a poet, a humorist and an educationalist. Certain principles have been associated, traditionally, with the names of outstanding men and, in some respects, this is unfortunate, because people are liable to forget that Ampere and Volta were men, not units, that Osborne Reynolds was not a number and that Carnot and Rankine were not cycles. It seems very probable, however, that their scientific and technical achievements were due, in large measure, to their personal characteristics and early training.

From medieval times, it had been considered that philosophy and reasoning, classed among the liberal arts, was studied by the intellectual elite; and that thinking based on observation and practical experience was associated with the less reputable classes.

This attitude, not yet extinct, fostered an intellectual snobbery which delayed, for many years, the recognition of engineering as an academic course of study. It was criticized by Rankine in his inaugural lecture as Regius Professor of Civil Engineering at the University of Glasgow, in 1855. Traditionally, the inaugural lecture was given in Latin: the title chosen by the new Professor was *De concordia inter scientiarum machinalium contemplationem et usum*, which may be freely translated *On the agreement between theory and practice in the mechanical sciences*. He pointed out that "In theoretical science the question is 'What are we to think?', but in practical science the question is 'What are we to do?'."

Actually, the differences between the two are those of degree, rather than kind. The so-called pure sciences are concerned with the search for ultimate truth, irrespective of the factors of time, space, facilities or budgets. Engineers, on the other hand, are concerned with the best approximation to ultimate truth that can be achieved with the facilities obtainable, in the time allowed and with the amount of money that can be extracted from a reluctant administration. Rankine's persistence in this line of thought and the cogency of his arguments finally persuaded the Senate of the University of Glasgow that he was right and the degree of B.Sc. was established in 1872, the year of his death.

The poet

But what kind of man was Rankine? Probably the best description of his personal characteristics is given in a memoir by Professor P. G. Tait, who was a close personal friend and was the first Professor of Physics at the University of Edinburgh:

> 'His appearance was striking and prepossessing in the extreme and his courtesy resembled almost that of a gentleman of the old school. His musical tastes had been highly cultivated and it was always exceedingly pleasnt to see him take his seat at the piano to accompany himself as he sang some humorous or grotesquely plaintive song . . . words and music alike being generally of his own composition'.

The following sample of his serious verses reflects his early associations with Ayrshire, but when he wrote them his eyesight had failed, so that he could not read. The bridal tour described therein was strictly imaginary, for Rankine never married.

Come busk ye braw, my bonnie bride,
And hap ye in my guid gray plaid,
And ower the Brig o' Doon we'll ride
Awa' to Carrick Hills, love.

For there's flowery braes in Carrick land,
There's wimplin' burns in Carrick land,
And beauty beams on ilka hand
Amang the Carrick Hills, love.

There dwalt my auld forefathers lang,
Their hearts were leal, their arms were strang,
To thee my heart and arm belang
Amang the Carrick Hills, love.

I'll bear thee to our auld gray tower,
And there we'll busk a blythesome bower,
Where thou shat bloom, the fairest flower
Amang the Carrick Hills, love.

In spring we'll watch the lammies play,
In summer ted the new mown hay,
In harvest we'll sport the lee-lang day
Amang the Carrick Hills, love.

When winter comes wi' frost and snaw,
We'll beet the bleeze, and light the ha',
While dance and sang drive care awa'
Amang the Carrick Hills, love.

His attitude toward love, however, was not always sentimental; sometimes his treatment was humorous and analytical. This approach is exemplified in the poem which follows, called *The Mathematician in Love*:

A mathematician fell madly in love
With a lady, young, handsome and charming;
By angles and ratios harmonic he strove
Her curves and proportions all faultless to prove,
As he scrawled hieroglyphics alarming.

He measured with care from the ends of a vase,
The arcs which her features subtended;
Then he framed transcendental equations to trace
The flowing outlines of her figure and face,
And thought the result very splendid.

He studied (since music has charms for the fair)
The theory of fiddles and whistles,
Then composed by acoustic equations an air
Which, when t'was performed, made the lady's long hair
Stand on end like a porcupine's bristles.

The lady loved dancing . . . he therefore applied
To the polka and waltz an equation;
But when to rotate on his axis he tried,
His centre of gravity swayed to one side
And he fell by the earth's gravitation. etc. etc.

It is interesting, but perhaps not profitable, to speculate on the possibility that these verses may have been inspired by some abortive love affair of which we have no record.

Rankine was a good conversationalist; it mattered not whether his audience were comprised of children or scientific pundits, he could entertain any of them. For friends who needed help, he would draw willingly on his vast fund of experience, particularly in those cases where he had some definite and specialized knowledge of the matter at issue.

His practical outlook is suggested by Clerk Maxwell, who wrote:

> 'Whatever he imagined about molecular vortices with their nuclei and atmospheres was so clearly imagined in his mind's eye that he, as a practical engineer, could see how it would work . . . there was no danger of his going on to explain natural phenomena by any mode of action of the machinery which was not consistent with the general laws of mechanism'.

Early background

William John Macquorn Rankine was born in Edinburgh on July 5th, 1820. He was profoundly attached to his parents and one of the most touching notes in his journal is the brief record of his lasting obligations to them for early instruction in the fundamental principles of the Christian religion and the character of its Founder.

His father, David Rankine, was a lieutenant in the Rifle Brigade and was said to be descended from Robert the Bruce, Earl of Carrick. Ultimately, he became a railway engineer and Secretary of the Caledonian Railway Company. William's mother, Barbara Graham, was related to Thomas Graham, who did significant work on diffusion in fluids.

As a boy, young Rankine briefly attended Ayr Academy and Glasgow High School but, on account of poor health, much of his early instruction in mathematics and physics was received at home, from his father. When he was fourteen years old, his uncle gave him a copy of Newton's *Principia*, which he read in the original Latin and which had a profound influence on his mental development.

He entered Edinburgh University to study the sciences at the early age of sixteen and, in his first year, was awarded the gold medal for an essay on *The Undulatory Theory of Light*. In the following year, he received a special prize for another essay on *Methods in Physical Investigation*. For some years he assisted his father in railway work and, in 1838, he became a pupil of John McNeil, and did water, harbour and railway work for him in Ireland. His experience on railways led him to contrive a method of setting out curves which was named after him. His first published pamphlet is on railway wheels, but his most outstanding contribution was a paper on the *Fracture of Axles* in which the propagation of cracks by concentration of stresses at sharp changes of section was first explored.

He also studied molecular physics and, in 1849, he wrote a paper on *The Theory of Molecular Vortices* which placed him among the founders of the science of thermodynamics. In the following year, he was Secretary of Section A of the British Association and he presented a paper on the *Laws of the Elasticity of Solid Bodies*. Two years after that, he speculated on *The Reconstruction of the Mechanical Energy of the Universe* (. . . a theoretical investigation regarding the possible action of natural forces in preventing the 'running down' of the universe).

In the year 1842, Queen Victoria paid her first visit to Scotland and Rankine was able to put his practical experience with combustion to good use in designing and building a large bonfire on Arthur's Seat, which overlooked Holyrood Palace. He is reported to have used radial passages under the fire to distribute the air supply.

Thermodynamics

Most engineers, particularly engineering students, associate the name Rankine with the thermodynamic cycle that bears his name. Regnault had shown already that, when steam expanded without doing work, it became dryer. Rankine predicted that when steam expanded adiabatically it would become wetter and, if initially dry, would follow the law $PV^{1.11}$ equals a constant. He also prepared for his students the first practical table of the densities of saturated steam and, using Joule's work, he made a remarkably accurate assessment of the specific heat of air.

Carnot had already established the criteria for a perfect heat engine cycle, which is represented by the area *ABCD* in the illustration. The thermal efficiency of this cycle is the ratio of the work done to the heat used and this is represented by *ABCD*/*EBCF*. However, in the case of a vapour, it is impracticable to compress adiabatically the mixture of liquid and vapour at *A* to produce saturated liquid at *B*. The Rankine cycle recognizes this fact and condenses all the vapour to a saturated liquid at *G*.

Neglecting the work done in the feed pump, the simple Rankine cycle is represented by the area *GBCD* and the heat supplied is represented by the area *HGBCF*. Therefore, the thermal efficiency of the simple Rankine cycle is *GBCD*/*HGBCF*. Thus, although the work done per pound of fluid in the Rankine cycle is greater than that in the Carnot cycle, the efficiency of the former is less, because the heat lost in the process has been increased to a greater extent. The additional work done is indicated by the triangular area *GBA*, but the heat lost to the cooling water has been increased by the rectangular area *GAEH*.

For many years, this was considered to be the standard of comparison for steam engine cycles, but it has ceased to be the criterion for modern steam plants, most of which employ progressive feed water heating, which increases the possible thermal efficiency. Rankine also pointed out that, theoretically, the same standard could be used for simple and compound engines.

It is recorded that, in collaboration with James R. Napier, he designed and patented an air engine in 1853 but apparently this was never built. It seems likely that the project was inspired by the pioneer work of Ericsson who, in the previous year, had installed a 300 h.p. air engine in a ship, and who had designed many such engines for other purposes.

Hydraulics

Glasgow was, and still is, an important shipping centre and it is only natural that Rankine's talents and experience should have been used extensively in the solution of marine problems. He investigated the forms of waves and the motion of water flowing past ships. He estimated the power necessary to propel ships of various sizes and speeds and deduced a general formula which, because of its confidential nature, he communicated to Napier in the form of an anagram. Its interpretation was:

> 'The resistance of a sharp ended ship exceeds the resistance of a current of water of the same velocity in a channel of the same length and mean girth, by a quantity proportional to the square of the greatest breadth, divided by the square of the length of the bow and stern.'

This explanation is faintly reminiscent of *A Mathematician in Love*.

Among his published works are *A Treatise on Shipbuilding*, which was published in 1866 and a memoir on James Elder, which he finished in 1872. He joined the Institution of Naval Architects in 1862 and contributed several papers.

His civil engineering experiences had impressed upon Rankine the importance of an abundant and pure water supply for civic use and, accordingly, he proposed to supply the city of Edinburgh with water from a lake in the hills but, in face of opposition from the local water company, the project was abandoned. Nevertheless, after he moved to Glasgow and founded the firm of Rankine and Thomson, he took over and completed an existing scheme for supplying that city with water from Loch Katrine.

All scientists and research workers, in common with engineers, are interested in questions of measurement and, therefore, it is not strange that Rankine should be concerned with the relative merits of the British and Metric systems. This motivated a satirical poem *The Three Foot Rule*, which he sang to his own accompaniment at a British Association meeting in Bath, in 1864. It is not on the record as to whether or not he also wrote the air.

When I was bound apprentice, and learned to use my hands,
Folks never talked of measures that came from foreign lands;
Now I'm a British Workman, too old to go to school,
So whether the chisel or file I hold, I'll stick to my three-foot rule.

Some talk of millimetres and some of kilograms,
And some of decilitres, to measure beer and drams;
But I'm a British Workman, too old to go to school,
So by pounds I'll eat and by quarts I'll drink and I'll work by my
three-foot rule.

A party of astronomers went measuring of the earth,
And forty million metres they took to be its girth;
Five hundred million inches, tho', go through from pole to pole;
So let's stick to inches, feet and yards and the good old three-foot rule.

The great Egyptian pyramid's a thousand yards about;
And, when the masons finished it, they raised a joyful shout;
The chap that planned that building, I'm bound he was no fool,
And now't is proved beyond a doubt he used a three-foot rule.

Here's a health to every learned man, that goes by common sense,
And would not plague the workman by any vain pretence;
But, as for those philanthropists who'd send us back to school,
Oh! bless their eyes, if they ever tries to put down the three-foot rule

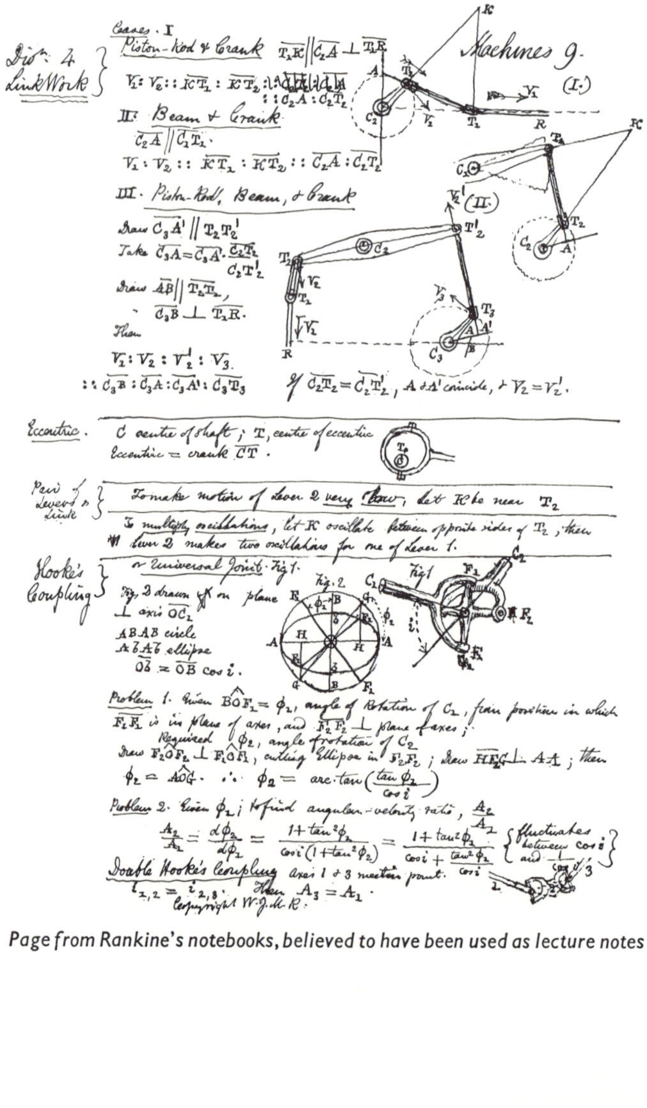

Page from Rankine's notebooks, believed to have been used as lecture notes

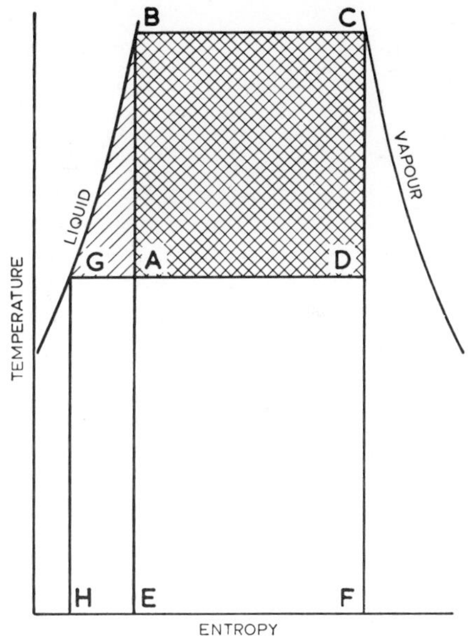

Carnot had established the criteria for an ideal heat engine cycle but Rankine's cycle takes into account the impracticability of the adiabatic compression of liquid and vapour mixtures. The simple Rankine cycle is represented by the area GBCD

In 1869, it is recorded in Rankine's journal that he applied to the Chancellor of the Exchequer for appointment as Master of the Mint. This followed the death of the Master, Thomas Graham, who was a distant relative of Rankine's mother. The appointment was not made because, subsequently, the office was virtually abolished, being conjoined with the Chancellorship of the Exchequer.

Engineering education

The influence of Rankine on scientific and technical education is described in detail in a monograph by Professor J. Small.[2] The first professorship of Engineering at Glasgow University was established in 1840 and the Regius Chair was then occupied by Professor Lewis Gordon. The establishment of a chair in Engineering was attacked, both on the ground that it was not a scholarly pursuit, and because, at that time, the system of Royal patronage was under fire. In 1855, Rankine became deputy to Professor Gordon and lectured on Applied Mechanics and on the Theory of the Steam Engine. The atmosphere then existing in academic circles is indicated in a motion placed before the University Senate, which is quoted in Professor Small's paper as follows:

> 'That, notwithstanding the high talents and acknowledged ability of the late Professor, the number of students attending the class of Civil Engineering had been small and that the endowment might, with propriety, be devoted to purposes of greater and more immediate importance in connection with the Higher branches of University Education'.

However, in 1855, Rankine was appointed to the chair, without dissent.

It is possible that this happy outcome was facilitated by the prestige of the Thomson brothers for, in 1846, William had already been appointed as Professor of Natural Philosophy. Ultimately he became Lord Kelvin. His brother James also was a well-known engineer; between them, they probably assisted materially in getting recognition for the profession in academic circles. Some opposition was expressed by other engineers, who considered that proficiency in engineering could only be achieved by practical apprenticeship.

At that time, Engineering was attached to the Faculty of Arts, but was not recognized as a degree course, being classed among the supplementary studies. Rankine dissented sharply from this view and his acknowledged eminence in the scientific field was probably a large factor in obtaining the approval of the Senate for the award of a Certificate of Proficiency in Engineering Science. The first of these certificates was awarded in 1863 and, nine years later, largely through Rankine's influence, the degree of Bachelor of Science was established.

Rankine received many honours and awards. In 1853, he was made a Fellow of the Royal Society and, later, he received the honorary degree of Doctor of Laws from the University of Dublin. From 1860 to 1862, he was President of the Philosophical Society of Glasgow and, when the British Association met in that city, he was President of the Engineering Section.

He was a prolific writer; he presented over 150 papers to various scientific societies. Four of his books have become classics in their respective fields. They are *Applied Mechanics* (1858), *The Steam Engine* (1859), *Civil Engineering* (1862) and *Machinery and Mill Work* (1869).

Because of the variety of his interests and his father's connection with the army, Rankine took a prominent part in the Volunteer movement, which was organized (circa 1859), to supplement the regular army and the militia, in case of invasion from the Continent. Its motto, accordingly, was "Defence, not Defiance" and it continued to exist until the early years of the twentieth century, when it was replaced by the Territorial Army. Volunteer training corps were formed at the Universities and Rankine became Captain of the First Company at Glasgow University. In 1860, this became part of the First Lanarkshire Volunteers. The Captain of the Second Company was William Thomson. At a later date, Rankine became Major, but ultimately, he had to resign because of the pressure of other work.

Tributes

Perhaps, the most important proof of merit in any professional man is the esteem of his contemporaries and successors and there is abundant evidence of this in the case of Rankine. Clerk Maxwell classed Rankine with Clausius and Kelvin as one of the three founders of theoretical thermodynamics and he stated further that Rankine had availed himself to the greatest extent of the scientific use of imagination. Professor Lewis Gordon, whom he succeeded at Glasgow, also paid tribute, both to his standing as an engineer in Britain and to his high reputation in other countries. Tait said that it was not easy to speak of the man himself in terms which, to a stranger, would appear unexaggerated and that he was the soul of honour in giving credit to others.

Sir Richard Southwell quotes Karl Pearson in praise of Rankine's contributions to terminology by his introduction of precise definitions of such useful names as 'stress', 'strain' and 'potential energy'. He also repeats Lord Rayleigh's opinion to the effect that Rankine's paper *On the Thermodynamic Theory of Waves of Finite Longitudinal Disturbance* was a very remarkable investigation, which had been much neglected.

His death at the comparatively early age of 52 was a great loss to science and to the community; but into the short space of 36 years he had crammed a degree and variety of achievement which it would be difficult for any other man to equal in twice the time. Of him, it may be said with truth, 'Age cannot wither him, nor custom stale his infinite variety".

REFERENCES

1. *Miscellaneous Scientific Papers of Mcquorn Rankine* by Tail and Miller (Chas. Griffin & Co., 1881).
2. J. SMALL, The Institution's First President, William John Mcquorn Rankine, *Trans. Instn Engrs Shipb. Scot.*, 1957.
3. Ten British Physicists by A. Macfarlane, John Wiley & Sons, Ltd.
4. *ASME News*, American Society of Mech. Engineers.

THE GREAT MASTERS

Greatly influenced by his self-taught, single-minded father, young Thomson spent almost a lifetime as a university professor but his heart was in applied science and engineering. Fortunately he was also a pioneer of boardroom boffins, eventually forming his own firms to pursue his ideas. Many of his contributions to science and engineering have proved to be of permanent value.

Kelvin of Largs—Natural Philosopher

by James A. McCash, B.Sc.(Glas.), M.I.C.E., M.I.E.E. (Member)

[Courtesy: J. M. Dentand Co.]

Professor William Thomson at the age of 22

In 1903, the University of London conferred its first Honorary Degrees. The four recipients were the future King George V and Queen Mary, Lord Lister and Lord Kelvin. Of Lord Kelvin, the Dean of the Faculty of Science said:

> . . . for more than half-a-century [he] filled the office of Professor of Natural Philosophy in the ancient University of Glasgow. He has promoted true ideas concerning the conservation of energy, the laws of thermodynamics and their application to the mechanics and physics of the universe. He has enquired into problems interesting to the chemist and geologist and important to the physicist and engineer. He has calculated the probable size of atoms; he has studied the structure of crystals; he has estimated the age of the earth. The world knows him best as the man who has shown how practically to measure electric and magnetic quantities. A greater philosopher than Democritus, in him are united the qualities of Archimedes and Aristotle.

His engineering memorials include his 1851 papers *On the Dynamical Theory of Heat*, the second and third 1866 Atlantic Telegraph Cables using his mirror galvanometer and later his siphon recorder; the nautical 'flying' sounding-machine and depth gauge; the tidal harmonic analyser and tide predictor; together with some seventy patents and his still extant firm of scientific instrument makers, Kelvin and Hughes Ltd.

At his funeral in Westminster Abbey his pall-bearers included Lord Rayleigh, President of the Royal Society, and the Presidents of the Royal Society of Edinburgh, the Geological Society, the Institution of Civil Engineers, the immediate Past-President of the Institution of Electrical Engineers (of which Kelvin had been President for the third time when he died), and M. Gaston Darboux, Perpetual Secretary of the *Académie des Sciences*.

Erudite family

William Thomson (1824–1907) was the fourth of seven children (three of them girls), born to Dr James Thomson of Belfast and his wife Margaret Gardner of Glasgow. Besides the father, Professor of Mathematics for 17 years, his brothers James, F.R.S., Professor of Engineering for 16 years, William, F.R.S., John, and his nephew James Thomson Bottomley, F.R.S., were to achieve distinction in the University of Glasgow, an association which lasted over 125 years.

Dr James Thomson was the youngest son of a farmer of small means but of granitic Scottish Covenanting ancestry, settled in Ireland since about 1650. James was born in 1786 and died in 1849. Beginning as a labourer on his father's farm and largely self-taught, he was probably unable to find the cost of class fees at the 'normal' matriculating age of 14. But, after becoming a teacher at the local Presbyterian Church school where he had been a pupil, he did matriculate at Glasgow University. Until 1814 he taught eight hours daily each summer to earn his keep and to pay for his winter expenses.

At this date Queen's College, Belfast, did not exist; a conveniently near and, to many Ulstermen a most congenial, university was Glasgow's. A *studium generale* for nearly 400 years, it had been a centre of learning with a relatively liberal constitution: it had almost no religious test for its graduates, unlike Oxford, Cambridge and Dublin. During the 18th century it had become highly regarded, both for its pragmatic teaching and its low fees.

When James was a student at Glasgow, the university's former instrument-maker, James Watt, was still at the height of his fame. Transport from Belfast was cheap, though perhaps tedious. Kelvin related how his father would travel in lime-carrying sailing smacks which sometimes took three days, and would walk the 25 miles from the port of Greenock.

In 1815, shortly after graduating M.A., James was appointed Professor of Mathematics and Lecturer in Geography in the Royal Academical Institution, Belfast, quickly building for himself a national reputation. By 1820 he was living in a house he had built in College Square, Belfast, and was working each day from 4 a.m., the evenings being dedicated to his family. He published much and his labours were rewarded by an honorary LL.D. of Glasgow in 1829, followed in 1832 by an invitation to become Professor of Mathematics there.

Infant prodigy

Both young James and William were born in the new house and their initial education was undertaken there by their father, who imposed an exceptionally severe discipline, even by early 19th century standards. When James was five and William only three years old they received a daily course of instruction during their constitutional walk before breakfast, when 'darling papa' would invent interesting topics of conversation carried on through successive mornings, and each evening after dinner they joined a pair of terrestrial and celestial globes on the table to study geography and astronomy. William is recorded as having had a fair competence in both subjects by 1830, when he was six.

There is today much curiosity about the early childhood of people who have been in the public eye. The psychologist demands of the nursery a full explanation of character, and subsequent behaviour. Whether or not this bears fruit, Kelvin's childhood merits some detailed description.

Kelvin's much-loved mother died when he was in his sixth year. She had borne three more children after him, two of them invalids. Her only sister, Aunt Agnes, ran the household for the family, from the time Kelvin was four years old until some years after his father's death. There was an interval of about three years when she married and went abroad with her dissolute husband whom she soon deserted, returning to Dr James's home and resuming as mistress of the house when Kelvin's sisters married. When he himself married, Kelvin gave her a house at Largs and she lived there happily enough, surrounded by her own and Kelvin's grand-nieces and grand-nephews until she died: Kelvin built a house for himself and his second wife nearby some 12 years later.

The obvious and clearly dominant family influence in Kelvin's childhood and adolescence was his father's. The two seem to have had an unusual affinity for each other. Willie usually lay on his father's knees when the family was gathered round the fire for an hour's reading or story-telling by the father, before going up to bed. When his father came home from his lectures, Willie would run to him and jump about him like a little dog, and then kiss him.

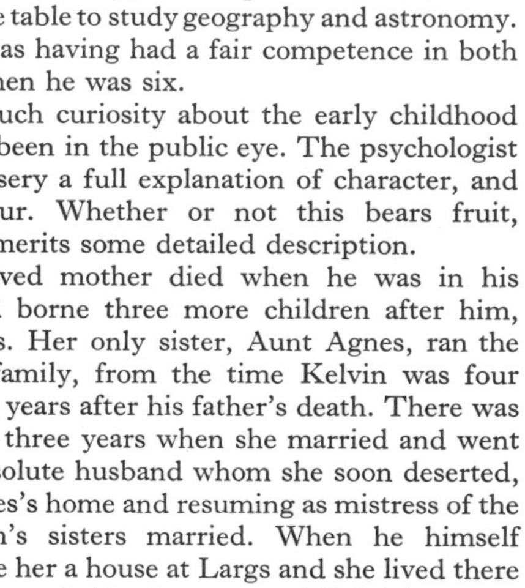

Kelvin's birthplace in College Square, Belfast: the creeper-covered houses were built by his father and at one time stood alone, facing the hills. The artist was Kelvin's niece

[Courtesy: Macmillan and Co.]

As a child he was easily irritated if crossed, but the irritation never lasted very long. After the death of his mother he developed a tendency to sleepwalking. The greatly-sorrowing father took the two boys into his bedroom to sleep and the sleepwalking soon passed.

Apart from about a year they spent at the Institution in Belfast, he educated them himself, adding Latin and Greek to their curriculum. They had history lessons from Aunt Agnes and read Voltaire's *Peter the Great* and *Charles XII* in the original. For light reading they had Maria Edgeworth's *Moral Tales* and Caesar's *Commentaries!* After dinner the family had a short mathematics lesson, followed, for example, by Pope's *Iliad* and *Odyssey*, plays of Shakespeare, Goldsmith or Sheridan, or selections from various poets.

In 1832, when James became a member of the university senate, he took up official residence there and the family moved to Glasgow when young James was ten and William eight, they were allowed to attend their father's lectures and, on one occasion, William begged his father's permission to reply to a question which no matriculated student present could answer.

Two years later, in 1834, the two boys matriculated, William being all of ten years old.

In accordance with the contemporary procedure for the degree of *Artes Liberales*, they received a solid grounding in classical languages and in the physical and mathematical sciences. In later life, Kelvin insisted that the study of logic was essential to any scientist. Natural Philosophy was included in Divinity courses in Glasgow and he attempted also to enact that every Arts Degree must include Natural Philosophy.

Holiday reading

He had acquired a reading knowledge of French and German by the age of twelve. During the summer of 1839, his father took the four boys to Paris where he left them for two months to improve their knowledge of conversational French.

In the course of a fortnight's study in the library of the *Bibliothèque Royale*, William read Laplace's *Méchanique céleste*, the five volume treatise on the nebular theory of the solar system. Young Thomson used the insight gained from it in an essay on *The Figure of the Earth* which won him a University Medal; he was to return to the subject intermittently for nearly 50 years.

In 1840 the family went to Frankfurt-am-Main. Despite his father's instruction that all other work should be left behind and all available time devoted to learning German, young William took with him Fourier's *Théorie Analytique de la Chaleur*. Later he recorded,

> I had become filled with the utmost admiration for its splendour and poetry . . . On the 1st of May, 1840, I took Fourier out of the University Library and in a fortnight I had mastered it—gone right through it.

He re-read it surreptitiously with a view to refuting some adverse criticism of it. His first paper was thus written before his 16th birthday.

Having completed his Glasgow A.B. course before the age of 17, Thomson entered as a student of St Peter's College, Cambridge, in April 1841. The journey there took three days by mail coach. The funds for the Cambridge course were raised with difficulty but in 1845 he was elected to a Fellowship of £200 p.a. and acquired several pupils. The same year brought him a Second Wranglership, a First Smith's Prizemanship, the Colquhoun Silver Sculls, and recognition as the best mathematician of his year.

He formed a life-long friendship with Sir George Gabriel Stokes, shortly to become Lucasian Professor of Mathematics. He also became a founder member, in 1843, of the Cambridge University Musical Society, his instrument being the French horn, and his 'gods', Beethoven, Bach, Mozart, Mendelssohn and Weber.

During his four years at Cambridge, Thomson published sixteen papers, mostly in the *Cambridge Mathematical Journal*, of which he became editor. The subjects included electrical theory, which led to much correspondence with Faraday. They met at the 1845 conference of the British Association.

During his sojourn in Cambridge William kept a mathematical notebook in which he recorded ideas for later consideration. A typical example reads:

> Aug. 14 (1844) Gourock—This morning the following theorem has occurred to me. I hope it has not been proved by Gauss, and if not, it is highly improbable that anyone else has. . .

His work as a teacher

In May 1846 the Chair of Natural Philosophy at Glasgow fell vacant. The electing Faculty had young Thomson in mind for the position but were apprehensive that his lectures might be above the heads of his students. There were six applications and the decision was to be made in September.

Thomson had spent the spring of 1845 in Paris, attending meetings at the *Académie des Sciences* and working under young Victor Regnault, then Professor of Physics at the *Collège de France* on the thermal properties of steam, the density of gases and the determination of latent heat. Regnault joined G. G. Stokes and George Boole, among many others, in recommending Thomson to the Glasgow Faculty and he was unanimously elected, shortly after his 22nd birthday. His father did not live long to enjoy William's success for he died of cholera in January 1849.

When he was twenty-eight, Thomson married Margaret Crum of Thornliebank, Glasgow, his cousin and childhood friend. His wife, an invalid at the time of their marriage, died childless in 1870.

He married again in 1874, following his acquaintance with the family of Mr C. R. Blandy of Madeira at whose villa he stayed during the laying of the Western and Brazilian Co.'s cable in 1872. Frances Anna Blandy, later Lady Kelvin, is described by her niece as of "kindly qualities and a sympathetic nature". She survived Kelvin but there were no children of this marriage either.

Thomson occupied the Glasgow Chair for fifty-three years, refusing on three occasions the, at that time, less well endowed and equipped Cavendish Chair of Experimental Physics at Cambridge. Eventually he resigned to concentrate on his theories of light. From 1904 till his death in 1907 he was Chancellor of the University of Glasgow.

Young Thomson felt very strongly that science teaching meant laboratory work. At that time there was no physical laboratory in any university for the systematic teaching of students. Victor Regnault at the *Collège de France* had a private *cabinet de physique* where he maintained apparatus for personal research.

Thomson visited Paris in 1845 and went to see Cauchy, the greatest of contemporary French mathematicians. The initial meeting was cordial but a failure, as Cauchy spent much of the time trying to convert Thomson to Roman Catholicism. This proved difficult for, as a youth, Thomson caught laughing in church, had been personally warned by his Presbyterian minister, "Ye'll no lauch when ye're in hell!"

The inner court of Old College at Glasgow: the Natural Philosophy rooms were in the corner to the right o, the turret and Kelvin's laboratory to the left on the ground floor

[Courtesy: J. M. Dent and Co.

Thomson was allowed by Regnault to join him daily for three months as one of the few research assistants in his laboratory. In later years William said that Regnault had taught him "a love of precision in all things, and that highest virtue of the experimenter—patience".

The result of the difficulties he had encountered in obtaining experience in experimental physics was that he established, in an annexed professorial wine-cellar, the Glasgow Physical Laboratory which he claimed to be the earliest physical laboratory anywhere for use by students.

In 1846 Thomson gave his introductory lecture on Natural Philosophy, beginning with a prayer, which was to remain his practice throughout his teaching life. The lecture, too, retained the same outline for 50 years thereafter; he began by indicating the procedure adopted in deriving systematic knowledge of natural science by deduction from the observation and classification of data—the Natural History stage—followed by the Natural Philosophy stage, until the assembled information satisfied a general law, to be reinforced by further facts found in systematic experiments and observations. The lecture ranged from astronomy through dynamics, to optics, geology, heat, electricity and magnetism.

In his early days, Thomson was embarrassed by having students older than himself but he was none the less popular with them, probably because of his fast-rising fame, rather than the assistance he gave them in passing their examinations. Half a dozen medical ancestors of mine gained an elementary knowledge of physics at his feet and their concensus was that they were glad to have been his students but that they could write their conjoint knowledge of Nat. Phil. on the back of a postage stamp. He frequently made discoveries while lecturing and pursued his ideas after class was over.

Students, particularly engineers, can imagine the joy of their professor illustrating acoustics with the aid of his own French horn (with a duster in the bell); and ballistics by firing his own Volunteer's rifle into a lead pendulum. The latter was still a demonstrator's adjunct in the 1920's when I was a student at Glasgow.

Similarly, his classes in mathematics for engineering students were something of a trial to his audience. "Do you know what a mathematician is?" he once asked and wrote on the blackboard

$$\int_{-\infty}^{+\infty} e^{-x^2}\,dx = \sqrt{\pi}$$

Turning to the class he said, "To any mathematician that is as obvious as that twice two makes four is to you!" Apparently he expected his engineers to know their mathematics before they came to his junior Maths class—"Mere sublimated common sense" he used to call it.

Ranking (with his close friends Clerk Maxwell and the Irishman Sir George Gabriel Stokes) as one of the three greatest mathematical physicists of the 19th century, he was also an exceptionally competent experimenter, who insisted on his students having practical experience. He wrote forty years later:

> I did all in my power to find for them in the Electro-dynamic Properties of Metals, Moduluses of Elasticity of Metals, Elastic Fatigue, Atmospheric Electricity, etc. I then had an ordinary class of a hundred students—including intending theological students. A surprised, notable German professor said
>
> "What! do the theologians learn physics?"
>
> I said, "Yes! they all do, I believe . . . their theology suffers not at all from their having learned something of mathematics and experimental physics.

Nowadays we might consider it a gross waste of scientific genius to make Kelvin lecture when he might have been much better employed on pure research but without his Chair he would have had no laboratory. With the aid of this he demonstrated that scientific knowledge was not something fixed once and for all by Aristotle or even Newton but an expanding universe; and its acquisition an essential component of a liberal education, like classical languages and logic.

[Courtesy: Glasgow University]

The Senate in strict order of precedence, leaving the Old College in 1870: Rankine (top) and Thomson are shown ringed

An enthusiast for wider education, he recommended in 1875 the enlargement of Owens College into Manchester University, and of Mason College into Birmingham University, adding, "I utterly differ from those who object to any moderate multiplication of universities throughout the British dominions, such as we have in Scotland." This has a topical ring to it.

The scientist

Silvanus P. Thompson considered Kelvin the greatest scientist of that golden age of science, the 19th century. Others consider Maxwell or Faraday superior physicists. Many of Thomson's mathematical papers, written before he had his own lab where he could verify his theories, nevertheless inspired and suggested the work of his contemporaries, including Faraday and Maxwell. The latter, shortly after graduating, wrote to Thomson for guidance in his reading of Ampère, Faraday and Thomson, correctly guessing that the latter had important unpublished work awaiting experimental proof. He received unstinting advice and remarked, "Thomson is glad that I should poach on his electrical preserves, including his method of Electric Images."

When Thomson, on the other hand, shortly after graduating, wrote to Faraday on electro-optic stress and on the physical state of dielectrics—anticipating Kerr's and Röntgen's work 30 years later—he received the non-committal reply that, he, Faraday, was proposing to resume experiments on these subjects: a further timid proposal from Thomson that he should call at Faraday's laboratory in the Royal Institution to learn more of his experiments was met by a short note that Faraday would be out of London.

They did meet frequently thereafter, but, though Faraday often said he could do with a colleague, he never asked Thomson—or anyone else—if he were interested. Faraday was in fact a lone worker and was possibly a bit put off by, if not actually mildly apprehensive of, Thomson's mathematical expertise. His own was an impirical approach.

Nevertheless, while by far the greater part of the specialized work of today's scientists is out-dated within a decade and the quality of their research seldom matches its quantity, Kelvin's massive output remains, like 19th century physics generally, the essential basis on which today's scientific and engineering achievements are built. To convey an idea of the width of his range, a few of his publications, selected from over 700, are listed in Table 1, not necessarily for their intrinsic value but for their topicality today, in both engineering and physics.

Kelvin missed being first with the Second Law of Thermodynamics by about a year, and his absolute scale of temperature was probably his most important contribution to thermodynamics. It derived from Carnot's concepts of an ideal cycle. Carnot himself had remarked in 1824 that the quantity of work done in his cycle by one unit of heat between any two temperatures was constant for all working substances, being a function of temperature on some absolute 'permanent' gas scale. Thomson later called this 'Carnot's Function'.

In 1848, Thomson reasoned that, since the work done per unit of heat per degree drop in temperature in a Carnot cycle was independent of the working substance, equal, calculable amounts of 'mechanical effect' could be used as uniform intervals in an absolute scale of temperature.

This was illustrated by Rankine in 1849 in his statement of the Second Law of Thermodynamics when he redefined temperature in terms of the work done in perfect engines.

Joule had proposed a definition but Thomson was determined that his proposed scale of temperature should be absolute in the scientific sense, i.e., independent of the physical properties of any substance.

For this he had to have the numerical values of the Carnot Function at different temperatures and the source of these was found in Regnault's recently-published memoirs. Using Regnault's experimental data for the specific heat of air at constant pressure, Thomson calculated and tabulated the Carnot Function at different temperatures. Thereafter he reverted to the Joule-Clausius proposal that the absolute temperature should be redefined in terms of the absolute zero based on a gas thermometer. The feasibility of this was demonstrated by the well-known Joule-Thomson 'porous plug' experiment, where the slight cooling effect due to molecular diffusion was shown to be negligible in most gases. The results of their work on the new scale were published in 1854 by Thomson and Joule.

In later years Kelvin was more interested in electrical problems and in the practical application of science.

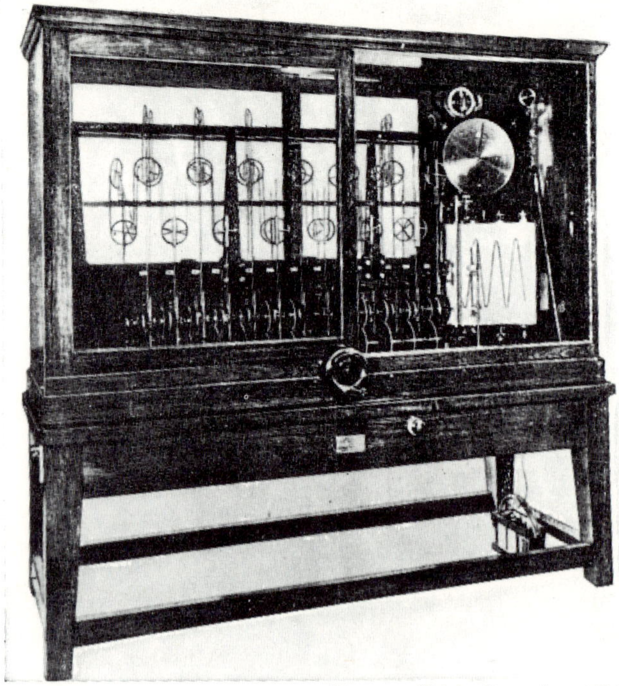

[Courtesy: Kelvin, Bottomly and Baird]

Kelvin's analogue computer for predicting the tides: the machine, designed in part by his brother, followed Kelvin's theory

The engineer

In 1902, the leading American electrical engineers, including Edison, Westinghouse, Nikola Tesla, and Elihu Thomson, on behalf of the American Institute of Electrical Engineers, gave a reception for Lord Kelvin at Columbia University. These men were primarily engineering industrialists, not given to lavish praise of pure scientists working in ivory towers: but Kelvin was a scientist at one and the same time pure, applied and approachable.

President of the Royal Societies of London and Edinburgh, and of the British Association, he was also President of the Institution of Electrical Engineers, of the Institute of Marine Engineers, Honorary Member of the Institutions of Civil Engineers and of Naval Architects, the Verband deutscher Elektrotechniker and some eighty other societies, about sixty of them foreign. He read papers to this Institution on several occasions.

In this eclectic approach to engineering and pure science Kelvin was—and probably remains—unique. He was ever insistent that all his scientific work should have immediate relevance to the social needs of the period; saying, "I see no use in a new discovery until it is applied to the use of mankind." He did not leave the translation of processes from test-tube into pilot plant to designers and craftsmen with the silly implication that 'rude mechanicks' could easily handle such development. He did both superlatively well himself as, for example, in the testing, laying and commissioning of the transatlantic cable. Yet it is for just this characteristic that he has been reproached: for not concentrating exclusively on fundamental work which might have enabled him, for example, to anticipate Rankine and Clausius.

[Courtesy: Illustrated London News]

Shipping the Atlantic Cable on board the 'Great Eastern' in 1865

Because it led Thomson away from scientific research and teaching to a career as a consulting engineer, this cable laying enterprise merits some detail. By 1850, electric telegraphy was very much in the news and, with Faraday's timely adaptation of gutta-percha to marine insulation in 1848, the stage was set for a transatlantic cable. The first submarine cable was laid in 1850 between England and France but parted shortly after. A second and eventually successful one followed in 1851.

In 1854 the American, Cyrus W. Field, obtained a concession for his Newfoundland Telegraph Co., and then joined with C. T. Bright, who had laid cables successfully in Ireland, and Dr O. E. W. Whitehouse, a retired physician and amateur electrician, to found the Atlantic Telegraph Co. of Great Britain in 1857. They obtained guarantees from the British and American governments and the loan of naval ships for cable laying. But the specification for the cable proved inadequate: the copper varied in quality and the insulation was too thin and often eccentric.

Thomson had previously proposed stranded conductors to avoid retardation of signals due to resistance combined with capacitance which, he showed, varied with the square of the cable length. The accuracy of this was queried by Whitehouse,

Table 1—Subjects of Topical Interest Investigated by Thomson

On the Economy of the Heating or Cooling of Buildings by means of Currents of Air	1852
On the Efficiency of Clothing for Maintaining Temperature	1884
On the Effects of Mechanical Strain on the Thermo-electric Qualities of Metals	1855
On Transient Electric Currents	1853
Electromagnetism as a Motive Power	1857
Electrical Traction on Railways	1902
Transmission of Energy by Direct Current	1907
On the Age of the Sun's Heat	1862
On the Internal Fluidity of the Earth	1872
The Age of the Earth	1895
On Safety Fuses for Electric Circuits	1884
Electrical Measuring Instruments to Measure Currents from 10^{-6} Milliampere to 10^{3} Amperes and to Measure Potentials up to 40,000 Volts	1887
Electrical Units of Measurement	1883
The Wave Theory of Light	1884
On the Possibility of Life on a Meteoric Stone falling on the Earth	1877
On the Living Cell	1904
On Electrically impelled and Electrically controlled Clocks	1866
On a New Form of Astronomical Clock with Free Pendulum and Independently Governed Uniform Motion for Escapement	1876
The Automatic Steering of Ships	1865
A Tide Calculating Machine, Tide-Gauge and Tide Predictor	1875
Problems relating to Underground Temperature	1860
Polar Ice-caps and their Influence in Changing Sea-levels	1888
The Microphone	1878
Towards the Efficiency of Screw-propellers in Aeroplanes	1894
The Size of Atoms	1883
Models of Radium Atoms to give out α and β Rays respectively	1904
An Attempt to Explain the Radio-activity of Radium	1907
The Problem of a Spherical Gaseous Nebula	1907
University of Wales; College of Bangor Laboratories—an Independent University of Itself	1885

who had been appointed the Company's chief electrician. Thomson was called in and his logic prevailed. He urged that the section should be four times as large and the insulation half as thick again.

This first cable parted during laying, which stimulated Thomson to write his paper *On Machinery for Laying Submarine Cables*. He designed his 'iron-clad' mirror galvanometer and paying out brakes for the cable. He also tested samples of the broken cable, laying down a new specification as a result of these. For developing signalling apparatus he

[Courtesy: Macmillan and Co.]

His last lecture at Glasgow was given in 1899: he was a popular teacher though he expected rather too much of his students

eventually received £500 and that only against a guarantee of a gruelling speed of three words per minute.

The cable was ultimately successful, though after great difficulty. Thomson had been on both cable laying expeditions but was obstructed by Whitehouse who defied the board and replaced Thomson's equipment by his own. Eventually Thomson was placed in complete control and Whitehouse sacked, but only after he had ruined the cable insulation by using induction coils at 2000 V. The cable was abandoned but the project continued.

In 1864 new capital was subscribed for a fresh attempt, Thomson serving on an advisory, scientific committee. The entire cable was loaded on Brunel's *Great Eastern*. After one cable parting in 1865, a third 1800 miles was laid and commissioned in July 1866. Signal speed was 14 words per minute, and Thomson received a knighthood for his part in the work.

Navigational aids

Thomson's engineering patents were largely electrical, as were his instruments, but there were very important mechanical and civil engineering applications. An example is his improved mariner's compass of 1876, for use in iron and steel ships. The previous version of the paper 'card', mounted above four needle magnets, was about fifteen times as heavy as Kelvin's, and it had a relatively short period of vibration. He doubled this to about forty seconds by replacing the solid card by a narrow engraved aluminium ring of about 10 in. diameter and relatively high moment of inertia which weighed less than 0·4 oz. Captain (later Admiral) Lord Fisher was responsible for the adoption of this compass by the Royal Navy as standard in 1889. It remained in universal use for about 40 years.

Another marine invention was the improved sounding machine of 1874. The traditional knotted rope held by hand was replaced by a heavy depth gauge attached to piano wire, motor-wound and automatically braked on touching bottom. The sinker was an air cylinder on which maximum piston travel was recorded by a pointer. By this means Kelvin—himself a competent yachtsman—was able to take soundings to 2700 fathoms, while travelling up to 14 knots.

Of great interest is his work on self-registering tide gauges, tidal analysers, tide-predicting machines and tidal phenomena themselves; the mechanization of the first three and elucidation of the fourth occupied him between 1867 and 1881. His harmonic method of tide analysis and prediction was partly based on theories by Newton, Laplace and Sir G. B. Airy.

Tide predictions had been fairly satisfactorily based on empirical observations for each port separately. Thomson, jointly with Prof. Tait, described a machine which was what we would now call an analogue computer. From accurate data on tidal heights, obtained by a self-registering tide-gauge, it would compute curves of tide tables for any future period and at any one of 140 ports.

Individual simple harmonic motions were generated by the uniform rotation of an adjustable crank, the vertical motion only being taken off the crankpin. Each tidal component and its characteristic amplitude were represented by their own gear and pulley, the pulleys all lying in the same vertical plane.

A cord, fixed at one end and carrying at the other end a stylus attached to a weight, passed round each pulley in turn. This arrangement of linked simple harmonic motions corresponded to the interaction of the tidal components, the result being recorded by the stylus on a moving drum.

Beliefs and character

Although his father had been a life-long radical and Republican, Kelvin himself died a Liberal Unionist. But he refused an invitation from his University to stand as a Liberal in 1869 because of the time involved.

Throughout his life, after adolescence, Kelvin held moderate, non-sectarian views. Yet, while attending church regularly on Sundays, Kelvin had no compunction about occupying Sunday afternoons with scientific calculation.

He considered Darwin's theory of evolution incapable of explaining creation and held it impossible "that atoms of dead matter should come together so as to make life". But he also held that, should such a thing be proved possible "we must not invoke an abnormal act of Creative Power to explain it".

Several friends remarked upon Thomson's gentle nature. He was always kind to younger men and fellow-workers, and his splendid modesty was widely admired.

Kelvin, for his time, travelled widely, from Canada to Russia and from the Bosporus to Brazil, everywhere demonstrating an unusual aptitude for making and keeping friends and greatly encouraging the cross-fertilization of ideas. In 1892 he received a peerage, the first scientist to be so honoured.

He died in 1907, completing a paper on his death-bed. Not long before he had written:

> One word characterises the efforts that I have made during fifty-five years; that word is failure. I know no more of electric and magnetic force, or of the relation between ether, electricity and ponderable matter than I knew and tried to teach to my students of natural philosophy fifty years ago in my first session as Professor.

Kelvin's fame will endure as a result of his work, both scientific and commercial, in thermodynamics and concise measurements. Probably his own choice of memorial, if any, would be today's world-wide trend in establishing industrial research establishments in which theory and practice are united.

He is commemorated by numerous institutions, including the new Kelvin Chair of Natural Philosophy at Glasgow; but perhaps most significant is a simple memorial inscription in Westminster Abbey — adjacent to Newton's.

[Courtesy: Hodder and Stoughton]

Kelvin at 81 was still very interested in experiment: here he is improving his compass

REFERENCES FOR FURTHER READING

1. Gray, Andrew, *Lord Kelvin. An Account of his Scientific Life and Work*. J. M. Dent and Co. (London): 1908.
2. Thompson, Silvanus P., *The Life of William Thomson: Baron Kelvin of Largs*. Macmillan and Co. Ltd, London. 1910.
3. Crowther, J. G., *William Thomson: Lord Kelvin. British Scientists of the Nineteenth Century*. Kegan Paul, Trench Trubner and Co. Ltd, London. 1935. (Paperback by Routledge and Kegan Paul Ltd, 1962).

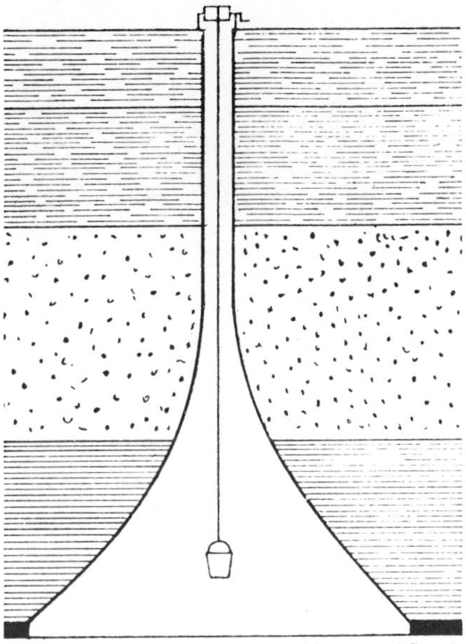

[Courtesy: Sir Isaac Pitman and Sons]

Fig. 1. Early mines were bell-shaped, widening within the ore or coal-bearing layer

Mining is one of the earliest technologies and its history is closely linked with that of mechanical engineering. It is from a 16th century mining manual that we derive most of our knowledge of mechanical engineering during the long period between the Roman Empire and the Industrial Revolution; and it is doubtful whether the latter would have taken its course, had not the steam engine been given its chance by the need for draining the ever-deepening mines.

Mechanical Engineering in Mining

by John Sinclair, M.Eng., Ph.D.

When tracing the history of the application of engineering to mining it is necessary to go back to the period of pre-history when dates are only approximate although progress has been made by the use of the Carbon-14 dating technique to measure the age of hoards.

One is inclined to wonder at the incredible achievements of the ancient predecessors of our mining engineers, considering their primitive knowledge and facilities. Excavations in Southern Turkey have revealed man as a city dweller in 7000 B.C., some three centuries before civilization began in Egypt. Technical competence in the manufacture of tools, weapons, textiles and jewellery was so high that it bears witness to a standard of living, seldom equalled in present-day Anatolia.

Metal mining preceded coal mining but the practices, problems and processes of both are similar.

Flint mining

The earliest miners were interested in flints from which they made their tools. They probably belonged to the middle stone age (Mesolithic) about 10 000 B.C. while the later pits were sunk during the Neolithic which ended about 2000 B.C. The early primitive pits were smaller and less symmetrical than the later Neolithic pits, and no galleries were driven from them. They were extended in diameter when they reached the superior third flint layer to allow the flint to be mined, as shown in Fig. 1.

The early coal pits some 9000 years later were of the same type. At Hexham in Northumberland, for example, the shallow bell pits were 8 to 10 yards apart with only a very small pillar of coal left between them at the bottom.

The picks and wedges of the early miners were of red deer antlers, the shovels the shoulder blades of deer or wild oxen. In the later pits as many as two dozen galleries would be driven laterally outwards to increase the area which could be worked.

The floor stone was hoisted to the surface in baskets by means of leather thongs. Dish-shaped lamps of chalk with rush wicks in animal fat were used for lighting. The freshly extracted moist flint could be worked much more easily than the dry flint on the surface so that it was made into weapons at the mines, using hard quartzite pebbles.

The flint mines of Europe were on or near the chalk outcrops in northern France, Belgium, and southern England. In Norfolk the sites of such mines extend over 34 acres. Sites also remain at Cissbury in Sussex, Worthing, Harrow Hill and Chichester.

Base metals

In the succeeding Bronze Age, lead, tin and copper were worked, and finally iron, which displaced bronze, but only with great difficulty because of its weight and liability to rust. Until man learned to control the carbon content of iron, bronze was a better material for tools and weapons.

Tin has been worked in Cornwall from the Stone Age to the present, down to a depth of 1000 yards, alluvial tin being won by streamers before the deeper workings began. The Phoenicians were customers of the early Cornish tin-miners.

Copper was probably the first metal used by man. It was used more than 15 000 years ago for colouring beads and glazes. It was also mined in Britain in the Bronze Age, and in Egypt before 4000 B.C. A contemporary Greek account indicates that the chief tool was the hammer. Discipline was extremely severe and child labour was used for transport of ore along the galleries. When the ore was very hard, fire was used to break it. This method, commonly adopted by miners before explosives became available, consisted in building a fire against the working and then quenching the heated ore with water, so that it cracked.

Gunpowder was first used for blasting minerals in Hungary in the sixteenth century, A.D., and paved the way for concentrated mining which had defeated the ancients.

The Romans worked copper in North Wales and Anglesey, iron in the Forest of Dean and they also revived Cornish mining and worked coal in many places. It was the mineral wealth of Britain which largely prompted the Claudian invasion, according to Tacitus. The Romans worked lead extensively in Durham, the Mendips and in Wales. Mining was carried on by freemen and slaves in the Roman Empire, the former occupying supervisory and administrative posts.

When the Roman Empire started to crumble, slave labour became scarce and freemen had to be retained in the mines by force. The recession in mining which followed the collapse of the Empire extended well into the Middle Ages.

Roman mining engineering was of a high standard, the Rio Tinto Mines in Spain being used as a training school. Unfortunately little is known about their engineering techniques.

Precious metals and gems

The precious metals were also worked in antiquity. Ireland with its gold wares was the El Dorado and gold implements and ornaments of Irish origin have been discovered right across Europe to Asia Minor. The Romans worked gold in North and South Wales and these workings have been continued spasmodically until the present day.

From the dawn of civilization, the recovery of nuggets washed out and transported by water has been a widespread art. Placer and dredger washings still supply much of the gold outside South Africa and Canada. Most lode gold is native but occasionally gold occurs combined with arsenic and antimony sulphides as telluride.

Ornaments of silver have been found in the royal tombs of Chaldaea, built in the fourth millenium B.C. and, throughout the ages, its pleasing colour which it retains untarnished in atmospheres free from sulphur, its malleability and ductility, made it, next to gold, the favourite metal for ornamental purposes. In ancient Arabia and Germany, silver was more valuable than gold and as recently as the seventeenth century in Japan the two were equal in price.

In the ancient world the Greeks were great miners of silver; in fact Athens owed its rise as a military and economic power to the 2000 Laurian mines, the richest in the Eastern Mediterranean. At first, about 500 B.C., labour was free but later slave labour was largely used and discipline was strict, the most difficult slaves worked in fetters and conditions were bad. In 413 B.C. 20 000 mine workers left the mines and went over to the enemy during the Peloponnesian War. This weakened Athens and contributed to its defeat. Again, three centuries later, the Acropolis was seized by revolting miners who had killed their overseers. These revolts were put down savagely, without improvements of conditions. Roadways were only 2 to 3 ft high; five face men and twenty carriers worked together, only one man working at the face at any one time. Ventilation was bad and conditions hot. Furnaces were used at the bottoms of main shafts for ventilation, a method first used in British coal mines about 1650 A.D.

Fig. 2. Mechanical engineering methods employed by German miners in Agricola's days (middle of the 16th century) when they were the leading exponents of the art

In ancient times and in the Middle Ages, India was the only source of diamonds from placer deposits in river gravels. In 1725 diamonds were discovered in Brazilian rivers and the market was flooded and prices fell, which led to Government control and taxation of output. In 1867 children picked up a diamond on a Boer farm and the find was not recognized for some time. Diamonds were next found in gravel banks of the Vaal River and a diamond rush began. More stones were discovered on de Beer's farm on an arid plateau and the mining town of Kimberley sprang up almost overnight. All these early finds were in loose soil and had to be exploited by dry-digging, the absence of water proving a great handicap.

Diamonds were next found quite unexpectedly and more

abundantly in the friable yellow sub-soil. Later the volcanic rock below was weathered on the surface and revealed the presence of diamonds and from this the present-day supplies are obtained.

Iron mining

Iron and aluminium are the most widely distributed of the metallic elements but to be economic, the percentage of iron must be high and the associated compounds must be reactive and readily fluxed away in the course of smelting. The minimum varies from 50 to 65 per cent iron down to 20 per cent in a readily accessible ore. Haematite contains up to ·05 per cent of phosphorus and sulphur and is used for acid steel.

About 1000 B.C., iron was being worked extensively in Central Europe at the head of the Adriatic Sea. The economic and political forces which this invoked soon began to affect Britain and with them began the series of invasions of the early Iron Age Britain.

The industry was, of course, of the cottage type with small hearths serving each settlement. At first charcoal from the forests was used for smelting but later coal was used and, finally, coke.

Blackband or clay iron-stone, containing some 30 per cent of iron associated with the coal measures, provided the iron ore of the industrial revolution until the middle of the nineteenth century.

With the decay and fall of the Roman Empire, much mining technique fell into disuse, but around 800 A.D. a revival set in when Charlemagne used prisoners of war for labour in Hungarian mines and in Spain, the Arab invaders reopened the gold and silver mines.

Improvement in international trade in the eleventh century expanded the demand for silver and gold, particularly in Bohemia, Austria and Saxony. The last reached a high standard of technical competence described in the first textbook of metal mining, *De Re Metallica* by Georgius Bauer (Agricola)[2] and published in 1556. This consists of twelve volumes and deals exhaustively with all branches of mining, including smelting and refining. It is illustrated by interesting woodcuts,* such as Fig. 2, showing the use of four-wheeled carts, manually operated windlasses (one provided with a flywheel), and the use of gearing.

A wide variety of pumps were available to drain the relatively shallow mines. Chain pumps with buckets or ball-and-chain pumps, driven by waterwheels, raised water 210 ft; ram and suction pumps were also used. A primitive form of fan is shown in Fig. 3.

Coal mining

The Chinese are credited with first working coal about 1000 B.C. In 371 B.C., Theophrastus refers to fossil materials that are called coles being used by smiths at Ellis in the Peloponnese.

Before the Roman invasion, the Britons are credited with some slight knowledge of coal since excavations of prehistoric sites in Monmouthshire and Derbyshire have revealed flint axes embedded in layers of coal, which are said to indicate workings during the Stone Age.

Coal cinders discovered in proximity to baths and other Roman remains indicate the use of coal for domestic heating but production was on a small scale, the workings being mainly open-cast at outcrops. The Saxons and Danes, whose invasions succeeded the Romans, appear to have disregarded coal entirely, using wood and peat to warm their dwellings, even near coal outcrops. Records show that, in 852 A.D., 12 loads of coal were to be sent each year to Peterborough Monastery. However, this may refer to charcoal.

No mention is made of coal mining in the Domesday Book (A.D. 1085), but by the time of Magna Carta in 1215 the industry had certainly got started on the shores of the Tyne, Wear and the Firth of Forth, with the gathering of sea-coal eroded by the tide from coal outcrops.

The monks of Tynemouth were probably the first to mine coal in the North of England and by 1269 coal was shipped from that port. The working of coal and the opening of small pits became general by 1307. The use of the new fuel was largely confined to the poorer classes as the hearth in those days was in the centre of the room, the fumes going out through a hole in the roof; and the fumes from coal were much more unpleasant than those from wood or charcoal.

The first access to coal seams was by inclined roads and headings and the coal was carried in baskets or 'corves' on people's backs (often women's) up the incline. This was followed by shallow vertical pits, bell pits (Fig. 1), and nearly horizontal adits, watergates or aqueducts.

Fig. 3. A 16th century waterwheel-driven centrifugal fan used for ventilating mines

A—hollow drum
B—hole for blowing
C—fan axle
D—lantern pinion
E—beam axles
F—spur gear
G—water wheel

Coal was carried up or hoisted by windlass or jackrail. The adit served to drain the workings down to the lowest outlet to the surface.

During the 15th century technique gradually improved, domestic consumption increased and round the Firth of Forth and Middlesbrough the land was studded with coal pits and salt pans. The chief centres of coal export were on the Tyne and the Forth. Fear of the exhaustion of coal reserves provoked the

**Several have appeared in these volumes; eg: page 145, pages 17 and 18, Engineering Heritage, vol. 1.*

Act of Parliament of 1563 prohibiting the export of coal from the realm!

With the greater depth of collieries, water difficulties increased as pits were worked below the limit of surface drainage and attempts were made to lift the water by machinery. The monks of Finchale Priory in 1486 invested "on the new ordnance of a pump" and on horses to work it at one of their Wearside pits while, in 1492, the Bishopric of Durham installed two great chains to draw coals and water out of their Whickham Colliery, the pumps being similar in design to the Chinese chain pump which is still used in agriculture.

In the sixteenth century wages were *3d.* to *4d.* per day and materials correspondingly cheap, but iron was scarce and picks so dear that they were often hired annually.

The mining of coal developed slowly but surely during the sixteenth century and men and women in Scotland entered the coal mines in early childhood and worked underground for the remainder of their lives. While the men were employed in getting the coal, women and children transported it on their backs to the pit-bottom and sometimes up spiral stairs to the pit top. Muscular development was the criterion of female beauty, hence the Scots saying "She's like the collier's dochter, better than she's bonnie". By an Act of the Scottish Parliament in 1606 if the son or daughter of a collier went to work in a pit, they were 'thirled' to it for life.

The colliers were regarded as inferior to the craftsmen and lived in remote isolated communities, to some extent ostracized by others. The relation between colliery owner and miner very closely resembled that between the lord of the manor and his serf, and in Scotland serfdom lingered on until 1799.

The ladies of the land supported Elizabeth's dislike of sea-coal but her successor, James I, had used it in Scotland and after he had burned it at Westminster coal became established as the normal domestic fuel. This helped to prevent deforestation of the South of England through excessive charcoal burning for the production of iron, and preserved a supply of oak for the Navy.

Ventilation

The gradual improvements in the technique of mining during the eighteenth century were largely the result of the increasing application of mechanical engineering. As the workings got deeper, mining difficulties increased. There were no safety lamps nor mechanical ventilation and the battle against the miners' enemy, firedamp, has been one of the main concerns of the coal mining industry for three centuries. Roads were driven in the coal from the bottom of the shaft and the coal pillars between were robbed until the strata above and below the seam showed signs of collapse. This method became systematized in the pillar and stall and post and stall systems (Fig. 4). In the seams of Shropshire a method known as longwall mining was used which is shown in Fig. 5. All the coal was removed without first cutting it up into pillars. This method is now used to win 92 per cent of British coal. The first ventilation systems used fires and furnaces to cause convection currents. This principle was only prohibited by Parliament in 1954, the last furnace having been taken out a few years previously. For the ventilation of single roads, a bellows or a small hand-driven fan was used.

The seventeenth century saw the introduction of the practice of firing the gas (hence the underground safety official known as a fireman). The fireman, covered with wet cloth entered before the miners. When he got near the coal face he crept on his belly, with a lighted candle at the end of a long pole. By creeping on the ground and keeping his face close to it he escaped the blast of any explosion. Fig. 6 is an old print illustrating this dangerous work. Single shaft mines with the shaft divided into two or more compartments by wooden brattice for the inlet and outlet of the ventilating air were common until a fire in the shaft of New Hartley Colliery led to the loss of 204 lives and the prohibition of single shaft pits.

Furnaces were placed underground, the first at Wallsend pit. The upcast shaft at Marley Hill reached 168 deg. F. with furnace ventilation. Serious colliery explosions averaged forty a year and eventually stirred the public conscience. The South Shields Committee was set up to investigate the problems and the Sunderland Society in 1813 persuaded Sir Humphrey Davy to investigate the safe lighting of coal mines. The only alternative to the candle was the Spedding flint and steel mill, invented in 1740, which provided a shower of sparks.

Davy, George Stephenson, and Dr Clanny invented safety lamps practically simultaneously and although Davy received the kudos, the lamps of Stephenson and Clanny were probably more successful.

A celebrated mining engineer, John Buddle, installed in 1807 a large air pump at Hebburn pit on the suggestion of an Irishman named Ryan. It had a square piston of 5 ft side and a stroke of 8 ft, made 20 strokes a minute and drew 8000 ft^3 of air/min through the pit workings. Buddle used a steam jet for ventilation in 1811 and in 1827 the forerunner of the modern rotary fan was installed at a pit near Paisley. The prototypes, such as the Guibal, were large, 40–60 ft diameter, and slow-running; but smaller diameter, faster fans came in by 1850.

Power supply and transmission

The application of power underground in British collieries began with the lapsing of the Watt patents in 1800. Initially it was very tentative but by 1850 mechanical methods were adopted for hauling, winding, pumping and for ventilation

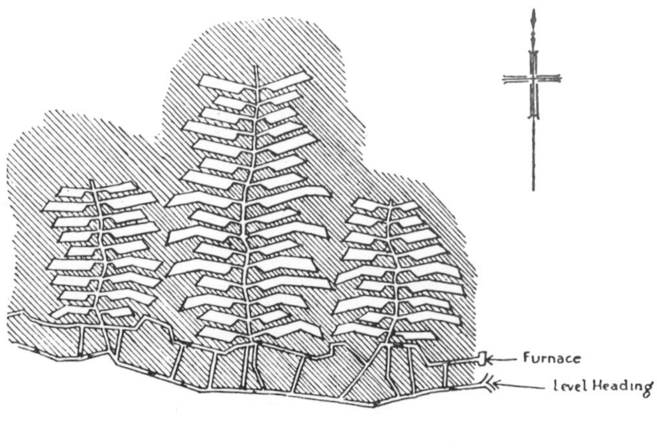

[Courtesy: The Mining Association]

Fig. 4. Plan of the post-and-stall workings at a South Wales colliery during the 17th century: the pillars of coal were robbed until the strata showed signs of collapse

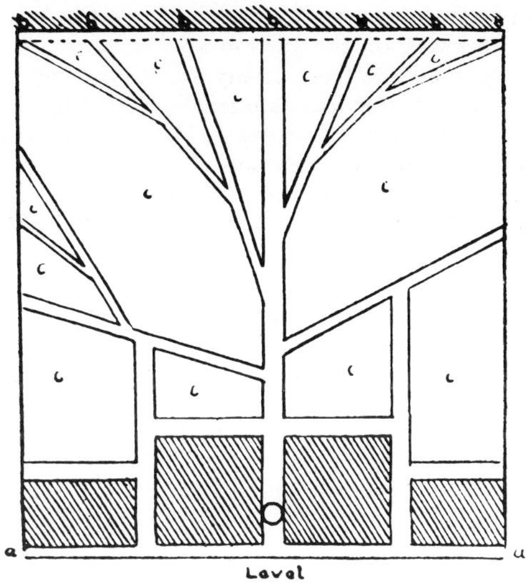

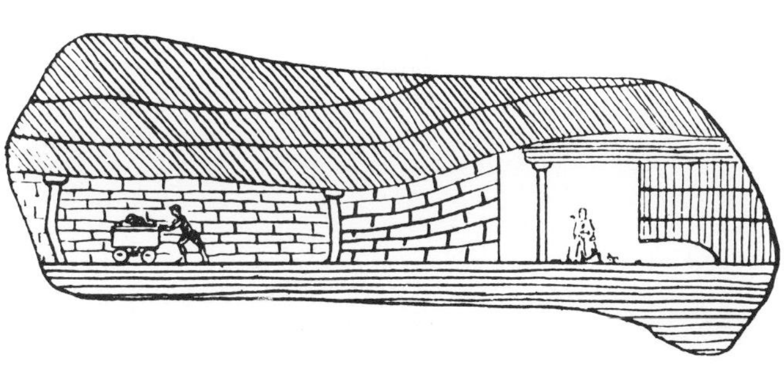

[Courtesy: The Mining Association]

Fig. 5. The method of longwall working first adopted in Shropshire in the late 17th century is now used to mine 92 per cent of British coal: all the coal can be removed without cutting it up into pillars since the wall supports the roof

Fig. 6. The only way of clearing the firedamp from the workings before efficient ventilation became available was this rather hazardous one of firing it; the 'firemen' being protected by damp rags and by lying low

on a fairly wide front. However, no suitable power system was yet available. Steam was generated at the surface, in some instances, and carried down the shaft for limited distances. In the majority of cases, however, boilers were installed underground in spite of the danger of explosion. In other cases haulage engines were installed on the surface and the ropes taken down the shafts and also used to drive pumps in dip workings. Even in 1882 the opinion was still expressed that rope transmission was superior to compressed air.

Since the transmission of power by steam had obvious limitations and rope transmission proved inefficient and expensive in maintenance, compressed air, and later electricity, was used. The first case of the former was in 1830 when Admiral Cockrane took out a patent for the sinking of shafts by the caisson method, using compressed air. This was also successfully adopted by Triger in France. It was used for lifting water in 1845. At Govan Colliery near Glasgow in 1849 compressed air was used to raise water and for haulage, the pressure being 20 lb/in^2. But not till 1865 was it suggested that air should replace steam underground. It was then used to drive early longwall coal cutters.

The early air compressors of 1850 were of the beam type with two vertical steam cylinders and two vertical air cylinders with brass ball valves in special cages. By 1870 the wet compressor had been developed in Belgium. The use of compressed air in driving the St Cenis and St Gothard tunnels led to improvements in drilling and other machines driven by this medium and in compressors.

In 1850 the majority of steam engines in collieries were still of the Newcomen, Watt or Cornish types but a few horizontal engines had been installed; for example, a two-cylinder winding engine was installed at Old Shildon Colliery, Durham by Hackworth about 1840, embodying his valve gear with single eccentric. By 1860, however, the slow-speed horizontal engines were in general use in mines and, with the introduction of electricity at collieries in 1882, they were used to drive dynamos through belts or ropes. Because of the cheapness of fuel at collieries, slide valve gear was used longer there than elsewhere but ultimately improved valve-gear and expansion gear were adopted.

Transport

At the end of the 15th century wheelbarrows had been used in the North of England to convey the coal to the shaft bottom where it was loaded into wicker-work corves for hoisting. The earliest wagon-way, the forerunner of railways, was used at the surface of collieries to reduce wear on roads, and wooden railways were used underground in Shropshire in 1620.

Early in the eighteenth century horses were used at Tyneside collieries. Both sledges and trains with small wheels were in use by 1789. In 1753, cast iron replaced wood for wheels and in 1767 plates of cast iron were used on top of the wooden rails; by 1794 light bars of malleable iron of oblong section were used at Walbottle Colliery.

Adopting a quite different approach in 1758,, James Brindley, engineer to the Duke of Bridgewater, constructed a canal from Worsley to Manchester, eventually extended to Leigh and to Liverpool. The Duke then decided to install an underground canal system to take coal by boat to Manchester from the inclined seams sloping to the South. At Worsley the main underground canals were constructed whose entrances are still visible. They extended for several miles and intersected 12 workable seams. In each seam canals were driven east and west and small boats were taken along these to be loaded with coal. The main canal was connected to the surface by shallow pits for ventilation and to provide access for workmen. The total length of underground canals was 33 miles.

Coal was carried in skips or baskets and emptied into the boats; later small boxes with 'pancake' or 'penny' wheels were used which ran on angle irons.

At the beginning of the nineteenth century, women and girls were still employed in many British coalfields in hauling the coal from the coal-face, and in Yorkshire they were also employed as miners. In 1840, on the instigation of Lord Ashley, afterwards Earl of Shaftesbury, a Royal Commission was set up to examine the employment of women and children in mines.

Their report, which was published in 1842, revealed appalling conditions: boys and girls of all ages from 8 to 18 were working underground for as long as 16 hours daily, crawling on all fours, dragging loaded tubs along roadways deep in mud which were only 30 inches high; and women stood up to their knees in water or worked bent double where men had refused to go.

Lord Ashley's Act excluding all women and girls and boys under 10 from underground work came into force in March 1843. It had a hostile reception in the coalfields from both coal owners and workers!

Until 1840 corves were still in use for coal transport from the face and up the shaft but since about 1760 they had been pulled by horses. As distances of workings from the shaft increased, the expense of horse haulage increased.

George Stephenson was one of the first to employ the steam engine in hauling coal from the face to the shaft-bottom; after his appointment as engine-wright at Killingworth Colliery in 1812, he altered an underground pumping engine so that it could haul coal up an incline to the shaft. Extensions to this system were made by the addition of two engines so that a haul of two miles was achieved. One engine was 1500 yards from the shaft bottom and the smoke from the boilers adjacent to this engine was conveyed in flues along the roadways for this distance.

Winding

In 1650 came the first improvement on the hand-winch in hoisting coal up the shaft: this was the cog and ring gin (Fig. 7) which consisted of a winding drum immediately over the pit. Horses yoked under a long horizontal bar, worked the drum through lantern gearing as shown. The drums were frequently blown off the pits by explosions! Reversal of winding meant reversing the horses.

About 1680 the whin gin was used with the drum placed vertically at some distance from the pit, the ropes being guided over pulleys at the pit-head. The diameter of the drum and the number of horses and levers could thus be increased for faster winding. The corve carried about $4\frac{1}{2}$ cwt and about 100 tons was considered a good day's winding from a pit 120 yards deep. The first pit cages were invented by Carr of Sheffield in 1790.

Waterwheels were also used for winding as shown in Fig. 8. They had two sets of opposed buckets so that the direction of winding could be reversed, according to which set was supplied with water; this was controlled by a simple sluice operated by hand levers. Such a wheel was equivalent to 16 horses and four men.

It is recorded that in 1763 coal was wound by a 'fire engine' at Hartley Colliery, presumably a Newcomen engine. Until 1780 the great difficulty with steam engines for winding was to obtain rotary motion. Steam, however, was certainly used to pump water to the surface where it was used to work waterwheel winding drums. Watt's first rotative engine was put into commission at Walker Colliery in 1784. In shallow Welsh pits a water-balance system was extensively used, a tank being filled at the surface to counter balance the coal, and emptied underground into a drainage adit.

The first quarter of the 20th century saw the introduction of electric winding gear, both a.c. and d.c. Ward Leonard systems. At present, of the 1410 winding engines in use in Britain, 860 are electric.

In the early part of the 19th century the search for more reliable methods of bringing men and minerals to the surface was engaging the attention of mining engineers. The descent into deep shafts by ladders and 'man-engines' with reciprocating platforms was a great waste of time and energy, as well as being dangerous, and hemp ropes were not too reliable. In 1830, George Wright Binks, a foreman rope maker at Woolwich Dockyard, had the idea of forming a rope of twisted iron wire. Simultaneously, Albert at Clausthal in Germany had the same idea and the first wire rope was given its successful trial at the Caroline pit there in 1834.

Drainage

The first steam engines, Savorys' in 1695 and the more successful Newcomen engine in 1713 at Griff, Warwickshire, were employed in pumping water from collieries, using low pressure steam and utilizing atmospheric pressure. The vertical motion of the piston rod was transferred by means of a heavy beam to the pump rods, whose weight, assisted by the slight upward pressure of the steam, raised the piston for the next stroke. The valve mechanism was at first hand-operated: subsequently modified by Beighton, it became more or less automatic.

Wooden pipes were first used for drainage but iron pipes were introduced in 1744.

The Cornish engine, which held the field for nearly a century, was a development of Newcomen's and Watt's ideas. It utilized a positive steam pressure on the top side of the piston in a vertical cylinder, with a vacuum created below the piston.

[Courtesy: The Mining Association]

Fig. 7. For winding, the first improvement on the hand-winch was this horse gin, dated about 1650

Fig. 8. Mechanical winding came about 1760 with the double-sided waterwheel: by pulling a lever, the operator could direct the water into one or the other wheel, thus determining the direction of winding

[Courtesy: The Mining Association]

Very reliable and cheap to maintain, its initial cost was high and it was heavy and large.

These engines were displaced by direct-driven steam pumps and the first use of electricity underground at collieries was for pumping at Trafalgar Colliery, Forest of Dean, in 1882. The supply was d.c., a.c. being first used at Denaby Main Colliery in Yorkshire for driving a haulage; and at Ackton Hall Colliery for supplying a coal cutter. Centrifugal pumps in turn displaced reciprocating pumps for many colliery duties and, in recent years, borehole and submersible pumps have found increasing employment.

Coal-face mechanization

The first coal cutters were no exception to a general trend of early mechanization in that they aimed at reproducing mechanically, the motions previously performed manually. They were in effect mechanically-operated picks, at first driven by hand and later by ponies.

In 1768 Willie Brown's Iron Man, driven by two men, was introduced at a Northumberland colliery and over the next thirty years many other machines of similar design were tried out but all failed owing to the lack of power supplies available underground.

Early in the nineteenth century, a horse-driven machine was tried out in a thick Yorkshire seam, this being the only type of seam in which it could be accommodated. The horse was harnessed to a frame in the base of which was a treadmill formed of an endless band of wooden slats, mounted on rollers, which turned a sprocket wheel when the horse began to walk; this wheel, through a series of levers, operated a swinging pick. The whole machine, complete with horse, was mounted on wheels running on rails laid along the coal face. This also was not a success and coal-face mechanization still awaited the advent of better underground power supplies.

The first power-driven machine, equipped with a disc at the edge of which were picks for undercutting the coal, was invented by Walker of Wigan in 1859. A bar machine, studded with small picks, rotated and drawn along the face by a ratchet and pawl mechanism was introduced about the same time.

It was not until 1870 that a really successful coal cutter was evolved, a disc machine. These machines were then more successful than the chain-type coal cutters of which the Baird of 1874 was an early example, consisting of a projecting arm or jib carrying an endless chain, the links of which carried boxes armed with cutters or picks. Trouble was experienced with the jib and chains at first but metallurgical developments eventually enabled chain machines to replace others early in the twentieth century. Their success in the U.S.A. no doubt influenced their ultimate success in Britain. Small mobile 'punching' machines on wheels or screw columns were also used successfully both in the U.S.A. and in this country, but have now fallen into disuse.

So-called conventional mining, that is, using coal cutters and conveyors at the face, reached its zenith a few years before the 1939–45 war and at this time the first power loader for simultaneously cutting and loading the coal, the Meco-Moore, came into successful operation. Since then a great many mechanical and automatic devices have been introduced.

At the same time as increased mechanization of cutting and loading equipment there has been a corresponding development in the support of the roof, both at the face and on the roadways. Wooden props and bars were replaced by steel in the early 1920's, first as rigid props and arches, later in the form of adjustable hydraulic pit props.

Conclusion

Although by the beginning of the eighteenth century, the coal mining industry of Britain had a considerable history of technical development behind it some of which was adapted from earlier metal mining techniques, its subsequent rapid growth was both the result and cause of the Industrial Revolution.

During the latter half of the eighteenth century the industrial areas of Britain were connected by canals to the navigable portion of rivers, thus making possible the shipment of coal from inland mines to the sea ports.

The acceptance of the steam engine as a power producer, the invention of locomotives, railways and steamships, all of which were big coal consumers, together with the use of coke in iron smelting, gave an unprecedented impetus to coal mining during this period. Further, the coking of coal was carried out quite early in this century for the manufacture of illuminating gas. It is not surprising that in the middle of the 17th century the British annual coal output was only $2\frac{1}{2}$ m. tons; by 1913 it had reached 287 m. tons, of which 88 m. were exported.

It is, therefore, fair to say that the growth of British mechanical engineering was partnered by the growth of the British coal mining industry. This partnership has continued to the present day when some 68 per cent of the annual output of 200 m. tons is won from fully mechanized faces. This has permitted an increase in productivity of nearly 20 per cent.

Two prototype, fully automated coal-face installations at Newstead and Ormonde Collieries, controlled from the roadway and requiring no miners on the working face, offer a prospect of further advances in productivity, and another result of a happy marriage of mechanical and mining engineering.

[Courtesy: National Portrait Gallery]

THE GREAT MASTERS

After Watt, his friend Murdoch was probably the most important contributor to the mechanical development of the steam engine, besides being the first to apply gas lighting, pneumatic power transfer and a host of minor inventions.
But to this day—as in his own age—he has been overshadowed by Watt.
Far less interested in wealth and fame than in pursuing his inventions, he remained a valued servant to the end, resisting every temptation to make himself independent.

pencil sketch by Sir Francis Chantrey, RA

William Murdoch—Faithful Servant

by James A. McCash, BSc (Member)

In London in 1824 when the friends of James Watt met to support the erection of a suitable monument to him in Westminster Abbey, M. Charles Dupin, a French engineer who was present, commented:

> There was to be remarked *among the spectators* a venerable old man, whose intrepid services I . . . wished had also been rewarded by some . . . marks of public gratitude: Mr Murdock directed the application of the new steam engines . . . showed all the resources of genius, . . . to triumph over every difficulty . . . [was] insensible to any temptation [and] withstood all menaces.

In 1839 Murdoch died in Birmingham, his death almost unremarked elsewhere. He had come to Soho to look for work 60 years earlier and there he died and was buried alongside Watt and Boulton in the Parish Church at Handsworth.

Background

The Murdochs were a prolific and versatile family, settled in Nithsdale, Ayrshire, since the 16th Century or earlier. William Murdoch's ancestors were millers and farmers in the Kyle Baillierie. An Auchinleck baptismal register records that all entries are "Attested by me David Murdoch, Session Clerk, 19th January 1746", and this register includes the "marrage contract" (sic) of William's parents John Murdoch and Anna Bruce in 1747. William, the third child and the eldest of the three boys, was born at Bello Mill on 21st August 1754.

The father, described as millwright and joiner, was fairly well-to-do and had some money to spend on the education of his children. One of William's many cousins became a mechanical engineer in Glasgow whose company, Murdoch and Aitken, built steam locomotives and colliery winding engines.

We know that for some years, young William received instruction in the 3 Rs at a cost of about 30*s*. pa for $5\frac{1}{2}$ days' a week at nearby Old Cumnock's already ancient parish school.

He was taught there together with some 35 other children and developed a clear handwriting and good style. The school dominie, William Rae, was under the powerful direction of the Reverend Dr George Muir, Minister of Cumnock Parish. Muir was a bold denouncer of insobriety and a reclaimer of spiritual and moral backsliders. He was even more notable for his encouragement of children who showed promise of mental powers.

Murdoch's notably temperate life, unshakable loyalty to his friends and his resolution stem largely from his father but also in considerable part from the example of the greatly-revered Muir, whose name has remained a household word in Cumnock for 200 years.

Following the precedent of his own father, John Murdoch joined the Woolwich Company of the Royal Regiment of Artillery as a 'mattross' (apprentice) in 1744, becoming ultimately a First Gunner. After service at Fontenoy and Mons he was back in Scotland in 1745 and it is almost certain that he was present as a combatant in April 1746 at Culloden Moor.

Parental example

After he returned to his father's Bello Corn Mill, John Murdoch did his own millwrighting and stone-dressing. Having made the patterns and coreboxes himself he had Dr John Roebuck of Carron Ironworks cast what William later claimed to be the earliest cast-iron millgearing in Britain. Suitably mounted, it was eventually to decorate William's front lawn at Sycamore House, Soho.

Some time before 1777, John and young William Murdoch together constructed a wooden 'horse' (probably an early tricycle). The machine is referred to by a local versifier who wrote in 1817:

> Murdoch, (the World should know thy scientific fame). . . .
> Sprung from a curious deep designing sire
> Who rode a horse no mortal e'er could tire,
> Improved upon his predecessor's laws,
> And wrung from darkness bright refulgent gas.

This proto-tricycle is believed to have resembled in its tiller, steering and wheel arrangement Murdoch's first locomotive design built between 1781 and 1784, now in the City of Birmingham Museum. Peter Macmillan, a blacksmith born in 1813, who is considered the inventor of the bicycle, lived less than 40 miles away and could well have seen or at least heard of the Murdochs' 'horse'.

As a young man, William experimented with coal-gas illumination in a large rock-hewn cave at Bello Mill, using the excellent local coal. He wrote later:

> At the time I commenced my experiments, I was certainly unacquainted with the circumstances of the gas from coal having been observed by others to be capable of combustion; but I am since informed, that the current of gas escaping from Lord Dundonald's tar oven had been frequently fired . . . but . . . I believe I may . . . claim both the first idea of applying, and the first actual application of this gas to economical purposes.

However, he did not seriously pursue this work until many years later. His other engineering, woodworking, stonemasonry and millwrighting was no doubt fostered by the needs for maintenance of his father's corn mill but Murdoch had, like his father, an innate liking for mechanical engineering.

He therefore decided to seek his fortune as a mechanical engineer; and where better than at the leading establishment of his day?

Murdoch's house in Redruth was lighted by gas. It was here, too, that he built his first steam locomotive

Greener pastures

In 1777, just before his 24th birthday, Murdoch set off on foot for the famous Soho Manufactory of Boulton and Watt. Samuel Smiles tells a probably apocryphal story of how, during an interview with Boulton, Murdoch accidentally dropped his oval stove-pipe hat; the resulting unexpected 'clunk' attracted Boulton's attention. Murdoch then confessed that he had turned his hat from a bit of timber "on a lathie of my ain contriving". And this got him his job—or so the story goes.

Throughout their long and intimate association, Watt's pre-eminent ability and immense prestige, jointly with Boulton's talent for commerce and publicity, combined to overshadow Murdoch. Most people know of him dimly as Watt's reserved, if energetic and competent, employee in Cornwall. He worked as a weekly-paid senior engine-erector in the water-logged mines of Cornwall for many years. Less well-known are his frequently unacknowledged contributions to the basic design and practical advance of steam, pneumatic, hydraulic and gas engineering.

Murdoch maintained a gentlemanly, grave and modest reserve to the end of his life. He was apparently without worldly ambitions. Consequently, much of his early inventiveness was for many years credited unchallenged to Watt or other contemporary inventors, and others usually derived the financial benefits from his work. A patent cost its inventor about £80 when enrolled—18 months' basic wages for Murdoch in those days!

Examples of designs known to be at least as much due to Murdoch as to Watt, are the piston slide valve, the 'sun and planet' gear of Watt's Patent of 1781 and Item 7 of his Patent No. 1432 of 1784, covering the motion of wheel-carriages. On 6th August 1784 Boulton wrote to Watt specifically asking him to include the latter in one of his steam engine specifications to save the expense of a separate patent, 'and by way of taking possesion'.

In fairness to Watt and Boulton, Murdoch had probably signed an agreement to make over to the Company any inventions of his.

The oscillating cylinder engine

Possibly the most notable example of unaccountable neglect was the case of Murdoch's unpatented, oscillating-cylinder engine. Rotation was obtained by direct connection of the piston-rod to the crank, the cylinder being mounted on trunnions. Murdoch's original demonstration model, designed to run on compressed air, is now in the Science Museum, Kensington. This design was successful, yet many years were to pass before its merits were tacitly recognised by its adoption in widely different applications. It incorporated in one adaptable, robust, yet light, design the 'puffer' engine—using high-pressure steam distributed by a square-section piston valve—convertible, if desired, to a condensing unit for stationary use. At the same time it obviated the overhead beam, a great advantage in confined spaces, and it pioneered the Y-bar gear. The model has an integral pump, easily adaptable to condensate service, which is operated by an eccentric block on the output shaft. The double-acting

Murdoch's own model of the oscillating steam engine: it did away with the clumsy overhead beam and was later widely adopted for transport applications ◀

[Courtesy: Science Museum, London]

The first locomotive which he demonstrated to Trevithick in model form ▶

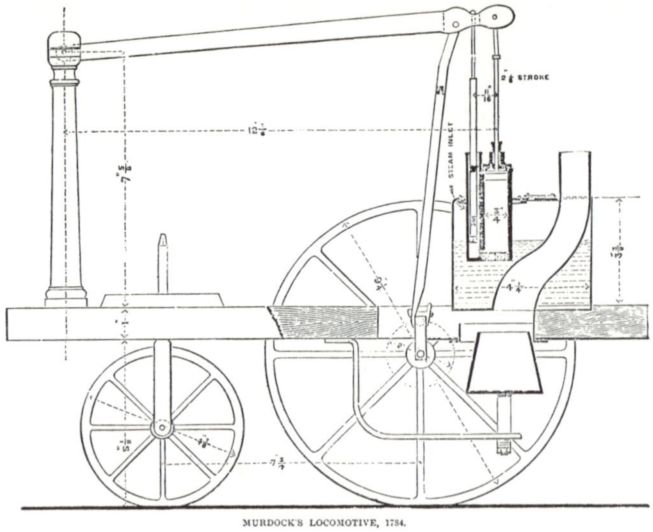

[Courtesy: *The Engineer*]

cylinder is slung on hollow trunnions, one of which is used as a steam port.

This engine, equally suitable for horizontal and vertical mounting, was to prove as portentious to subsequent development as was Watt's first condenser and air pump; and, taken jointly with Murdoch's locomotive and his later 'long-D' slide valve, it places Murdoch second only to Watt in the annals of the reciprocating steam engine.

Thus, Trevithick and Vivian's Patent of 1802, for instance, shows an oscillating cylinder engine and Joseph Maudsley used this design in his famous oscillating marine engine for the *Endeavour*. In 1858 it was used in the paddle-driven *Great Eastern* of 18 900 tons. Another adaptation of the oscillating cylinder, used in conjunction with his highly successful boiler, was utilised by that most successful of the early road locomotive builders, Walter Hancock of London in 1827.

Murdoch survived for many years after these notable patentees had been widely and publicly acclaimed for their successful originality in design and it no doubt gave him satisfaction (if little advantage) to see his earlier pioneering work so well received many years later.

The occasion for his own design is not definitely known, but, as a simple, lightweight alternative to Watt's 'rotative' engine, it was possibly designed for transport duties.

The locomotive

In 1780, a prosperous engine supervisor and notable Cornish engineer-inventor, John Budge, who had a small engine-building foundry at Luckingmill, had offered Murdoch an equal partnership. But Murdoch would not be tempted away from Watt, although working on occasion with Budge and, it is recorded, making use of his foundry for his later locomotive developments. Thus, he deliberately relected the opportunity of teaming-up with 'a Matthew Boulton of his own'. But he retained Budge's friendship and the use of his workshop for many years.

Apart from his lack of greed and vanity, the only explanation of Murdoch's touching faithfulness to his first and only employers in the face of frequent neglect in the early years of his career—was his veneration for Watt. They were similar in temperament, constantly experimenting and innovating; interested in solving the problem, rather than collecting the reward, though, of the two, Watt was probably the better businessman.

To Murdoch, probably the greatest disappointment endured during his service as a young, underpaid engine-erector, was the deliberate though, in the long run, ineffective, opposition of Boulton and Watt to the development of a high-pressure non-condensing locomotive. Boulton wrote to Watt in July 1784 that Murdoch and his father had planned a steam carriage and also a hot-air engine years before.

> Brown [Murdoch] and his father were about a steam carriage some years ago, [he also] mentions a hot air engine as the thing he had thought of before he came to England.

Murdoch designed and built the first known working model of a steam locomotive at Redruth between 1781 and 1784 when he demonstrated it to interested friends and neighbours, drawing a model wagon at 6–8 mile/h. That same year, he discussed with Boulton a more sophisticated, gear-driven, condensing locomotive, suitable for operation on various gradients. This led to Watt's specification of the design characteristics of such a locomotive, sent in a letter to Boulton, together with a recommendation that Boulton should dissuade Murdoch from further pursuit of his locomotive proposals. Watt wisely added:

> Rather than lose him . . . let him have an advance to the extent of £100, to enable him to prosecute his experiments, [for otherwise] if he succeeds tolerably, he will attach himself totally to it, and we shall lose the benefit of his knowledge and experience . . . still we must not prevent him from going on with his scheme.

Murdoch, confident of success, listened to Boulton's advice but continued his work in private for the next two years. Watt had guessed William's probable reactions accurately; Boulton never understood him.

The spare time design and final construction of this model had taken Murdoch well over two years. He now hoped to demonstrate it to Watt himself and to obtain patent protection either in the Company's or in his own name. For Watt's superior engineering ability he had the same veneration as for that of his own father. These were the two men who had trained him in engineering design and undoubtedly had the greatest influence on his own development both as a man and as an engineer.

In the absence of a reply from Watt himself, Murdoch set off with his model for the London Patent Office in 1786. Soho was doubtless advised of Murdoch's intention and Boulton intercepted him on the open road near Exeter and somehow persuaded him to return home. Watt quickly came to see Murdoch and shortly thereafter wrote to Boulton ". . . William [now] applies himself to his business".

This interim comment by Watt is occasionally adduced as evidence that Murdoch thereupon desisted from further locomotive development, but this is incorrect. Despite the active discouragement Murdoch had now received from both of his employers, he built a second, very similar, model locomotive with a 1 in diameter cylinder in 1792. His son John later wrote on Murdoch's behalf that Trevithick and his cousin A. Vivian had called at their house in Redruth about 1794 and asked him to show his model. Trevithick's own locomotive first ran in 1801 and was greatly influenced by Murdoch's work. However, he used higher boiler pressures.

In a letter of Trevithick's there is an oblique reference to a large steam carriage made by Murdoch. His grandson also said that Murdoch had constructed for himself a steam carriage between 1792–1794, sufficiently large for going to and from the mines; and lighted with gas taken from a bladder-type reservoir!

Pneumatics

Murdoch's least-known locomotive experiments were on air motors. In Alexander Gordon's *Treatise on Elemental Locomotives* (1832), we read

> Mr Murdoch of Soho, made several calculations and some experiments for a locomotive engine, in conjunction with the late Mr D. Gordon, in the year 1819, for the purpose of introducing compressed air as the immediate motive power for carriages. . . . Their object was to store up power in the elastic air, compressed highly, and to let it down into a piston engine by degrees, instead of steam.

These proposals, far ahead of their time, came to nothing. But for over 30 years, Murdoch showed a marked fondness for the engineering possibilities of compressed air; he was probably the first to demonstrate its use in several applications which are important today.

On 25th January 1788, a letter from Boulton to Watt describes the use by Murdoch of compressed air for raising water without pumps or beam. Their agent, Wilson, elaborates on this by writing to Watt:

> Murdock let me see his new method of raising water; it is by compressing air at the surface which is forced down small pipes to the bottom of the shaft. . . . There is no bucket, but there is a clack to prevent the water falling back . . . from the side is a very small tube, which goes from the top of his house to his well 40 feet and another larger tube by the side, up which the water ascends and runs off continually in an equal stream. . . .

He did not patent the invention, nor did anyone after him, until 1865.

With compressed air supplied by the foundry cupola blowing-machine, he used an air motor which successfully drove the pattern-makers' lathe at Soho for some 35 years. This was a reciprocating engine, 12 in diameter × 18 in stroke, geared to run the lathe at about 225 rev/min.

In his machine shop he eliminated counter-shafting and belt-drives for machine tools, using instead a central exhauster pump to maintain a vacuum. Each machine was driven independently by its own vacuum piston engine connected to the pump system. This is said to be the first successful application of an earlier proposal by Papin of Blois (1647–1712), who had been defeated by lack of suitable mechanical equipment in his day.

Other pneumatic ideas of Murdoch's included a large air-driven elevator for castings in his foundry; a pneumatic doorbell at his house; and the atmospheric pressure/vacuum system for transmitting light articles through an exhausted tube, later developed by the London Pneumatic Dispatch Company about 1853 and itself the progenitor of the atmospheric railway.

In 1785 Murdoch had married Mary Painter, daughter of a mine-captain of Redruth, and brought her to his home in Cross Street, where all their four children were born; William in 1788 and John in 1790, the two others dying in infancy. His wife too died in 1790, possibly at the time of John's birth, and in 1801 the two boys were taken to their grandfather's home at Bello Mill. About 1805, they joined William in Soho, William Jr becoming an employee of the Boulton and Watt. Here they remained until the father retired in 1830. Murdoch never remarried.

Coal gas

In 1791, he was living in his small home-cum-office at Cross Street, Redruth, probably with his mother-in-law looking after his two young sons. Alone and in his spare time he experimented in his workshop with domestic kettlefuls of various gas feedstocks, both vegetable and mineral, burning

[Courtesy: The Gas Council

Murdoch's original gasholder at Soho was the forerunner of today's water-sealed designs

At 'Sycamore House' Murdoch installed a pneumatic bell push; mounted on the lawn is the first cast-iron, toothed gear wheel, ordered by his father

the issuing gas. He decided that, of wood, peat and coal, the latter was the most suitable raw stock from the points of view of light brilliancy and cost. His work was to became the basis of the powerful new gas and chemical engineering industry.

He designed, and built in his backyard, prototype generating, water-washing, storing and distributing plants and designed both fixed and portable burners. Then he publicly demonstrated the efficiency of coal gas as an illuminant by using it in his own house and offices in 1792; the farthest burner being about 70 ft from his vertical, cast-iron retort. This installation is the first practical application in the history of coal gas lighting. He also devised a 'gas torch'—a kind of bagpipe filled with gas, with the aid of which he lighted himself home over the moors at night.

About 1794 or 1795, Murdoch informed Boulton and Watt of his success in the use of coal gas as an illuminant, confirming that very considerable economies could be made by its use in lighting as a substitute for sperm whale oil-burning lamps and tallow candles. He proposed that a patent be sought. James Watt, Jr, on the Company's behalf, refused to consider a patent in view of an unexpired patent of 1781, held by Lord Dundonald, and because of continuing expensive litigation over open infringements of James Watt's own patents. Murdoch was recommended to cease experimenting for the present.

Recognition at last

Disappointed with the Company's attitude to all his inventiveness, Murdoch did stop experimenting at Redruth, and returned home to Bello Mill in 1797 where, both at Lugar and at nearby Muirkirk, efficient foundries and a forge were available. Now, for the first time since his appointment, his valuable services to Boulton and Watt were openly recognised, probably as a result of this gesture. In 1798 Murdoch was back on the permanent staff at Soho, working on the equipment—including the boring-mill and lathes—for the new Soho Foundry.

The first patent in his own name was taken out in 1799, and in 1800 he was urgently offered, and accepted with pleasure, a new contract from the new partners of Boulton, Watt and Co., as Engineer and Superintendent. This appointment coincided with Watt's retirement from active business and brought Murdoch's salary to £300 *pa* plus 1 per cent commission.

Thereafter he became the mainstay of the firm, doing considerable work as a production engineer in addition to his designing and inventing. Although he made very little out of his own inventions, he eventually left some £14 000—a considerable fortune in those days—so that his 1 per cent commission from the firm must have been remunerative.

In 1798 one of the new foundry buildings became the first works to be permanently lighted by gas. Again Murdoch pleaded that a patent be applied for but was met by a second refusal.

In 1799 and 1801 Philippe Lebon of Brachay, France, was granted patents for the distillation of coal (and wood) and for a spark ignition gas engine too! Three years later Lebon was dead, murdered in the Champs Elysées. But with the publication of Lebon's work, including a proposal to introduce gas street-lighting in Paris, the Soho Company decided at last to put the development of gas-making equipment in hand.

The first public result was the well-known exhibition of lighting of their works facade.

Murdoch's fight for recognition as the originator of the application of coal-gas to illumination was ultimately successful, as shewn by the award to him of the Royal Society's Rumford gold medal.

But he had no patents and others, particularly Winsor, were able to cash in, though Boulton and Watt continued to make the equipment commercially. Eventually Samuel Clegg, trained by Murdoch at Soho saved the Gas Light and Coke company from bankruptcy by becoming its chief engineer.

In 1809, Matthew Boulton had died and Watt was occupying his retirement in minor inventive pursuits. He involved Murdoch in these, probably as much for the mutual pleasure they had in each other's company as for Murdoch's now recognised ability to select the best answer to a design problem from the flood of solutions which usually presented themselves to Watt.

Murdoch was now guiding the hands of the young directors of the Company but, before Watt died in 1819 in his 83rd year, the two friends and colleagues were to see the Company surpassed in many detailed aspects of steam and general engineering by new, eager and competent competitors who sprang up throughout England and Scotland. After Watt's death, Murdoch continued with the Company until 1830 though his powers were failing when he was retired at the age of 76. His sons and grandson left the Company at the same time.

A venerable octagenarian, largely forgotten by a generation remembering only Watt's work, Murdoch died on 15th November, 1839, in his 86th year, at his home, Sycamore House, Handsworth, in Staffordshire.

REFERENCES FOR FURTHER READING

1. SMILES, SAMUEL, *Lives of The Engineers: IV Boulton and Watt.* John Murray, London, 1861.
2. DICKENSON and JENKINS, *James Watt and the Steam Engine.* Clarendon Press, Oxford, 1927.
3. MCCASH, JAMES, 'William Murdoch—Man of Little Showing'. 1966, *The College Courant*, vol. 19 *et seq*, Glasgow University.

Frederick Winsor giving one of his lectures on gas lighting at the Lyceum Theatre; from a diorama in the Science Museum. The story of gas engineering begins overleaf

[Diorama in the Science Museum]

Fig. 1. Pall Mall in 1808: the first street in the world lit by gas

THE EVOLUTION OF ENGINEERING

Gas Engineering

by W. T. K. Braunholtz, OBE, MA, PhD

A whole new supply industry sprang into existance as a result of William Murdoch's pioneering. Originally used largely for lighting, gas was overtaken by the more efficient electric lamp but it has held its own for heating and cooking and the techniques developed by gas engineers have laid the foundation for a vast chemical industry.

More than a thousand years ago the Chinese knew of the occurrence in nature of combustible gas, which issued from 'fire wells' and was collected in skins or piped through bamboo tubes and used for illumination. Its early use for heating furnaces is also recorded. In the West the eternal flame in the Roman Temple of Vesta is supposed to have been fed continuously with natural gas.[1] The observation that an inflammable and illuminating gas is produced by the heating of coal, came many centuries later and was not exploited on a commercial scale until the beginning of the 19th century.

At that time the only sources of domestic and public lighting were tallow candles and oil lamps, and movement in the streets at night was a hazardous undertaking, as the light afforded by the street lamps hardly enabled the passer-by to distinguish a watchman from a thief.[2] The proposal to light the streets of London with gas manufactured from coal was at first a subject of ridicule (Sir Walter Scott referred to it as a mad suggestion to light London with smoke!) which, however, turned to admiration with the first public demonstration of gas lighting in 1802, to celebrate the Peace of Amiens, and the first public exhibition of street lighting by gas in Pall Mall in late 1807 or early 1808 which is shown in Fig. 1.

For the greater part of the 19th century, gas held undisputed pre-eminence as an illuminant, but gradually the emphasis changed and the important position of gas in modern society is now based primarily upon its properties as a generator of heat. To provide for its production and distribution, as well as for its countless fields of application, is the function of gas engineering.

The pioneers

Scientific progress has owed much to the early investigations conducted privately by leisured individuals who communicated their findings to one or other of the learned societies. Thus, the first authentic record of the production of a combustible gas by the heating of coal is usually ascribed to the Rev. John Clayton, who reported his observations to the Royal Society in 1688. There followed numerous experimenters, notably George Dixon who, in the second half of the 18th century, was interested in the extraction of tar from coal by heating, but who also investigated the production of coal gas and its use as an illuminant. His experimental apparatus was, in crude embryo, the classical process of gas production by coal carbonisation. Others were engaged in a similar exploration, notably Philippe Lebon in France and Jan Minckelers in Holland. These investigations which, at that time, had no commercial significance, provided the background for the subsequent development by William Murdoch of the controlled manufacture of gas from coal.

Murdoch has been described as the greatest mechanically inventive genius that Scotland has produced.[3] His pioneer work in the field of gas was incidental to his main career as a brilliant mechanical engineer with the firm of Boulton and Watt. While working in Cornwall, in the closing years of the 18th century, he installed in his house at Redruth a small gas-making plant and piped the gas to light his office there. In 1798, he was transferred to the firm's Soho (Birmingham) steam engine factory, where he continued these investigations. In 1808, he reported his work to the Royal Society, in a paper (for which he was awarded the Society's Rumford Gold Medal), as follows:

> My apparatus consisted of an iron retort, with tinned copper and iron tubes through which the gas was conducted to a considerable distance and there was burned through apertures of varied forms and dimensions . . . The gas was also washed with water and other means were employed to purify it.

After experimenting with iron vessels of various shapes, in which a charge of some 10–15 lb of coal was heated, Murdoch was led to the design used at the works of Phillips and Lee at Salford in 1805. A cage containing 15 cwt of coal was lowered into an iron pot, $3\frac{1}{2}$ft diameter and $5\frac{1}{2}$ ft deep, set in a furnace and externally heated as shown in Fig. 2; after carbonisation, the cage was removed and the coke emptied out.[4] Coal is a poor conductor of heat, however, and in order to reduce the time required for the heat to penetrate the charge, Murdoch reverted to a horizontal retort of D-section (with the flat side downwards), and closed at one end.

Commercial production

It was Frederick Winsor, a layman "of but little practical experience but possessed of unbounded imagination", who first proposed the distribution of illuminating gas on a commercial scale and from a central source. His idea led to the formation, several years before the first public railway, of the Chartered Gas Light and Coke Co., the first commercial gas undertaking in the world. This was in 1812, the year of Napoleon's retreat from Moscow. Other countries were not slow to follow suit: in Europe, France (c. 1816), Austria (1817), Belgium (1819), Germany (1825), Holland (1826) and in the New World, the USA (1816) and Canada (1836). Most of the early Continental gas works were built and operated by British engineers.

The science and profession of gas engineering may be considered as beginning with Samuel Clegg, who had been an apprentice with Boulton and Watt in Murdoch's time and, in 1812, had installed a private gas works in the Strand which is illustrated in Fig. 3. Clegg was appointed engineer of the new Chartered Company in 1813, at the age of 32. The first chemist of the Company was Frederick Accum, who published in 1815 the earliest authoritative treatise on gas manufacture.[5] But it was Clegg who introduced gas purification by means of lime and whose inventions included, in 1815, the first gas meter and gas governor and, in 1816, a horizontal rotary retort (Fig. 4), rotary gas meter and self-activating governor. He developed also the form of horizontal retort set that remained the standard until the mid-century, comprising groups of five iron retorts, 7 ft long, arranged inside a single coal-fired (and later coke-fired) furnace.

Design of retorts

Indeed, the horizontal retort remained the sole means of coal carbonisation by the gas industry until nearly the end of the century, although its form and setting were progressively improved. Thus, George Malam, who succeeded Clegg as engineer of the Great Peter Street works of the Chartered Company, introduced the elliptical retort shown in Fig. 5 to carbonise coal rapidly in a thin layer, and George Lowe

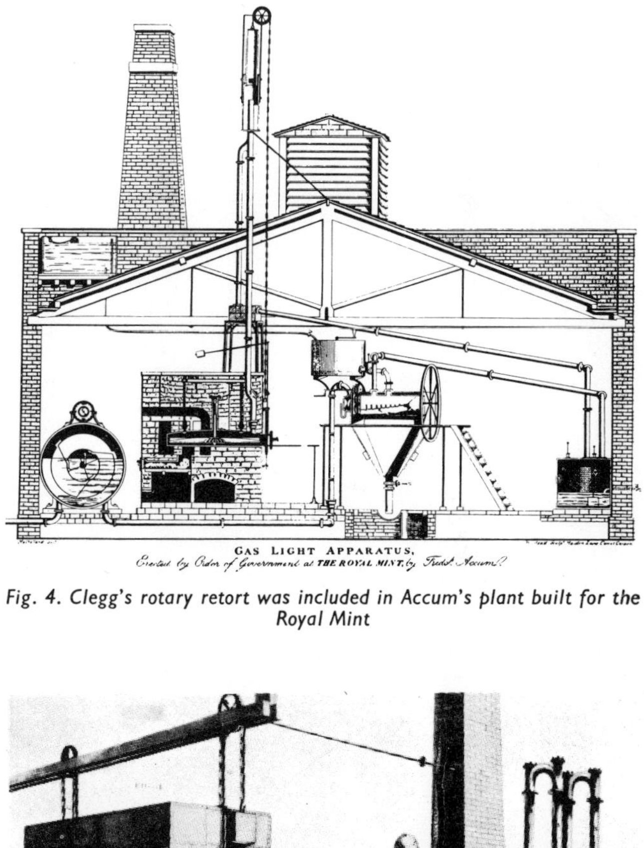

Fig. 4. Clegg's rotary retort was included in Accum's plant built for the Royal Mint

Fig. 2. Model of Murdoch's cage and furnace design: heat took too long to penetrate the coal

Fig. 3. By 1812 Clegg had installed this gas-making plant in the Strand: lime was used for purification

devised, in 1831, the through retort, comprising a tunnel 20 ft long with a door at each end. This increased the load per charge and effected economies in fuel.

Improvements were made also in the materials of which retorts were manufactured. The old cast iron retorts had the advantage of being non-porous and good conductors of heat, but they underwent distortion with prolonged heating and had a useful life of only about 200 days. Retorts made of fired clay, on the other hand, showed less distortion, could operate at higher temperatures (giving increased gas yields) and had a useful life some three to five times that of iron retorts. Clay retorts were therefore generally introduced after about 1850, in improved settings and with the later refinements of self-sealing doors, built-in producers and heat recuperators.

All horizontal retorts, however, present the physical difficulty of charging coal evenly into a long, narrow, horizontal tube at red heat. At first this had to be done manually, with long scoops as shown in Fig. 6, but a great deal of engineering ingenuity was expended in an effort to operate the retorts mechanically. One of the best-known and most successful machines, shown in Fig. 7, is that developed by de Brouwer, in 1900, in which coal is passed round the deep rim of a rapidly rotating wheel and projected from it into the retort.

Another, very important, method of improving the mechanical operation of gas retorts was to invoke the help of gravity by setting the retorts at an angle or vertically, charges of coal being introduced intermittently at the top and the coke falling from the lower end when an iron stop was removed. Inclined retorts were usually set at an angle of 32° to the horizontal and were widely used between about 1890 and 1930.

The earliest proposal for an intermittent vertical retort dates from 1828, but this type of gas-making plant was not introduced commercially until 1902. Of far greater importance was the vertical type into which coal is charged continuously, passing slowly down the heated retort by gravity and being removed continuously as coke at the bottom. A variety of designs were patented and when, by the Gas Regulation Act of 1920, gas quality became based on calorific value rather than illuminating power, the continuous vertical retort really came into its own. Steam was introduced at the base to cool the issuing coke and, at the same time, increase gas production by generating water gas. By 1960, more than 60 per cent of town gas in Britain was made in such retorts, the advantages of which include saving in ground space and labour, coupled with improved efficiency and operating conditions.

Other processes

In coke oven gas production, the emphasis is normally on the efficient production of metallurgical coke rather than of town gas. Essentially, modern coke ovens are large, narrow horizontal chambers, set side by side with gas-fired heating flues between each pair, in a long 'battery'. They are charged intermittently with crushed coal and the coke is pushed out by a mechanical ram. Coke ovens have been built at some of the larger British gas works (the earliest was in 1912), but their use primarily for gas making is more common abroad. With later improved designs including waste heat recovery and firing by producer or blast furnace gas, the coke oven has become capable of supplying surplus coal gas to outside users and large quantities are taken by gas undertakings to supplement their own production.

No sooner had the production of town gas from coal became established commercially than attempts were made to introduce gas made from other organic materials, *eg* fish oils, fats, etc. The earliest patent, granted in 1815, involved the thermal decomposition on cracking of the oil in heated retorts, and both Covent Garden Theatre and the Post Office were lighted with oil gas. Other systems were devised but they could not compete economically with coal gas.

On the other hand, the production of water gas has been of great importance to gas undertakings as a means of supplementing coal gas supplies or of covering sudden daily or seasonal peak demands. There are two kinds of water gas. Blue water gas is an equal mixture of carbon monoxide and hydrogen with a calorific value around 300 Btu/ft^3, burns with a blue flame and is made by passing steam over incandescent coke; and carburetted water gas is enriched by addition of gas of high calorific value formed by the decomposition of a hydrocarbon oil in contact with hot brickwork. The introduction of blue water gas dates back to 1824, whilst Lowe's carburetted water gas plant was first used in England in 1890. Subsequent engineering improvements have included self-clinkering grates, automatic coke feed, and valve operation, and waste heat recovery.

Fig. 5. Malam's elliptical retort was meant to improve heat transfer

[Diorama lent to Science Museum by British Gas Council]

Fig. 6. All horizontal retorts presented the difficult problem of pushing coal into a long, narrow tube

[Crown copyright, Science Museum]

Fig. 7. To overcome the problem illustrated in Fig. 6, De Brouwer invented his rotating wheel charging machine

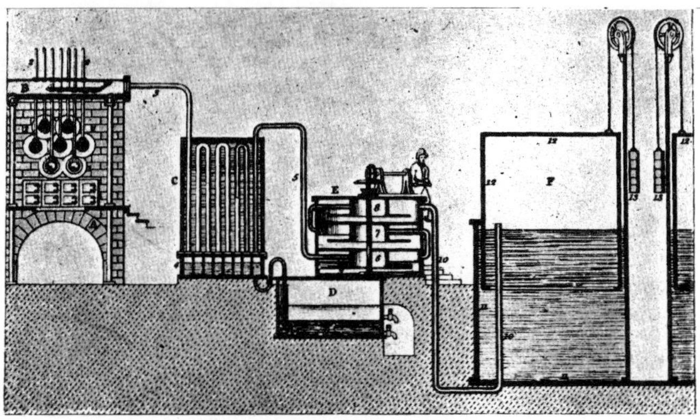

Fig. 8. Early purification methods (c. 1832)

Purification

Thermal decomposition of coal produces not only a mixture of combustible gases and a residue of solid coke, but a host of other chemical by-products which leave the retorts in company with the gas. Some of these (*eg* tar and water vapour) may give rise to stoppages by condensing in the supply pipes, others (*eg* hydrogen sulphide and organic sulphur compounds) form noxious and corrosive products when burned. Thus, the purification of the crude gas before its distribution was accepted from the outset as a vital problem of gas engineering, and even the earliest gasworks included apparatus for the removal, at least in part, of condensed tar and water and of sulphur compounds, as shown in Fig. 8.

Both physical and chemical methods were used. Simple air- or water-cooled condensers did not remove the persistent tar 'fog' and for this the earliest practice was to pass the gas through a bed of coke breeze or through a large number of small holes. Later, more efficient types of tar extractor were devised, notably the Cathels bubbler washer (1868), the Livesey washer, using perforated metal tubes submerged in water (1870), and the Pelouze and Ardouin impingement extractor. Other physical processes involved washing the gas with water to dissolve out the ammonia, for which purpose Lowe invented the tower scrubber in 1846 and Hills introduced a rotary scrubber filled with moist brushwood in 1848. Washing the gas with oil to remove naphthalene was not usual until the beginning of the present century, owing to the importance of naphthalene in increasing the illuminating power of the gas.

The recovery of benzole and toluole from coal gas by oil washing was introduced during 1914–18 as a matter of national importance, since benzole was urgently needed as the starting point for the manufacture of phenol and its derivatives for explosives and disinfectants, as well as providing a valuable motor fuel, toluole was essential for the manufacture of a variety of high explosives of which the best known is trinitrotoluene (TNT). Their recovery from indigenous sources released vital shipping that would otherwise have been required to import them.

Chemical methods of gas purification involve primarily the removal of constituents containing sulphur, which produce sulphur dioxide when burned. As early as 1805 Clegg, and in 1815 Malam, among others, were largely responsible for the use of lime to remove these compounds and, at the same time, the incombustible carbon dioxide.

The lime was first used as a sludge and later, owing to the difficulty of its disposal after use, in a dry state, spread out on shallow trays in enclosed boxes through which the gas passed. Other oxides were tried and when it was discovered, in 1849, that iron oxide, converted into sulphide by contact with hydrogen sulphide, could be revivified by atmospheric oxidisation, the use of this material, either as natural bog ore or chemically prepared, was rapidly introduced, still in conjunction with a proportion of lime to remove the organic sulphur compounds.

By about 1912, iron oxide had completely replaced lime in gas purification.

From the engineering point of view, the installation of large, shallow, rectangular, cast iron oxide purifier boxes had its drawbacks which led, in 1930, to the introduction of upright cylindrical tower purifiers in which the trays of oxide were arranged one above the other and could be removed mechanically for emptying and recharging. In recent years there has also been much investigation of methods for removing hydrogen sulphide by chemical solutions, and processes for removing organic sulphur compounds without the use of lime have been successfully developed.

Storage and distribution

The earliest apparatus for storing large volumes of gas, invented by Lavoisier in 1782, was a rectangular metal box with its open side freely suspended in a tank of water, as shown in Figs. 3 and 8. By 1816, the box or bell had become cylindrical and this has remained the basis of all water-sealed gasholders up to the present day. The first telescopic gasholder, with more than one moving section, was built in 1826 and since then there have been various developments in the methods of bracing, framing and guiding the 'lifts' as they rise and fall, and in methods and materials of construction.

The erection of these vast hollow vessels presents the engineer with many problems and the present century has seen the introduction of new types of gas storage, including the waterless or piston-type gasholder in which the gas is contained under a piston moving inside a fixed shell; spherical or cylindrical steel vessels under pressure; and underground storage in porous strata or in natural or artificial caverns.

The rapid and unco-ordinated expansion of the gas industry after 1812 led to intense competition between rival supply undertakings and a chaotic proliferation of gas distribution networks, particularly in the Metropolis. By 1850, there were 13 separate gas undertakings supplying London alone and G. Livesey has recorded that, during those first 40 years, it was not at all unusual for three, four or even five companies to have mains in the same street, and often on both sides. Consumers were constantly changing from one company to another and there was ample opportunity for clandestine tapping of a competitor's main. It was not until 1860 that each undertaking was given, by Act of Parliament, a monopoly of supply in a given area. At the same time statutory gas testing and penalties for non-compliance with standards of quality and pressure were introduced.

The continuous supply of purified gas, at a constant pressure, from a central works to thousands of individual consumers, both domestic and industrial, and to innumerable public street lamps, constitutes a special branch of gas engineering. A variety of non-ferrous materials have been tried for the supply pipes (wood, lead, stone and, in recent times, asbestos cement, copper and plastics), but cast iron pipes, with flanged or spigot-and-socket joints, were used almost exclusively for the first 100 years. Many of these old cast iron mains are still in perfect condition today. The smaller service pipes were, for a time, made from surplus gun barrels left from the Napoleonic wars or from lengths of wrought iron tube.

By 1850, cast iron mains up to 36 in diameter were being used and up to 48 in by 1870, but steel pipes began to be inserted where special tensile strength was required. Indeed, since the turn of the century, there has been a great development in the use of seamless, welded steel tube, particularly for pipelines for the transmission of gas over long distances (even thousands of miles) and at high pressures (up to 800 lb/in^2). Such pipelines require special protection against soil corrosion by suitable coatings and wrappings and against electrolytic action from stray earth currents by cathodic protection.

The distribution of gas in metallic cylinders or bottles, was used in London in the early years of the 19th century; also later for the lighting of railway carriages and, more recently, as an emergency supply of fuel for motor vehicles or in places where a piped supply is not available.

Town gas is normally distributed from gasholders to urban areas at a relatively low pressure of about 2–4 in wg, which is derived from the weight of the floating bell of the holder. To maintain a constant pressure at the consumer's meter, the gas is passed through a governor, a type of automatic valve invented by Clegg in 1815 and shown schematically in Fig. 9. It was subsequently improved by others and developed for a wide range of sizes and pressures.

At the time of the introduction of gas lighting on a commercial basis, no means existed for measuring the amount of gas used by a consumer. The method of charge was a fixed annual sum per burner of given dimensions, burning for a specific number of hours each evening after sunset. Clearly this system was open to all kinds of abuse and once again it was Clegg who, in 1815, put the matter on a scientific footing with the invention of the first gas meter. His earliest design and, indeed, those of many subsequent inventors, depended upon the rotation of a wheel divided into two or

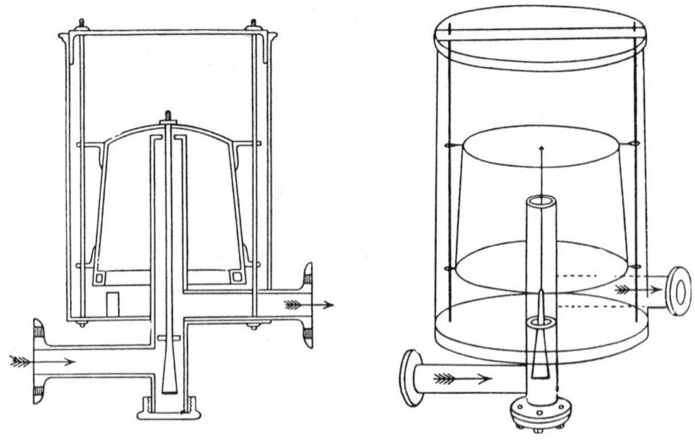

Fig. 9. Clegg's automatic valve was designed to maintain constant pressure at the consumer's meter

more chambers and enclosed in a sealed casing. Movement was due to the pressure of incoming gas which entered a hollow horizontal spindle. The casing was partially filled with water into which each chamber dipped in turn as the wheel rotated, and by this operation a measured volume of gas was expelled through the meter outlet.

In 1820, Malam invented the first apparatus for measuring gas without the use of a water seal by means of a series of bellows and, in 1838, Defries patented a dry gas meter constructed on the principle of the common kitchen bellows.

Both the wet and dry types of gas meter, in improved designs, have been in use ever since for measuring the passage of small and large volumes of gas. The prepayment or penny-in-the-slot meter was invented in 1889 by Thorpe and Marsh. A manually wound clockwork was actuated by the insertion of a coin and shut off the gas after the passage of a standard quantity of gas, which was measured by the revolutions of a wheel.

Lighting

The third major function of gas engineering, after the production of combustible gas and its distribution through networks of underground pipes, is to devise and perfect methods and appliances for its efficient utilisation.

William Murdoch, the pioneer of gas production from coal, experimented, as early as 1792, with different types of burners to produce light. In some the gas issued through a number of small holes, in others through slits which might be straight or circular. The latter type was called the Argand burner after a Genevese who, in 1784, had invented a circular wick for oil lamps.

The first commercial gas burner in its simplest form was known as the rat-tail, a metal tube closed at one end which was perforated with a single hole. Other types were known picturesquely as cockspur (1808), cockscomb (1808), batswing (1816; widely adopted for street lamps), fishtail (1820) etc., Burners using the Argand principle, with the gas passing through the annular space between two concentric tubes and burning at a circular row of apertures in a steel ring at the top, were introduced in 1809. In improved form, they were chosen in 1860 as the standard apparatus for the statutory testing of the illuminating power of gas; the 'candle power' being defined as the number of standard sperm candles burning at standard rate that gave an illumination equivalent to that produced by the gas when burned under standard conditions.

The Argand burner continued to be used extensively, in various designs, for domestic and public lighting throughout the century. In 1852, the new House of Commons was lit by gas, and the middle of the century saw a great extension of gas lighting in domestic premises. Improved burners with refractory (steatite) tops replaced corrodible iron and artistic invention ran riot in the varied and sometimes fantastic design of Victorian gas fittings.

An alternative method of using gas for illuminating purposes was based on an observation, made by Drummond in 1828, that a brilliant light was emitted by lime when heated to a high temperature (hence limelight). Cruickshanks, in 1839, tried to make a platinum mantle coated with lime and other refractory oxides.

Impetus was given to these experiments by the discovery, by Bunsen in 1855, that a hot non-luminous flame could be produced by mixing air with the gas before it issued from the burner. Making use of this principle Hogg, in 1868, invented an aerated burner with a mantle of platinum or refractory material, and in 1882, a Frenchman, Clamond, exhibited at the Crystal Palace a mantle consisting of a conical basket composed of threads of calcined magnesia. These early forms of incandescent mantle were, however, neither efficient nor mechanically strong, and it was the epoch-making investigations of the Austrian, Auer von Welsbach, beginning in 1884, that made incandescent gas lighting a commercial proposition. He discovered that a far higher incandescence is produced by using mixtures of certain oxides.

A few years earlier the electric bulb had been introduced and had begun to face the gas burners of that time with serious competition. The incandescent upright burner with the Welsbach mantle arrived just in time to give new life to gas lighting and, indeed, introduced a new era of domestic and public lighting. Many improvements were added from time to time, mainly with the object of increasing the efficiency of light production and its better distribution around the burner. Space allows only passing mention of the ancillary advantages of heating and ventilation accompanying interior gas lighting; of the introduction of the inverted indandescent burner in 1897, which threw most of its light downwards instead of upwards; and the increased efficiency due to preheating the air and gas by the hot products of combustion; (the regenerative burner, was introduced in 1900. By compressing the gas before combustion, the high-pressure gas burner of 1907 improved this efficiency still further.

Heating

Although in the first half of the last century gas was used mainly for illumination, the heating properties of the flame had also attracted many investigators. F. Winsor, in 1807, claimed to be the first to "cook, melt, boil and distil by gas fire, either in a kitchen or dining room", and many similar references may be found in contemporary literature. Soyer, the famous French chef of the Reform Club, described in detail the gas stoves he installed in the Pall Mall premises in 1841, and an exhibition of gas cooking stoves was held at the Polytechnic Institute in 1851.

The earliest gas cookers were merely cast iron boxes, with a door in front, gas burners inside and, perhaps a gas ring standing on top of the box. To this basic conception have been added, from time to time, many devices for increasing efficiency, ease of operation, and safety, until the modern gas cooker includes such refinements as thermostatic control, enamelled finishes, safety taps and self-lighting burners.

The earliest proposals for using gas for space heating were confined to the convector types of stove in which air passed over or through a tube heated by a gas flame. The first to be patented was by Barnes, in about 1832. Around 1850, crude attempts were made to produce a radiant fire by heating lumps of pumice or other incombustible material in a coal grate by a gas burner below the bars. The introduction of the Bunsen burner in 1855 should have given a new impetus to these developments but in fact progress was slow, because of objections to imperfectly purified gas burned in flueless appliances and their crude and inefficient designs.

The firebrick back was introduced in 1859 and the first radiant elements of woven wire in 1877. Radiants in the form of hollow balls were used in 1880, and brush-like tufts of asbestos fibre in 1882, but it was not until around 1905 that the more familiar refractory radiants were introduced. Modern developments have been directed towards improved radiant efficiency, coupled with circulation of convected heat, and self-lighting devices.

The direct heating of water by means of a gas burner placed beneath an iron bath was known in 1850, but the first instantaneous water heater (or geyser) was patented by Maugham in 1868. In it, the water to be heated trickled down over a series of baffles in direct contact with the hot combustion products. This was followed by the enclosed type, in which the water was not in contact with the heating gases. The modern design of instantaneous gas water heater, developed since the turn of the century, with its safety devices and self-acting valves, is the culmination of long and detailed research into the physical and engineering principles.

Conclusion

Gas engineering is compounded of a variety of sciences, both pure and applied, notably chemistry and physics, and civil, mechanical, chemical and electrical engineering. William Murdoch, the father of the gas industry, was an eminent mechanical engineer and the importance of mechanical engineering in providing the machinery, plant and appliances for its development will be apparent from what has been said.

Many gas engineers have also been qualified mechanical engineers and it was not fortuitous that the founder President of the Institution of Gas Engineers in 1863 was Thomas Hawksley, who went on to become President of the Civils in 1871 and this Institution in 1876.

This brief historical review breaks off at the turn of the century with the gas industry still using the old methods of producing town gas from coal and making the most of the various by-products resulting therefrom. The techniques developed by gas engineers were intimately related to the advances of mechanical engineering in general and have, in turn, helped to lay the foundations for the vast chemical engineering activities of today.

REFERENCES FOR FURTHER READING

1. Chandler, Dean, *Outline of History of Lighting by Gas*, 1936.
2. Chandler, Dean, and Lacey, A. Douglas, *The Rise of the Gas Industry in Britain*. 1949. British Gas Council.
3. Armstrong, E. F., *Murdoch Centenary Lecture*. 1939. Institution of Gas Engineers.
4. Stewart, E. G., *Town Gas: its Manufacture and Distribution*. 1958. Science Museum (HMSO).
5. Accum, Frederick, *A Practical Treatise on Gas-Light*. 1815.

THE GREAT MASTERS

Like many of his able contemporaries, the Rev. Edmund Cartwright had an extremely versatile mind but, unlike most, he took up an entirely new career in middle life which secured him his place in history. There is little to indicate why he took up engineering after being a successful clergyman and a keen writer, unless it was the result of the same enterprising and inventive mind which he demonstrated throughout his life.

Edmund Cartwright—Parson, Poet and Inventor

by W. English, M.Sc. (Tech)*

From an engraving in 'The History of Wool and Wool Combing' by James Burnley

Edmund Cartwright was born in 1743 at Marnham, in Nottinghamshire, the younger brother of Major John Cartwright, the well-known pamphleteer, who spent some fifty years promoting radical reforms of Parliament. Edmund was educated at Wakefield Grammar School and at University College, Oxford, and became a fellow of Magdalen College but he did not graduate as Doctor of Divinity until 1806.

His father, William Cartwright, had married a cousin, Ann Cartwright, in 1731, and they had five sons and five daughters. Edmund, from an early age, was intended for the church and his education was planned accordingly. He soon became distinguished for proficiency in his studies. Had he been allowed his own choice of a career, he would have entered the Navy. But two of his brothers were preparing for that service and it was decided that he should enter holy orders.

Clergyman

After a few years at Brampton, in Derbyshire, of which he was presented with the perpetual curacy, he became rector of Goadby Marwood, near Melton Mowbray in 1779 and, seven years later, was appointed a prebend in the cathedral of Lincoln.

Cartwright married Alice Whittaker, the daughter of an alderman of Doncaster, in 1772. She bore him eight children —only five of whom survived her—and died in 1785. The youngest daughter, an infant at the time, was adopted by Major and Mrs Cartwright.

Of the other children, Edmund became rector of Earnley, in Sussex, and died in 1833.

At the age of 41, Cartwright decided to take a holiday at Matlock. It is interesting to speculate whether he would ever have turned inventor if he had not taken that holiday.

For Matlock is not far from the village of Cromford, where Richard Arkwright† had already built a water-driven cotton mill and it was natural that this venture should be a frequent topic of conversation. So the rector was particularly intrigued when he got into conversation with a number of Manchester business men there. In his own words,

> One of the company observed, that as soon as Arkwright's patent expired, so many mills would be erected, and so much cotton spun, that hands never could be found to weave it. To this observation I replied, that Arkwright must then set his wits to work to invent a weaving mill. This brought on a conversation on the subject, in which the Manchester gentlemen unanimously agreed that the thing was impracticable; and, in defence of their opinion, they adduced arguments which I certainly was incompetent to answer, or even to comprehend, being totally ignorant of the subject, having never at the time seen a person weave. I controverted, however, the impracticability of the thing by remarking, that there had lately been exhibited in London an automaton figure which played at chess. Now you will not assert, gentlemen, said I, that it is more difficult to construct a machine that shall weave, than one which shall make all the variety of moves which are required in that complicated game.[1]

The conversation took place in 1784, a year after Arkwright's patent for his spinning machine had in fact expired. He held a second patent, however, covering a series of preparatory machines and this did not expire until 1789.

First power loom

Apart from some improvements he made on an agricultural machine at his father's home during a college vacation, Cartwright had never hitherto invented anything. But now he began to think of ways in which cloth could be woven on a power-driven machine and he was not the only pioneer who succeeded in what experts have described as impossible.

He began making his weaving machine, assisted by a carpenter and a smith. To his delight the machine wove a piece of cloth when complete. He took out a patent in 1785, but the drawing in the specification is meaningless, apart from a large handle to be turned by 'powerful men'. The description is vague; in his own words:

> The warp was placed perpendicularly, the reed fell with the weight of at least half a hundred-weight, and the springs which threw the shuttle were strong enough to have thrown a Congreve rocket. In short it required the strength of two powerful men to work the machine at a slow rate, and only for a short time.[1]

Then, and not before, he decided to find out how a handloom was worked, and discovered that only one person

† *cf. p. 36, Engineering Heritage, vol. 1.*

operated the loom, using far less effort than was required to turn his machine. But the discovery did not deter him. He now began to make a study of the hand-loom weaving operations, and was soon designing another loom for power drive, based on the principles of the hand-loom.

Problems and details

This he patented in 1786, adding improvements in 1787. The machine formed the basis, with many improvements by other inventors, of the early power looms.

It was probably the need for converting continuous rotary motion such as was generated by the water-wheel, into three different reciprocating movements at three different but accurately controlled times, which made the problem seem insoluble to those familiar with weaving; and, in fact, Cartwright found the propulsion of the shuttle to and fro across the loom to be his most difficult problem. His daughter Mary tells of the amusement it afforded his children to watch their father imitating the action of a weaver throwing his shuttle as he walked up and down the room.[2] In his second patent the shuttle was thrown by a spring actuating a 'picker stick', which was returned to the striking position by a cam, thus re-loading the spring.

His patents relating to weaving appear to have covered all kinds of details with a view to ensuring the successful working of the loom and the prevention of faults which the hand-loom weaver would normally anticipate.[3]

One of these details was a device which stopped the loom on the breakage of a warp thread. But in practice it was many years before it was considered necessary to apply such a stop motion to power looms; a weaver tending the usual four looms was expected to detect and repair such breakages before any defect appeared in the cloth. The idea, however, was applied at quite an early date to warping machines.

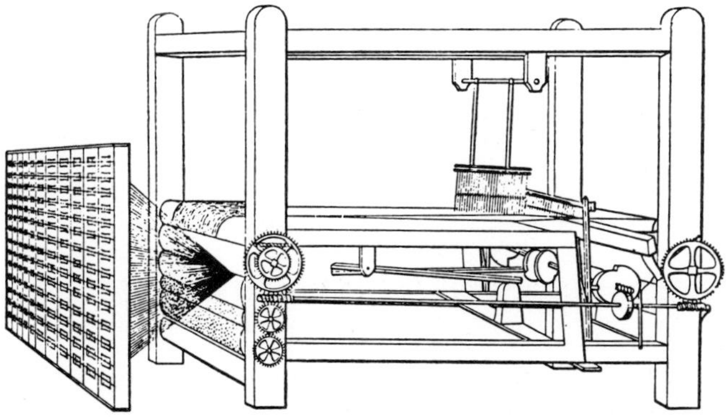

This drawing from the second power loom patent shows a change from the normal hand weaving procedure: the yarn came directly off the bobbins

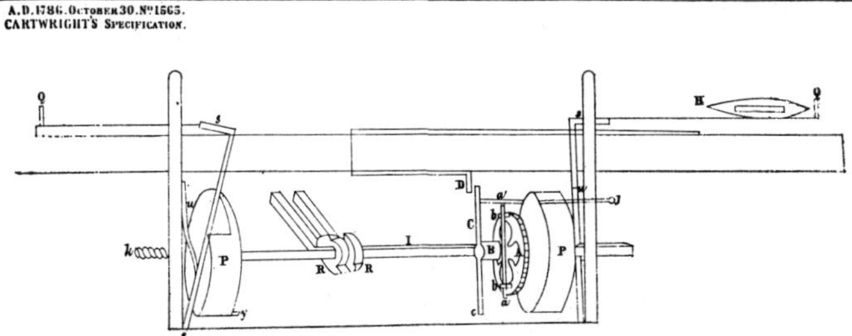

Picker stick details of the loom above

H — shuttle R — cams S — picker stick Q — sliding picker K — spring

A drawing in one of his specifications shows an idea which he had obviously not previously tried out—that of gearing together an eccentrically mounted wheel and an elliptical wheel, so as to cause the reed to strike the cloth more quickly, and to give more time for the shuttle to pass.

Other devices to be found in the specifications were weft tension and weft stop motions, 'temples' to keep the cloth taut laterally, a ratchet wheel take-up which regulated delivery of the cloth to ensure uniformly spaced weft threads; and a device for preventing the shuttle from re-bounding from its box. One outstanding improvement was that of movable shuttle-boxes so that two colours of weft could be used, e.g. to produce check patterns.

A drawing from the second patent specification shows a variation from the usual procedure in hand-loom weaving. As will be seen, the loom is arranged to weave direct from yarn on bobbins, thus obviating the warping, i.e. the filling of a roller with hundreds of threads. Except for certain specialities, this method has never been adopted. Another drawing in the same specification shows cam *P* which operates picker sticks *S*, *H* being the shuttle and *Q* the sliding picker with its cord attached to the stick. Cams *R* lift the healds through the levers shown.

An extremely important requirement in a power loom was that of a stopping motion, should the shuttle either fail to reach its box or rebound from it at the end of the throw. Otherwise the shuttle would be trapped between the reed and the cloth, resulting in the breakage of hundreds of warp threads. Cartwright developed a stopping device for this purpose.

The Doncaster failure

Upon completing his first loom he moved to Doncaster, where the workpeople were more skilful than in the village. In 1786 he discussed his loom with Manchester men, knowledgeable in weaving processes, and there he found even more skilful workmen to make his loom. But he failed in his other objective, that of interesting the manufacturers in his machine and obtaining financial support. Confident that his loom could be operated successfully, he set up a spinning and weaving mill in Doncaster and there struggled with the numerous problems, not only of weaving by power, but also of spinning, then still in an early stage of development. In fact, Cartwright introduced several improvements in spinning machines.

The venture was sufficiently successful to encourage Cartwright to erect a much larger factory, 84 ft by 25 ft and four storeys high in Fishergate. "Twenty looms were shortly set to work, ten for muslin or muslinette, eight for cotton, one for sail cloth and one for coloured check."[4]

The mill has been described by a contemporary writer, unfamiliar with industrial developments of this kind, as full of 'vast machines'. He further wrote,

> In the summer of 1787, my father and mother paid Dr. Cartwright a visit at Doncaster; but when she entered the vast building, full of engines thundering with resistless power, yet under the apparent management of children, the bare idea of the inevitable hazard attendant on such stupendous undertakings quite overcame her feelings and she burst into tears.

During this period, a Manchester firm contracted for the use of some four or five hundred Cartwright looms. They built a mill and installed a steam engine, but only twenty-four looms had been set to work, some of them weaving check patterns and one weaver tending two looms, when the place was gutted by a fire. Although there was no conclusive evidence, it was generally thought at the time that the fire was the work of hand-loom weavers who feared the venture would rob them of their livelihood.

Certainly the suspicion was so strong that any further projects of this kind were abandoned. But in Doncaster too, things were not going well for Cartwright and, although a steam engine (with a cylinder of 42 inches diameter) had been installed about 1788, he had to give up five years later, heavily in debt.

How successful, from a mechanical point of view, was Cartwright's loom? Opinions differ, and some authorities give more credit to other inventors for the development of the power loom. Yet the looms in the Manchester mill were said to be working satisfactorily before the fire, and although Cartwright failed to make a success of his Doncaster mill, most of the failure was probably due to his lack of business and manufacturing experience. There is no evidence that the looms as such were at fault. The fact that many of them wove muslin-type fabrics, requiring the use of relatively fine and delicate yarns, is an indication of their mechanical adequacy. His loom, in fact, became known throughout the industry as the Doncaster loom.

There is another factor which contributed to the failure of the mill. Looms weaving cotton had to be stopped after weaving a yard or so, in order to 'dress' the warp threads with a glutinous paste, and then to allow this to dry, before weaving was resumed. It was an operation requiring care and skill. The weaver had to spread the correct amount of paste uniformly, with a brush, and afterwards fan the threads to shorten the drying period. Cartwright provided automatic means for applying paste while the loom was working, but this was not successful since he could not dry the threads quickly enough. There was considerable loss of production time, a loss more serious in power-driven machines than in manual equipment.

But even in hand-loom weaving the loss in output due to dressing was economically unsound and in 1803 William Radcliffe of Stockport, with the help of Thomas Johnson (one of his assistants in whose name the patent was taken out) devised a machine which dressed and dried the warps before they were put in the loom. This scheme was a success, both for hand-loom and power weavers. As a result, Radcliffe's own business, along with those of other hand-loom weavers, was shortly faced with disaster.

It was estimated that by 1817 there were about 2000 power looms in Lancashire and this number had risen to 80 000 by 1832.[1]

Wool combing machine

In spite of his misfortunes, Cartwright continued to experiment with other mechanical devices. The most important of these was a wool-combing machine. Although some of the processes of converting wool into yarn had been mechanized, the combing operation continued to be manual. Cartwright took out his first combing patent in 1789; it refers to a relatively simple design, based on a cylindrical carding machine, but in his 1790 and 1792 patents it was obvious that he had attempted to imitate the actions of the hand-combers.

Hand-combing was a highly skilled and arduous operation, and gave employment to a large number of men in the wool industry. The two main operations were lashing to a fixed vertical comb a series of tufts of wool, arranged so that they hung down in a fringe; and, as illustrated, striking downwards into the wool fringe by means of a hand-comb held over the shoulder. Repetition of this gradually straightened out the long fibres and transferred the shorter ones to the hand-comb. The combs were kept hot by being placed, intermittently on a stove.

Subsequently the combed wool was joined to form a continuous length and wrapped in ball form.

In Cartwright's combing machine the frame *A*, 5 to 6 ft in diameter, carried radially arranged comb teeth as shown. At the lower end of the structure *B* a pair of rollers *C* fed the wool to the radial teeth. Operated by cranks and levers, *B*, known as the crank-slasher, moved up and down, the down movements forcing the wool into the comb.

At each revolution a smaller revolving comb *D* struck through the fringe of the wool, removing short hairs. The combed wool left in the radial teeth was then removed by the rollers *E*. Again copying the technique of the hand-comber, the comb table was caused to rub against the flue of a stove and, in a later specification, a stove was placed within the table so that the comb teeth passed through an aperture in its side.

The machine became known as Big Ben, because the violent downward blow of the crank-slasher was a reminder of the prowess of a pugilist known by that name. It appears to have had a moderate success, since even the earlier models did the work of twenty hand-combers.

Upwards of forty petitions for its suppression were presented to Parliament in one session, and Cartwright offered to limit the number of his machines to be installed in any one year.

A bill for the total suppression of the machine was rejected but since it was not entirely satisfactory it gradually went out of use.

Had it proved successful, some 50 000 wool combers in Great Britain alone would have been displaced, as they were ultimately by the introduction of more efficient combing machines. But unlike the later looms, these machines were not based on Cartwright's ideas.

Although Parliament extended his patent for a further fourteen years, Cartwright did not receive any substantial gain from his invention, largely due to a number of infringements, causing him considerable trouble and expense. As a

By 1832 some 80 000 power looms such as these had been installed in Great Britain

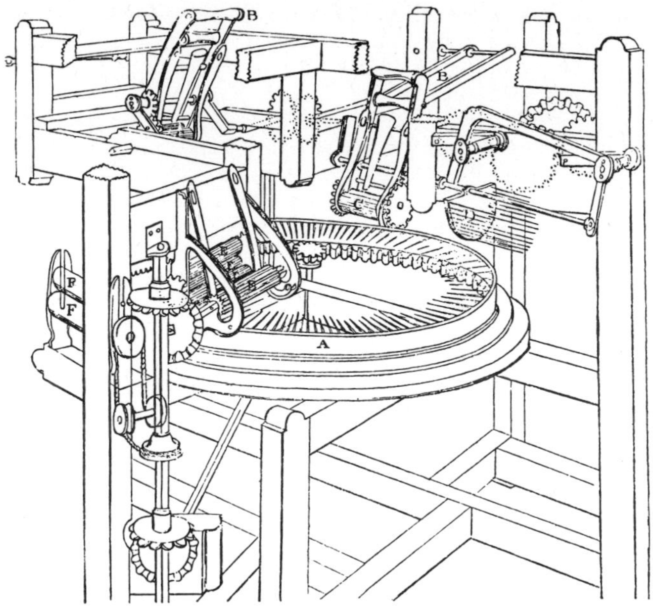

◀ *Cartwright's combing machine worked but was not a complete success and later machines were based on different ideas*

A—circular frame
B—crank-slasher
C—rollers
D—revolving comb
E—rollers for removing combed wool

The primitive methods used by 50 000 manual wool combers until long after Cartwright turned his attention to this process ▶

[Courtesy: Bankfield Museum, Halifax]

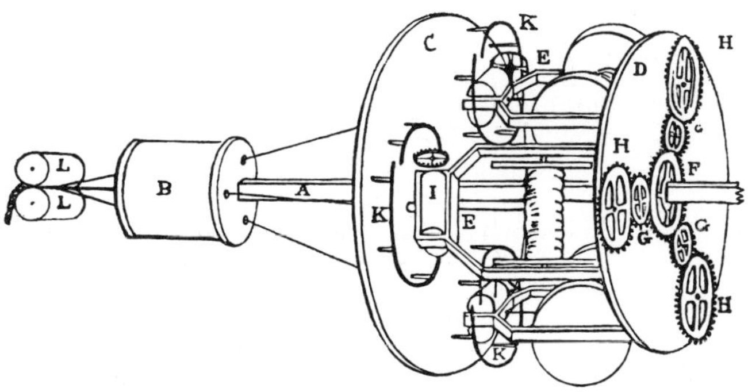

Rope-making machine: another child of his ever-fertile mind
A—axis B—drum C, D—pulleys L—delivery rollers

result, he had a very poor opinion of the value of patent protection and once wrote to a friend,

> A patent is a feeble protection against the rapacity, piracy and theft of too many of the manufacturing class. There is scarcely an instance, I believe, of a patent being granted for any invention of real value, against which attempts have not been made to overthrow or evade it.[2]

Other inventions

He married a second time in 1790; a Miss Kerney, of Portman Square, London. In 1792 he invented a rope-making machine. In the illustration, the large bobbins carrying the strands can be seen between the two discs or pulleys *C* and *D*. These strands pass through the holes in the drum or top *B* and the rotation of the entire assembly (with the exception of the delivery rollers *L*) about the axis *A* inserts the necessary twist.

Among his other inventions were bread-making and brick-making machines, moulded bricks for archways, various agricultural implements, metal packings for steam engines (he was the first to introduce completely metallic packing) and an engine driven by alcohol.

He had moved to London in 1793 and there became friendly with Robert Fulton,* the American marine engineer, who was here studying painting under the Anglo-American artist Benjamin West. He collaborated with Fulton in experiments on the application of steam to navigation and Fulton, in turn, may have been influenced by Cartwright when working on a flax-spinning and a rope-making machine which he patented about 1794. But he acknowledged Cartwright's as the superior machine.

Cartwright's interest in agricultural matters began during his earlier years at Goadby Marwood, where he made experiments while cultivating his glebe. It developed later when he worked on an experimental farm owned by the Duke of Bedford at Woburn.

Between 1800 and 1807, he was awarded a silver and a gold medal by the Board of Agriculture for essays on agriculture, the culture of potatoes, and on manures; and a silver medal by the Society of Arts for the invention of a three-furrow plough.

Financially, however, he was not able to recover the losses he had incurred through trying to develop his earlier inventions and in 1807 some fifty influential men in Manchester signed a memorial in his favour, which was presented to Parliament; and in the following year Cartwright put in a petition for financial assistance. In it he named the burning down of the Manchester mill as the main cause of his failure and estimated that he had spent between £30,000 and £40,000 on developing his inventions.

Among others who gave evidence before the committee appointed to deal with the petition was William Radcliffe who stated that the invention of Cartwright's loom was so important:

> that were it suppressed, or its application no further extended, the cotton trade would be chiefly confined to spinning, and the weavers on the continent, in the end become the sole manufacturers.

Mainly because of the benefits conferred on the country by the invention of the power loom, he was granted £10,000. He bought a small farm at Hollander, near Sevenoaks, where he spent the remaining years of his life.

* *cf. p 126: Robert Fulton—Man of Vision.*

The writer

There was another side to Cartwright's life: he wrote a number of book reviews, including one on Johnson's *Lives of the Poets* and, for a period of ten years, was a frequent contributor to the *Monthly Review*, whose editor considered him to be a reliable reviewer, impartial and objective. The scope of his interests is indicated by the fact that he once read a paper to the Manchester Literary and Philosophical Society on a new theory of the planetary system. But his main literary activity was in poetry. Even when he was in the midst of his studies of the hand-loom he issued a new edition of his poems. He had begun writing at an early age, and his friendship with George Crabbe who became his neighbour in 1772, no doubt proved a stimulus.

His poems have been described as mostly commonplace, yet Scott thought well of his *Armine and Elvira*, which passed through seven editions in just over a year. *The Prince of Peace*, an ode inspired by the American war of independence, described some of the horrors of that war. His daughter has told that the writing of poetry became to him in his later years a frequent solace under disappointment, and contributed with other mental exercises to promote that cheerfulness of spirit for which, even in his most advanced age, he was remarkable.[2]

In 1793, following the collapse of his Doncaster mill venture, he presumably underwent an experience common among those who have met with reverses after prosperity:

> When Fortune smil'd, my friends how kind and true!
> Those friends, when Fortune frown'd, alas, how few!
> Hurt tho' I am, yet why should I repine
> While Susan's unabating love is mine?
> Of this possessed I wealth nor friendship need;
> Of this deprived, I should be poor indeed!

Cartwright—the man

Long after Cartwright had left Doncaster, his friends there watched his career; the editor of the local paper noted his activities and frequently praised his philanthropy and, above all, his freedom from jealousy of rival inventors.

According to his friends, he was a man of "plain and simple manners, with a strong and enlightened understanding."

The benevolent, yet practical, side of his character is well illustrated by some of his non-clerical activities while he was at Brampton. The numerous inhabitants of that parish were mostly very poor and unable to pay for medical treatment. He therefore began to study medical books to be able to help and advise those who were ill.

On one occasion he attended a fourteen-year-old boy who was suffering from what was then known as 'putrid fever'. The remedies he applied, however, proved valueless. On one of his visits he saw in a corner of the room a tub or 'wort', (a fermentation infusion of malt for brewing beer) and this recalled an experiment where he had met with "a piece of putrid meat being made sweet by being suspended over a tub of wort". He subsequently wrote an account of the incident in a medical work published in 1799:

> The idea flashed through my mind that the yeast might correct the putrid nature of the disease, and I instantly gave him two large spoonfuls. I then told the mother, if she found her son better, to repeat this dose every two hours.

A few days later the boy had completely recovered, and there followed at intervals several more recoveries from the same fever with the same remedy, including that of one of his own servants, given twenty-four hours to live by the doctor; and of a farmer of over seventy, whom he found near dying, with blackened tongue and pulse scarcely perceptible.

The more one considers his early interests and experiences, the more difficult it is to understand the sudden application of his talents to engineering. Even to his daughter Mary:

> There was nothing in the pursuits of the first half of his life calculated to lead his mind to study the theory of mechanics, or in his habits to bring him acquainted with their practical application.

A Doncaster man who knew him well at the time he lived there, has described Cartwright as having an open countenance and fresh complexion. Crabbe's son wrote of him: "Few persons could tell a good story so well, no man could make more of a trite one." But although Cartwright was admired for the vigour of his inventive genius, he was also condemned for the rashness of his speculations.

He could be forthright too. Preaching in Mellor Church, near Stockport, condemning the rioting and machine breaking, widespread at the time and being interrupted by a member of the congregation, he denounced the interrupter even more thoroughly than the rioters.

He appears to have kept in close touch with his brother, Major Cartwright, communicating his various ideas and inventions; the Major was always keenly interested in Edmund's progress, even assisting him financially.

The last years

In his extreme old age Cartwright wrote of the fears and forebodings of this period of life under the title *Climacteric Thoughts*. After references to "horrid phantoms" such as fever, palsy, cancer and dotage, he ended:

> May then that Heaven to whose decree I kneel
> Teach me to bear what all are born to feel.
> Give me, as hitherto, thro' trials past
> A patient mind, unshaken to the last!

His prayer was granted, for he remained active and cheerful almost to the end. In 1822 he was working on an idea for propelling hand carriages by steam and, only three months before his death, he wrote to a friend, "I have luckily discovered a method of working an engine by explosion. . . . You will be surprised when I tell you it is by gunpowder." He continued working on the theories involved in this idea, but never reached the point of applying for a patent.

He died in 1823 at Hastings, where he had gone to recuperate, and is buried at Battle Church. He had lived up to the philosophy expressed less than a year before the end:

> With mind unwearied still will I engage
> In spite of failing vigour and of age
> Nor quit the conflict till I quit the stage.

Head of the statue in Bradford by H. C. Fehr

REFERENCES FOR FURTHER READING

1. *Encyclopaedia Britannica*, 7th Edition, Cotton Manufacture.
2. STRICKLAND, M., *A Memoir of Edmund Cartwright*, 1843, Saunders and Otley, London.
3. BARLOW, A., *The History and Principles of Weaving*, 3rd Edition. 1884, Sampson Low, Marston, Searle and Rivington, London.
4. HATFIELD, C. W., *Historical Notices of Doncaster*. Series One. 1886, Brooke, White and Hatfield, Doncaster.

THE GREAT MASTERS

A ruthless giant of the Industrial Revolution who twice married for money, quarrelled with his brother and openly lived with his mistress; a freethinker in religion and politics; a man who could be mean and generous in turns; an ironmaster of genius who is best remembered as a great engineer.

John Wilkinson—Man of Iron

by John Hoyland (Graduate)

The life of John Wilkinson is interwoven with the rise of Great Britain from a minor industrial power to the foremost manufacturing country in the world. He had a major role in the development of iron smelting and was recognised as a leading authority on the metallurgy of iron in many countries.

After a period of considerable activity during the 16th and 17th centuries, iron smelting in England declined rapidly. The reason for this was a fuel problem. Charcoal was extensively used in the production of iron and so rapidly were vast areas of woodland disappearing that Acts of Parliament were passed to prevent it; for it was thought that the supply of timber for shipbuilding and domestic purposes was in danger.

This was the state of affairs when John Wilkinson was born. He was tremendously successful and eventually became a millionaire, a very rare occurrence in those days, but his wealth and his enterprise did not outlast him for long.

The early years

In 1728, Isaac Wilkinson was employed as a workman or overlooker on one of the many haematite iron ore furnaces that abounded in the Workington area of Cumberland. He must have been a man of considerable energy for he and his wife also worked a small farm at Little Clifton. As was the custom on market days the farmers' wives took the produce to Workington in horse-drawn carts.

It was on such a market day that Isaac Wilkinson's wife gave birth to their first son John as she rumbled her way over the rough Cumberland roads in the market cart. The occurrence produced quite a sensation among the simple country folk and they prophesied that "John, sum tyme wod bee a girt man".

Isaac Wilkinson was a man of many talents and quick discernment, but he lacked the firmness and consistency which distinguished his son. Isaac often told the following little story about himself.

> I worked at a forge in the North. My masters gave me 12 shillings a week; I was content. They raised me to 14 shillings; I did not ask them for it. They went on to 16 and 18 shillings; I never asked for the advances. They gave me a guinea a week. Said I to myself, if I am worth a guinea a week to you, I am worth more to myself. I left them and began business on my own account—at first in a small way. I prospered.

About the year 1740 the Wilkinson family moved to Backbarrow in the Furness district where charcoal iron smelting had been carried out successfully for the past 25 years.

It was near one of these furnaces that Isaac took the house 'Bare Syke', and set up his business as 'potfounder' in adjoining buildings. Assisted by his son John, he bought molten iron by the ladle from the furnace and carried it across the street to cast it into his own moulds. Thus he made the common box smoothing iron by using a dry sand moulding technique with properly supported cores. This was his speciality.

In the space of a few years the business prospered to such an extent that Isaac dispatched his son to Dr Caleb Rotherham's academy at Kendal, where he obtained a good education. When he was 17 years old, John was apprenticed to a Liverpool ironmonger who, no doubt, was responsible for his basic business training.

In 1748, the potfounding business having become quite lucrative, Isaac Wilkinson acquired the iron furnace and forge at Wilson House, Lindale, on the River Winster. It was intended that the haematite iron ore should be smelted with peat moss, large tracts of which were to be found in that area. This was a move to break away from the use of charcoal. When John returned home, he and his father experimented with the preparation of the peat before smelting, but they were unsuccessful and returned to using charcoal. But first they cut a canal for transporting the peat to the furnace and tradition has it that an iron barge was used.

The two men were successful in some of their experiments. They made bricks from the local clay and the smoothing iron trade continued to prosper. A large water wheel was erected on Lindale Beck to provide power for grinding the bottoms of these irons.

Partnership with William

John Wilkinson pined after distinction and therefore decided to seek his fortune away from his father. He left for Wolverhampton in the year 1755 or 1756, at the age of about 27. This decision was probably prompted by conversations with the manager of the Backbarrow iron furnace who had been trained there. Being unsuccessful at Wolverhampton

Isaac Wilkinson's house at Lindale, now converted into stables; the iron pipes supporting the roof are said to have been cast by the Wilkinsons

and bent on obtaining employment, John moved to Bilston where he took a position with a Mr Hoo of Bradley.

On 17th November 1755 he married Ann Mawdsley, a woman of ample wealth. The following year a daughter, May, was born and, shortly afterwards, his wife died. The little girl was brought up by a Mr and Mrs Flint. She subsequently married a clergyman without consulting her father, who said he would never speak to her again.

His wife's money no doubt enabled John to buy the Bradley works at Bilston where he built a blast furnace. It was here that he experimented with mineral coal instead of charcoal for the smelting of iron. For four long years he was unsuccessful. A man of indomitable spirit and a practical experimenter rather than a theorist, he persevered and eventually achieved his purpose in 1772, doubling the output from the furnace.

At the time John went to Staffordshire, his father, desperate for success as an ironmaster, took on lease a charcoal furnace at Bersham near Wrexham. That he had an inventive mind is shown by his having taken out patents for methods of casting box irons and rolls in one piece, and for a bellows. Nevertheless, his Bersham enterprise was not successful and by the time John and his younger brother William took over the works in 1761 or 1762, Isaac was a poor, tired old man. He retired to Bristol where he died in 1784, after being dependent on his sons for many years.

The partnership of John and William Wilkinson prospered in the form of the New Bersham Company. They re-modelled the works and, in addition to their father's specialities of box irons and rolls, they manufactured cog wheels, pipes, cylinders, shells, grenades and guns.

The latter led to popular stories about John Wilkinson selling arms to the French during the Seven Years War. Recent research by Dr W. H. Challoner has revealed that Wilkinson did in fact export cannon to the French for use in the War of American Independence, but not during the Seven Years War. These exports were legal and were sanctioned by both French and English Courts.

Temperamentally, William Wilkinson was cast in almost the same mould as his brother. Both were individualists with forceful characters. William was educated at Nantwich in the school of Dr Priestly, the founder of modern chemistry and co-discoverer of oxygen, and later at Warrington Academy. Priestly was a man who tried to cultivate connections with people likely to be of use to him. Knowing that Isaac Wilkinson was a wealthy man, he married his daughter Mary in 1762. Later he wrote letters saying that "I had little fortune with her" and "I unexpectedly found great resource in her two brothers who had become wealthy".

It was unlikely that two such personalities as the Wilkinson brothers would get along together for ever. The characteristics of each seemed to get in the way of the other. The antagonism that persisted came to a head some time after 1793 when John purchased Brymbo Hall, an estate of about 500 acres, near Wrexham.

The soil was poor and the situation bleak, but the site was rich in coal and ironstone. John Wilkinson improved the land by "good tillage and manuring with lime at a rate of ten tons per acre". He also brought under cultivation about 150 acres of wild heath, installed a steam threshing machine and attempted to grow hemp. Industrially, Brymbo was developed by the sinking of pit shafts and the erection of blast furnaces.

It was over this that the two men quarrelled. William said that the Brymbo estate threatened the existence of the Bersham works, and in these circumstances he ought to receive a consideration from John. The story goes that, when John would not agree, William collected a great army of men and marched with them to the Bersham works where they began to break up the machinery. John, on hearing about this, collected a greater number of men and followed his brother's example. In a short time the Bersham works became a great wreck. The story goes on to say that each brother obtained as great a quantity of the spoils as possible.

That this account is true is scarcely credible. John and William were both keen, forthright businessmen and it seems most unlikely that they would choose this particular means of ending a partnership when the Bersham Iron Works was an extremely profitable concern. What really

[Courtesy: W. H. Sansom]

John assisted his father by carrying ladles full of molten iron from this furnace across the road to their pot foundry

[Crown copyright: courtesy of Science Museum, London]

Model of the water-driven multiple spindle cylinder boring mill (with the old trolley feed system in front) and (left) the original drawing of this machine. Firm support of the cylinder provided the accuracy which made Watt's engine feasible

happened was that the case went to arbitration and subsequently, in 1795, the stock and works were sold. It was virtually the end of the Bersham Works but the Brymbo Works continued with the now typical Wilkinson success.

In 1763 John Wilkinson had married for the second time. Mary Lee of Wroxeter brought John another fortune. It was again a marriage of convenience; shortly afterwards he took over the blast furnace and colliery at Willey, near Brosely, not far from Coalbrookdale. It was at Willey on 27th January, 1774, that John Wilkinson first described himself as an 'iron master', when he took out a patent for 'a new method of boring guns and cannon'.

Boring cylinders for Watt

It is difficult to overestimate this invention as it was really the introduction of the principle of slideways to large machine tools. Up to this time the object to be bored, already cast to nearly the intended size, was mounted on a trolley which was drawn on rails towards the rotating boring head. The cutters took the path of least resistance and although any section might be circular, the hole was not necessarily cylindrical. Wilkinson dispensed with the core and cast the workpiece solid to eliminate blowholes. It was then rotated in bearings and the cutter was fed into it along a straight, rigid bar by means of a screw. Almost before the patent was taken out, Wilkinson modified the mill for engine cylinders.

It was about this time that James Watt was developing his invention of the separate condenser. He required a closed cylinder with a stuffing box and cylinder cover and this necessitated more accurate boring than had hitherto been possible. Smeaton had doubted Watt's idea on the grounds that the accurate machining work required could not be done, although he saw the advantages of the new invention.

Wilkinson was commissioned to cast and bore a cylinder for Watt's experimental engine. The engine was successful, which Wilkinson was quick to perceive for he ordered one to supply blast at the Willey ironworks. This was the first engine to be built by Boulton and Watt at Soho. Wilkinson's manufacturing neighbours who were considering buying engines of the Newcomen type showed much interest in his purchase. They waited until they could see what the Boulton and Watt engine could do. Great pains were taken over its construction by Watt and it worked perfectly.

As soon as the blast engine was installed, Wilkinson began to worry Watt about a forging hammer. Watt's single-acting engine did not answer the purpose and it was not until Watt brought out his rotative engine that Wilkinson obtained a hammer to his satisfaction.

There was great trust between Boulton and Watt and John Wilkinson. In a letter to Watt, Boulton wrote—"I admire John Wilkinson for his decisive, clear and distinct character, which is, I think, a first rate one of its kind." Watt also admired Wilkinson for he would have no one else make the cylinders for his engines and eventually sent his son to the Bersham ironworks for a year, to be trained in business studies and mathematics. The boy was also to spend half of each day in the carpenter's shop to fit him for some employment, less precarious than his own.

Boulton and Watt placed orders with John Wilkinson to supply all cylinders for their engines. For this purpose Wilkinson redesigned the boring mill. The cylinder was now clamped securely on to a stationary work support and the boring bar passed through it. The bar was hollow and had a slot cut along its length. It revolved in bearings at both ends of the cylinder. The disc with the cutters was free to slide along the boring bar, but was fastened to one end of a rack in the centre of the hollow bar through the axial slot. The cutting head was fed into the work by a pinion engaging the other end of the rack.

Wilkinson's cylinders were so accurate that Watt could promise his customers that "a 72 in. cylinder would not be further from absolute truth than the thickness of a thin sixpence in the worst part".

These opinions and a long association between the three were destroyed when, in later years, William Wilkinson exposed John for manufacturing and selling several Boulton and Watt engines without permission of the inventors. Deprived of their royalties, Boulton and Watt sued and eventually compelled Wilkinson to pay substantial compensation. Thereafter they manufactured their own components at Soho. William's exposure of John indicated the bitterness and hatred which then existed between the two brothers.

Other inventions

Wilkinson often said "more is done by scheming than by working". It appears that, when pondering over a new project,

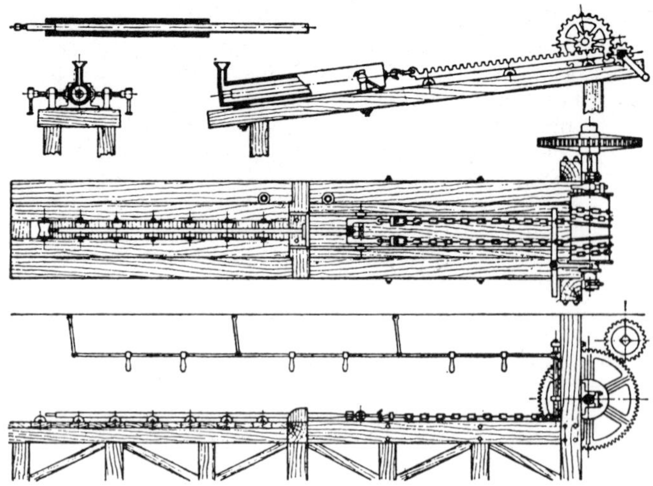

Wilkinson's lead pipe drawing technique of 1790 remained in common use until extrusion was introduced for this purpose a century later

[Courtesy: Radio Times Hulton Picture Library]
At the height of his power, John Wilkinson struck his own token money for payment of his workers who could use it in the neighbourhood

he would lie in bed holding an iron ball over a copper basin, the idea being that, if he accidentally fell asleep, the ball would clatter into the basin and wake him up. He was so enthusiastic that he would manage without sleep in order to continue his experiments. On one occasion he was so tired, not having rested for three days, that he slipped and was only saved by a workman from falling into a ladle of molten metal.

He took out patents on many of his inventions, some of which showed remarkable anticipation of things to come. One was a coal-cutting machine, known as the 'Iron Man'. When he attempted to introduce it into his own mines he was thwarted by the miners who said that if the Iron Men were going to cut the coal they could also set the props and roof timbers.

This mechanisation problem, causing labour troubles, was surely a foretaste of the future.

In 1789 Wilkinson took out a patent for a method of rifling ordnance, similar to the idea that Armstrong later proposed. The rifling of small arms had been carried out in Germany since the 17th Century but the difficulties in applying the principle to cannon and other large guns would have been insuperable; Wilkinson's idea was soon forgotten. The fact that he had considered the problem shows that his thinking was ahead of his day.

Being financially interested in lead mines and lead smelting, he took out a patent for a new method of making lead pipes in 1780. He practised it on a large scale and it proved very profitable to him. It is probable that this was his most important invention next to his boring mill. The old method of pipe making was to cast lead into a circular mould round a mandrel or core bar. Local defects could hardly be avoided and the pipe was, in many ways, unsatisfactory. Wilkinson's idea was an extension of the old method. After casting in the usual manner, the wall thickness of the pipe was reduced by rolling or drawing through dies. The reduced pipe was of much greater length and sounder quality. Perhaps Wilkinson may have heard of this idea for it was described as early as 1728 by M. Fayolle, but in the eyes of English Law John Wilkinson was 'true inventor' as he was the first to put the idea into practice.

In 1792, he patented an improved steam-driven rolling mill. A pitch chain from the beam of the engine was wrapped round the axle of the upper roll. The reciprocal motion of the engine gave partial rotation to the top roll. Only the top roll was driven, the lower one being rotated by friction. The drawing submitted with the patent showed no method of applying rolling pressure and it is probable that he intended the weight of the top roll to provide this.

This mill, with 5 ft diameter and 6 ft long rolls, each weighing some 8 tons, was built for rolling puddled bars at the Bradley Works but was not successful, as the rolling action used only a quarter of the circumference of the rolls.

Wilkinson's was not the first rolling mill, for Boulton and Watt had invented and built a model in 1781 which astonished all the ironmasters and Wilkinson immediately ordered a full-scale mill from them for his Bradley Works.

Wilkinson's patent for making pigiron was perhaps more successful than he realised. The drawing submitted with the patent showed a 10 ft high blast furnace, blown with three tuyeres and built up inside a casing of iron plates. This had a far-reaching influence in modern foundry practice. He had used this furnace for experiments in smelting iron from the cinders of refineries and reheating furnaces which, up to that time, had been thrown away. The furnace was very much like a cupola for remelting iron, which it could do without much alteration. The introduction of this cupola idea should be credited to John's brother William. In France, where William spent part of his life, the cupola was known for many years as the *Fourneau á la Wilkinson.*

Although it is possible that John Wilkinson obtained some of his ideas from his workmen who could not afford to take out patents, he was an inventor of no mean order. Unlike his father's, most of John's patents appear to have been put into production. His reputation had become so widespread that he was wrongly credited with many discoveries. Cochrane's distillation of coal to obtain tar was one, probably because Cochrane's boiler and apparatus were on a piece of land adjacent to one of his works.

Wilkinson had experimented with hot blast in 1801 and had scrapped the idea because the hot air charred the leather goosenecks, the pipes connecting the blast main to the tuyeres. A simple arrangement of concentric metal tubes eventually solved this problem and increased output. The reason why a man like Wilkinson was beaten by this difficulty is left to conjecture. It is also peculiar that this far-seeing man used his first steam engine as a blowing machine to replace the bellows, and his second to pump water from the tail race of a water wheel to the dam to provide power for cannon boring. It took quite some time before steam engines were used to drive rotating machinery.

Iron barges

Above all he believed in iron and in using it for every conceivable purpose. He presented a cast iron pulpit to the Wesleyan Chapel at Bilston, also window frames, pillars and so on. It became known as Wilkinson's Iron Chapel. He foresaw that some day houses and ships would be made of iron.

When the barge builders were experiencing difficulty in obtaining timber, they were able to take up an independent and disobliging attitude with Wilkinson who decided that he could well do without them. He set his blacksmith to work forging iron plates from cast bars for the barge *Trail.* On 6th July, 1787, in front of a large crowd and after a salute from 32-pounder guns, the *Trial* was launched. The following week he wrote one of his characteristically brusque letters to Mr James Stockdale—

> Yesterday week my boat was launched—it answers all my expectations and has convinced the unbelievers who were 999 in a 1000. It will be a nine day's wonder and afterwards be like Columbus's egg.

It appears from a contemporary notice that the boat was made from $\frac{5}{16}$ in. thick iron plates, riveted together on an elmwood framework, a total weight of about 8 tons. The boat was used for carrying castings between Brosely and Bridgenorth on the Severn and on the Birmingham canal. By October 1788, Wilkinson had launched another two boats of similar construction. But many years elapsed before ships were generally made of iron.

To commemorate the successful launching of the boat, he had medals and tokens struck and issued; this was the period when he issued his own coinage. A larger amount of copper coin needed to be minted because of great changes brought about by the factory system and the consequent sub-division of labour, necessitating many wage payments. The Royal Mint did not have the capacity to do this and several industrialists issued coins (or tokens) of their own. Wilkinson's tokens, struck at the Soho mint, were mainly of $\frac{1}{2}$*d* face value and carried such motifs as a tilt hammer, a brig under full sail, a female figure with a cog-wheel and a boring tool, and a smith working at his anvil. In all cases, Wilkinson's bust was on the obverse. In 1788 he issued his only silver coin of 3*s* 6*d* and, before the Government intervened and stopped the practice, he had issued notes to the value of one guinea.

It would appear that Wilkinson was too engrossed in his iron to become involved in public life. The only public office that he ever held was that of High Sheriff of Denbighshire in 1799. It was in that year that the first iron bridge in the world was built for the purpose of joining the fast-growing towns of Madeley and Borsely on opposite sides of the Severn. Abraham Darby III of Coalbrookdale was chief mover of the scheme but John Wilkinson was an active promoter and subscribed part of the cost. He urged that the bridge should be made wholly of iron and not partially, as suggested by the architect, and as usual he imposed his will. He was accused of being iron mad, but the bridge of over 100 ft span was erected and still stands.

Wilkinson took an interest in land and its improvement. He was extremely successful with Brymbo. Castlehead Rock was a different matter. The Rock rose out of the Winster inlet of Morecambe Bay and was almost an island at high tide. Wilkinson purchased this barren place in 1765 and proceeded to build a house on the north side. He had the bare rock covered with soil, which was carried in horse panniers, to make beautiful gardens. It was under a laurel tree in the gardens of Castlehead that he kept an iron coffin in readiness for his death. He took delight in showing this macabre item to his friends and also kept several more of different sizes which he offered gratis to them.

Character

The personality of John Wilkinson was very complex. This strongly built man with the deep sonorous voice and pockmarked face was what the Englishman calls a 'character'. He was perhaps eccentric, generous and mean, arbitrary and unscrupulous, masterful, passionate, perseverent and vigorous. He had the power to organise men and the ability to select those of the very highest calibre.

That he was a generous man cannot be denied. He gave to his old and trusted employees substantial pensions when they retired, a feature not contained in many industrial welfare schemes today. After Priestly's house in Birmingham had been wrecked by a mob who suspected him of being a revolutionist, Wilkinson provided him with money and shelter. When, eventually, Priestly went to America, it was with a large sum of John's money.

On the other hand Wilkinson could be extremely miserly. One of his notebooks contained details of the exact cost and the expected quantity of beer that should be consumed during a traditional ceremony of blowing in a new furnace. On another occasion, one of Wilkinson's managers found him arguing with his housekeeper about a discrepancy of threepence in the housekeeping accounts. Yet, the same manager's request for £20,000 for a lead mining scheme was answered by an immediate "well £20,000 you must have".

One of Wilkinson's biographers mentions his "ill regulated

[Courtesy: Science Museum, London]

The Brosely ironworks on the Severn from a painting in 1788

domestic life". During his second wife's life, while she was 'imprisoned' at Castlehead, he took a mistress by the name of Ann Lewis who bore him three children in rapid succession. He was about 80 years old at this time. When his second wife eventually died at the age of 83, he had the children declared legitimate and they took the name of Wilkinson.

It would also appear that Wilkinson had many other mistresses during his life as he set up various houses as alternative accommodation. He was certainly not puritanical in outlook. In the year 1785 he became acquainted with the *savants*, while working in France on the erection of one of Watt's engines, and he met them again later in connection with the Paris waterworks scheme, for which he supplied 40 miles of cast iron pipe. He acquired much of their doctrine, becoming a freethinker, somewhat tainted with their irreligion. Concerning the church and the monarchy he was always outspoken. This, not unnaturally, gained him quite a few enemies.

Although it would appear, from the bibliography, that ample documentation is available on the life of John Wilkinson, his biographers are rather sketchy about his youth.

The relationship between Isaac Wilkinson and his son must have been excellent. There can be no doubt that the father inspired the son, and passed on to him a deep love of iron; a love that developed into an obsession, for iron became John's whole life. How he met, and subsequently married wealthy women must be left to conjecture. Without doubt these marriages enabled him to get sufficient capital to buy other businesses. It was this money, coupled with his initiative and hard work, that eventually made him a millionaire.

He was a hard-headed businessman who would only carry on an enterprise if it proved profitable to him. When he bought a business he usually improved it until it was making a good return for the capital outlay.

Many of Wilkinson's inventions, including the 'cupola', showed great foresight and proved to be profitable. His lead pipe making machine was very successful and remained so until the process of extruding lead pipe was invented. It is unlikely that James Watt would be a household name but for Wilkinson's cylinders. But for Watt, Wilkinson might never have re-designed his cannon boring mill for cylinders. It is surprising that this boring machine was never patented by Wilkinson.

Obviously the quarrel between John and William Wilkinson went deeper than the Brymbo-Bersham affair, and there must have been other matters about which they differed. The clash of their similar personalities was the root cause of their differences and eventual break.

Wilkinson was concerned that there should be an heir to carry on the businesses. Having no legitimate grandchildren, he tried his nephew, Joseph Priestly junior but after three years discarded him as being unsuitable. Thomas Jones, son of another sister, was taken on subsequently.

John Wilkinson died at Bradley on the 14th July, 1808. He was a millionaire, his enterprises were established in several parts of England and Wales. Castlehead was left to Ann Lewis, his mistress, and the remainder of the estate was to be put into the hands of executors for the three children. On the decease of all three, Thomas Jones was to inherit.

Jones knew the background of the children and proceeded to claim the whole of the estate. He disputed the will in Chancery but the result of protracted proceedings was an upholding of the will. Meanwhile, most of the fortune had been wasted on faulty management of the enterprises, leaving no-one to benefit. The nephew eventually died a bankrupt.

Things also went wrong with the iron coffin; it was too small to take the wood and lead coffins in which the body had been transported to Castlehead. Eventually a larger one was cast and all three coffins were deposited. Over the whole was erected a large iron obelisk weighing over 20 tons. This had been ordered by Wilkinson and is shown in the illustration. It does not, however, bear the epitaph as he wrote it himself, which ran:—

> Delivered from persecution of malice and envy, here rests John Wilkinson, iron master, in certain hopes of a better state and heavenly mansion, as promulgated by Jesus Christ, in whose gospel he was a firm believer. His life was spent in action for the benefit of man, as he hopes, in some degree to the glory of God.

Obviously, he had returned to religion at the time of his death—at least outwardly.

John Wilkinson was a man of tremendous personality. So strongly did his character impress itself on the imagination of the people with whom he came into contact, that many legends about him were in circulation, both during his lifetime and afterwards. One of these was a belief that on the seventh anniversary of his death he would re-visit the works at Bradley, mounted as usual, on his grey horse. On that day several thousand people gathered on Monmore Green expecting him to put in an appearance.

This legend probably sprang from the concern that Wilkinson expressed over the running of his businesses. He had left written instructions as to how they should be managed and developed.

Was he a crank or an eccentric, or was he a genius? Certainly his ideas benefited the iron smelting industry, both in England and on the Continent, on which the whole structure of 19th century engineering was built up: and some of his achievements live on in modern times. One can only agree that "his life was spent in action for the benefit of man".

Acknowledgement

The author wishes to thank the Staff of the Library, Barnsley College of Technology and Mr R. Bassinder for their assistance during the preparation of this article.

REFERENCES FOR FURTHER READING

CHALLONER, W. H., *John Wilkinson (Ironmaster)*. History Today, 1951.

DICKINSON, H. W., *John Wilkinson—Ironmaster*. Hume Kitchin, Ulveston, 1914.

PALMER, A. N., *John Wilkinson and the Old Bersham Ironworks*. Transactions of the Honorable Society Cymmrodorion, 1899.

SANSOM, W. H., *Man of Iron*, Metallurgia, 1963.

SMILES, S., *Lives of Boulton and Watt, Metcalfe and Telford; Boulton and Watt;* John Murray, London, 1865; 1874; 1878.

The lower part of the cast iron pillar he had cast for his own monument; the epitaph finally inscribed was not quite the one he wrote for himself

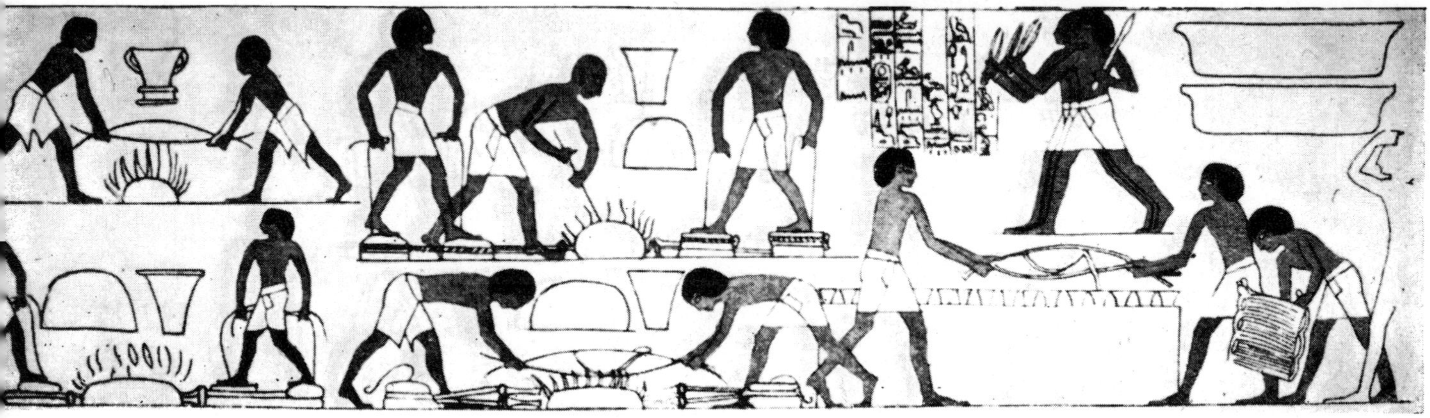

Fig. 1. This Egyptian tomb painting of the second millenium BC depicts the casting of the huge bronze doors for the temple of Ammon at Thebes: similar techniques were later used for iron smelting

Iron and Steel Making

by F. C. Thompson DMet, FIM

Iron and steel have been used in weapons, tools and fasteners for thousands of years. This is the story of how ferrous metallurgy developed from prehistoric puddling of spongy material in a hole in the ground to the processes which finally made cast iron and rolled steel generally available as basic engineering materials.

Even before the bronze age, iron was known and used by man. There exists a necklace found in a pre-dynastic Egyptian tomb in which iron beads alternate with precious stones, no other metal being present. Such iron was clearly esteemed above gold and was too rare to be used in anything but ornaments for the mighty.

Finds of the earliest iron objects are meteoric. Their high nickel content, (4–30 per cent), at once distinguishes them from later man-made metal, whilst in the natural state some, but not all, show the characteristic octohedral pattern which has never yet been completely imitated. It was at one time believed that meteoric irons were too brittle to be hammered but this is now known to be fallacious.

Another objection to the use of such iron by early peoples has been the belief that, with the tools then available, it would have been impossible to cut pieces from large meteoric masses. This is disproved by such facts as the finding of a copper chisel still embedded in a meteorite; and the use made by the Eskimos of such iron prior to their first contact with outsiders. At one period the ancient Egyptian name for iron signified "something hard from heaven".

This, the earliest known ferrous material, then, played little or no direct part in the development of civilisation, but it may well have initiated a belief which, for thousands of years, affected the worker in metals. A curious feature cropping up repeatedly in history is the reverence paid to the god-like smith and, later, the belief that he and the devil are closely related. From realising the celestial origin of iron it was no long step to the belief that the worker in such a heavenly material must needs himself be superhuman. Traces of human sacrifice, too, are not rare in ancient stories of the melting and working of iron, among which may be noted Needham's reference,[1] in his work on the history of iron in China, to the wife who threw herself into the furnace so that the melting of her husband's steel for the imperial sword might not fail.

Man-made iron

It was not until perhaps 2000 BC that iron was produced by man, at any rate in any significant amounts. Among the sources of our information are the discoveries of the archeologist and subsequent metallographic examination; tomb paintings from Egypt, such as that shown in Fig. 1 (which actually refers to bronze but shows the state of the metallurgical art) and ethnological studies of primitive peoples. We have, for instance, an eye-witness account only a hundred years old of the manufacture of Indian Wootz steel by methods which differed little from those used in prehistoric times.

There are sound grounds for the belief that metallurgy started somewhere on the southern slopes of the Caucasus. From this centre it spread slowly in two streams, east and west, the eastern development leading to the well-established iron production in India and China.

Wootz, a product of primitive Indian tribes, has been produced for some 3000 years. The initial product, reduced directly from the ore as a sponge, was repeatedly hammered down on a stone anvil andreheated in charcoal until sufficient carbon had been introduced to give a steel capable of being hardened by quenching in water. Why such steel should have had such outstanding qualities in cutting implements is still incompletely understood.

In India, until the coming of Europeans, iron and steel making techniques had not advanced for 1000 years. They had, however, produced forgings of wrought iron of a size which are the wonder of the world. Their Wootz steel was unequalled until the invention of the crucible process, and even after that was used for surgical instruments and similar purposes: it was in an attempt to reproduce it that Faraday started his pioneer work on alloy steels which, incidentally, led to the discovery of the very first stainless steel—containing some 50 per cent of platinum!

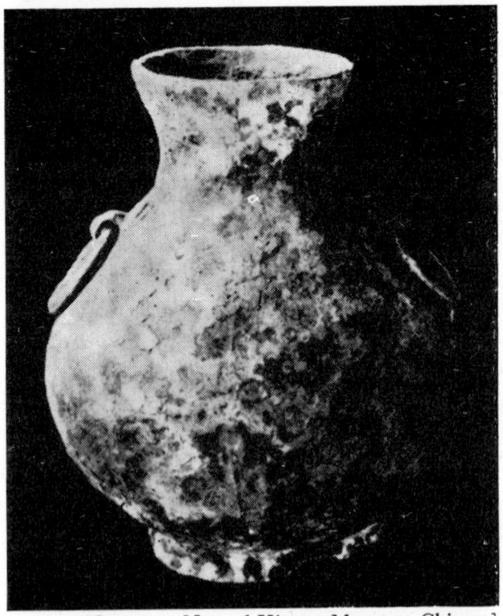

[Courtesy: Natural History Museum, Chicago]

Fig. 2. Chinese cast iron vase, dated between 200 BC and AD 200: cast iron was to remain unknown in Europe for another 14 centuries

It was in China, as the recent, extraordinary work of Needham[1] has revealed to the Occident, that the greatest ancient advances took place, molten iron being a fairly common metal there 1500 years earlier than was the case in Europe. Fig. 2 shows a cast iron vase, made possibly as early as 200 BC. In size, too, Chinese metallurgists produced a casting which is one of the wonders of metallurgy. In the gardens of the Kai-Yuan monastery stands a cast iron lion, made in AD 953, which is almost 20 ft high, 17 ft long, with a wall thickness of up to 8 in. The degree of organisation needed to pour such a casting, quite apart from the melting of the metal in backyard furnaces, is fantastic.

The western stream of metallurgical advance split up and became the source of mid-Eastern and European metallurgy. Diffusion of knowledge was slow: the Hallstadt civilisation, roughly mid-way between the late bronze and early iron ages, which flourished in western Austria somewhere around 1000 BC, did not reach this country till about 500 years later.

The conservatism of technology up to the last few centuries is remarkable. Even as late as AD 1550 there was no specialisation as we now know it. The same man might well be the original prospector for the ores, the mine-manager who supervised their extraction, the engineer who designed the furnaces and other equipment by which the metals were won and the hammers by which the product was worked into a useable form, the metallurgist in charge of the processes and even the financier who provided the necessary funds.

Although what we should now term a joint stock company was already known to Agricola[2] in the middle of the 16th century, it was more usual for the head of a family of smiths to combine all these functions. The 'mysteries' of this, like other arts, would be strictly preserved until the time came for him to transmit them to his son.

Ore mining

The red or reddish-brown, colour of the iron ores would attract attention to them and the weathered outcrops could be worked with suitable tools. In prehistoric times these were antlers used as picks and animal shoulder blades as shovels.

As time went on, the surface ores would be worked out and deeper mining became essential. Simple methods of raising small loads were available very early on but a much more formidable difficulty presented itself in the different character of the ore itself: instead of the soft, weathered material on the surface there was the extremely hard underground ore body. Such material could not be worked with the tools then available and by the first half of the second millenium BC fire-setting was introduced. Fig. 3 shows a surmise, based on the results of actual excavation, of what was then possible.[2] The soot from the fire is still to be seen on the walls, and the atmosphere in the confined space may be imagined. That slaves and convicts were predominantly employed in the old mines is some indication of the conditions which obtained.

Against the wall of ore a wood fire was built and the ore raised to a temperature at which considerable expansion occurred. It was then quenched by throwing water on the wall which resulted in lumps of the shrinking ore cracking off. For some unexplained reason, vinegar was at times substituted for water, being regarded as particularly effective for this purpose. It is said that Hannibal made the road across the Alps for his elephants by the same technique.

Slave labour, at times female, was later used for hand-picking and crushing and it is difficult to realise the extent of the human suffering involved in obtaining the metals used by Greeks and Romans.

Whether iron ores were calcined before smelting is uncertain, but the experience gained from weathered material must surely have suggested that reduction to the metal took place more readily when this had been done.

The earliest smelting furnaces were no more than holes in the ground some two feet or so in diameter and about

[Courtesy: ICI Ltd]

Fig. 3. In fire setting, the first non-manual method of underground mining, the ore was cracked by the heat. The sketch shows a reconstruction of a copper mine in the Tyrol, dated between 1600 and 800 BC

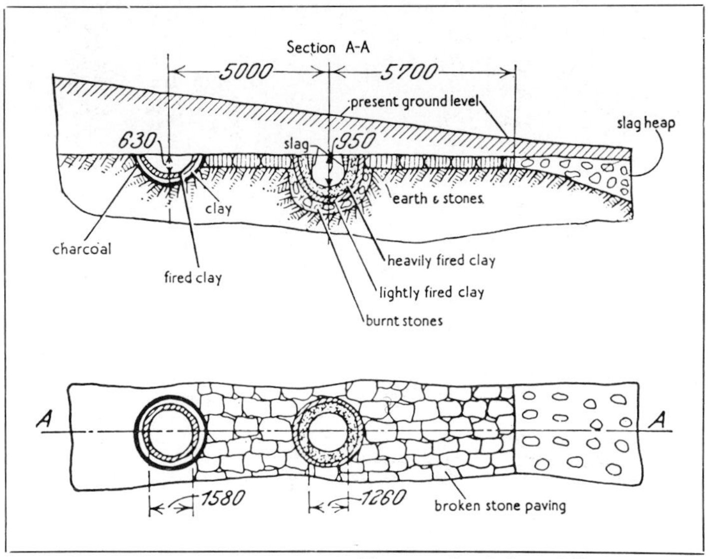

[Courtesy: Routledge and Kegan Paul Ltd]

Fig. 4. Pre-historic furnaces excavated in Austria, dimensions in mm

Fig. 5. At the time Agricola wrote in the 16th century, hole-in-the-ground furnaces were still in use, though now lined with brick

as deep. Lined with clay, they were dried out by burning wood fires in them, and to this fire a charge of wood or charcoal with a small amount of ore was added. After the furnace was full, further fuel was added from time to time until the reduction was judged complete. The product, a crumbly mass of iron, slag and unburnt carbon, was levered out, the 'bloom' being a few pounds or so in weight. A diagram of a furnace, one of the earliest subjected to careful investigation,[4] is shown in Fig. 4. How little change occurred during the next three thousand years or so is nowhere better illustrated than by a picture in Agricola's book, reproduced in Fig. 5, which shows an almost identical furnace, save that it was brick-lined.

Then as now, higher output, increased productivity and larger production units were the aims of the producer. Although, in principle, furnaces changed little for thousands of years, there were attempts to improve efficiency. The first of these appears to have been an effort to raise temperatures, either by building a simple dome to conserve heat, or by raising the height of the furnace. Weathering and deliberate destruction mean that we know less about furnace construction above ground than below.

Primitive African furnaces, starting from a shaft only a foot or two high, gradually developed to more than man-height. The effect of an air-blast was known from earliest times. Fig. 1, for instance, shows slaves working foot bellows. In addition, furnaces were sited against the windward slopes of hills—bole hills—even in comparatively recent times and there is still a district in Sheffield known as the Bolehills.

The early goat-skin bellows were later improved (Fig. 6) and in mediaeval times much larger ones were extensively operated by water-wheels as shown in Fig. 7. In China there were most ingenious hand-worked force pumps which enabled the producers to melt iron many centuries before this had become possible in the West.

A European trend which became marked in classical times was the great increase in the size of the producing unit. Thousands were employed at the same time in Greek and Roman mines and mints. Whether the production of iron was conducted on a similar scale is doubtful; it is not a metal which lends itself nearly so easily to mass production, nevertheless, Roman slag heaps containing some 100 000 tons have been found in Noricum.

At the temperature which could be reached in these early furnaces, iron remained a paste-like solid. The blooms produced, therefore, had to be manipulated with such tools as were available; at first, in all probability, wooden holes and a partial destruction of the furnace were necessary to extract the metal. The output per heat, even as late as the 16th century, was therefore only of the order of a hundredweight.

Wrought iron and steel

When iron ore is reduced by carbon at temperatures far below its melting point, the metal produced is soft and pure. A typical analysis is Carbon 0·06, Silicon 0·11,* Manganese 0·10, Sulphur 0·015, Phosphorus 0·69 %. Although this spongy mass can be consolidated by hammering on a stone anvil into a form sufficiently hard to cut corn or flesh, it is useless against armour. Neither as a tool, nor as a weapon is it as good as steel.

The heavy defeat inflicted by the Romans on the Celts at the battle of Addua in 223 BC was attributed by Polybius in the main to the poor quality of the Celtic iron which

> easily bent, and would only give one downward cut with any effect, but after this the edges got so turned and the blades so bent that, unless they had time to straighten them with the foot against the ground, they would not deliver a second blow.

It is clear from the Icelandic sagas that the same sort of difficulty existed there as late as AD 1000.

Because of this unreliability of the iron weapon in com-

* *Partly in the slag which is included.*

parison with bronze, the latter persisted long after iron was available; it also gave rise to the legends of miraculous swords such as King Arthur's *Excalibur*. Then, as now, the choice of alternatives was determined by cost and suitability.

The development of human civilisation is not as sharply divided into Copper, Bronze, and Iron Ages as some archaeologists would have us believe. Then, as now, the peasant used the implements which he could afford, the soldier the best available.

An enormous step forward was taken when it was first realised that the hardness of iron could be greatly enhanced by long heating in a carbonaceous material. There are many ancient objects known in which this had clearly been done, at first almost certainly by accident. Further, when a cutting instrument was required, a simple form of case-hardening was resorted to at an early date; or a fully carburised iron edge was welded to a soft iron backing.

Soft iron cannot be appreciably hardened by heat-treatment but this carburised iron could be improved enormously. Exactly when and where steel was first used is not certain but the Hittites in Anatolia have been credited with this achievement.

To the Greeks of Homer's day the advantages of quenching of steel were certainly known. This is much more than hinted at in those lines in the *Odyssey* which describe the blinding of the giant Polyphemus by Odysseus:

As when the smith an hatchet or large axe
Temp'ring with skill, plunges the hissing blade
Deep in cold water, (whence the strength of steel)
So hissed his eye around the olive wood.

Simple as these lines may at first sight appear, they are in fact full of illumination. By around 800 BC it was evidently well known, not only to the smith himself but also to the ordinary citizen, that steel could be vastly improved by quenching. How far the advantages of tempering after the quenching were known is by no means clear, but the curious nostrums used as quenching baths may well have provided this element of the treatment automatically. The symbolism implied in the use of the urine of a red-headed boy is not difficult to unravel in view of the connection of Mars, the red star, with the God of War; and such ideas have persisted almost to the present day. As a student, some 55 years ago, I was credibly informed that a Sheffield firm of toolmakers imported from Egypt barrels of camel's urine for quenching certain complicated tools.

The Romans, so far as one can see, made little advance in the metallurgy of iron. They were in essence manufacturers, applying on a bigger and better scale already known techniques. They increased output vastly, both in the mines and the works; they introduced water-power on a commercial scale to perform operations previously done by hand, but for weapons they still used, to a large extent the brass which was their main metallurgical contribution.

With the collapse of the western empire the so-called dark ages set in though, as recent work has shown, they were in fact far less dark than was once supposed, and the eastern empire carried on, more or less without a halt, until finally overthrown by the invasion of the Turks in the 15th century. When Rome fell the barbarians were at first dependent on the iron already available and on that procurable by plunder. They had, however, already produced small quantities themselves and the cessation of Roman supplies rendered them more and more dependent on home-produced metal.

[Courtesy: Macdonald and Evans Ltd]

Fig. 6. Hand-operated bellows for mine ventilation in the 16th century: note the rock lashed to the top as counterweight

Roman furnaces had produced semi-fused iron blooms up to about 100 lb in weight but, as civilisation recovered and expanded, the growing demand led gradually to improvements in furnace design which raised the temperatures attainable and thus the output obtained. Another effect, due to the increasing shortage of timber, was a continuous increase in the use of mineral coal. It had been used in furnaces by the Romans but on a comparatively small scale. During the middle ages it was used increasingly, though charcoal was still employed for many purposes.

The resultant increase of furnace temperature and the demands of fuel economy led to a variety of furnace designs until something not altogether unlike the modern blast-furnace was evolved. Fig. 7 is a reconstruction of a 16th century blast-furnace. It bears an obvious resemblance to the German *Stückofen* which consisted of two truncated cones, placed base to base and up to 16 ft high. The height of the earliest blast furnaces, which were generally square in cross section, was less than 20 ft and the hearth in which the molten iron collected was possibly less than 1 ft across.

Cast iron

Towards the end of the 15th century, a revolutionary advance was made which rendered it possible for the first time in Europe to melt iron as against producing it in the semi-liquid phase. This made iron castings possible, at first mainly for armaments and later as the basic engineering material, before the production of cheap malleable iron and steel.

On occasion small pieces of white cast iron have been recorded in much earlier excavations and these have, quite unfairly, been attributed to carelessness on the part of the digger in permitting modern material to contaminate his finds. But long before the iron-age, copper castings of high purity had been made, for which a temperature not far short of 1100°C is required. If soft iron is allowed to remain in its reduction furnace sufficiently long to take up some 2 per cent of carbon, it will begin to melt at 1150°C.

The temperature which could be reached in mediaeval furnaces was just too low for regular carburisation and fusion of the iron in economic amounts. Hitherto the only ferrous materials had been wrought iron, or wrought iron carburised in the solid state; this gave a product analogous to the old shear steel from which, for instance, the Damascus and Toledo blades were forged. Molten steel was not to make its appearance for another three hundred years. But cast iron, which could be made in relatively large quantities and cheaply, was now placed at the disposal of western mankind. In this country it was used for naval guns before 1543 and it came into increasing use for cauldrons, fire-backs, brackets and dozens of other purposes. As demand increased, furnaces increased in size and blast pressures were raised.

The earliest means of providing forced draught for furnaces was, as Egyptian paintings show, men blowing through wooden pipes. Hand and foot bellows made from skins were later employed and used by primitive peoples even down to very recent times. More ingenious means were used in later mediaeval times. It was the application of water power, however, which rendered possible the use of the relatively huge bellows which made the blast furnace feasible, as shown in Fig. 7. The water which drove the wooden wheel turned the wooden camshaft which operated the bellows. As usual, where water power was not available, similar plant was driven by tread-mills and at times worked by animals.

In the late Middle Ages, the deforestation resulting from the demand for fuel became more serious. An increasing tendency to look to coal instead of charcoal tended to move the main centres of production to the coal mining areas, and districts such as the Weald of Sussex, the earliest centre of English iron making, were driven out of production.

The steel age

It was a clockmaker, Huntsman, who, in AD 1740, first produced a cast of molten steel. Finding that the properties of the steel then available were too irregular for his springs, Huntsman had set about finding a more uniform material and, after many failures, succeeded in melting 50 lb charges of 'blister steel', i.e., wrought iron, preferably Swedish, which had been heated with charcoal for days at a temperature around 1100°C in large stone chests. This blister steel had a high carbon content and low melting point.

But first he had to discover a mixture of clays for his crucibles, sufficiently refractory to stand up to the much higher temperatures involved and then deoxidise the steel, presumably by the reduction of silicon from the walls of the clay crucible.

Steel, almost uniform in composition and ranging in carbon content from some 0·6 to 1·5 per cent, became an article of commerce. Such material, though suitable for cutting tools, wire, etc., was still too hard to be suitable for ordinary constructional purposes and far too costly. Matters remained in this state for more than a hundred years, until there came the greatest advance of all when, in 1856, Bessemer announced his 'pneumatic' process of making steel by blowing air into molten iron.

The first Bessemer furnace went into operation three years later and we could almost say that the modern world had been born. This is not the place to consider the claims made for William Kelly of Kentucky as the first inventor of the process—he may well have been—but that it was Bessemer who introduced the first constructional steel to the engineering world there can be no doubt.

The part played by Mushet in making the process practicable for ordinary irons is too often forgotten. The conversion of high carbon cast iron into a low carbon, malleable steel by blowing air through the molten charge resulted in the steel becoming highly oxidised and brittle unless produced from iron of high manganese content, such as certain Swedish irons. With most British ores the product proved far from satisfactory and Mushet realised both the cause of the trouble and means of overcoming it. By adding ferromanganese to the molten charge after blowing but before pouring, the iron oxide present in the metal is reduced and replaced by innocuous oxide of manganese. The steel then solidifies as a sound ingot, free from blow-holes.

In the original Bessemer converter, the lining was acid (i.e. siliceous) and unable to withstand the attack of the basic slag required for the removal of sulphur and phosphorus. The process was a success, therefore, only with iron low in these elements, a fact which made it useless for the large deposits of phosphoric ores throughout the world. This was later overcome by the invention of Thomas and Gilchrist. They gave the converter a basic lining of dolomite which could withstand attack from a slag high in lime, that was needed to absorb the phosphorus from the iron.

In the ordinary reverberatory furnace, the temperature was inadequate for making steel and it is to the genius of Siemens that we owe its modification into the regenerative open-hearth process. By using the waste heat from the furnace to heat the incoming producer gas and air, a temperature sufficiently high to melt and cast carbon-free iron is attainable.

The fundamental difference between the Bessemer and the open-hearth processes lies in the fuel. In the open-hearth it is the producer gas, whereas in the Bessemer process the charge carries its own fuel, mainly the silicon content in the acid process; and both silicon and the phosphorus in the basic version. It follows that the open-hearth process has a higher degree of flexibility as regards the type of iron used; it can deal with steel scrap much more easily.

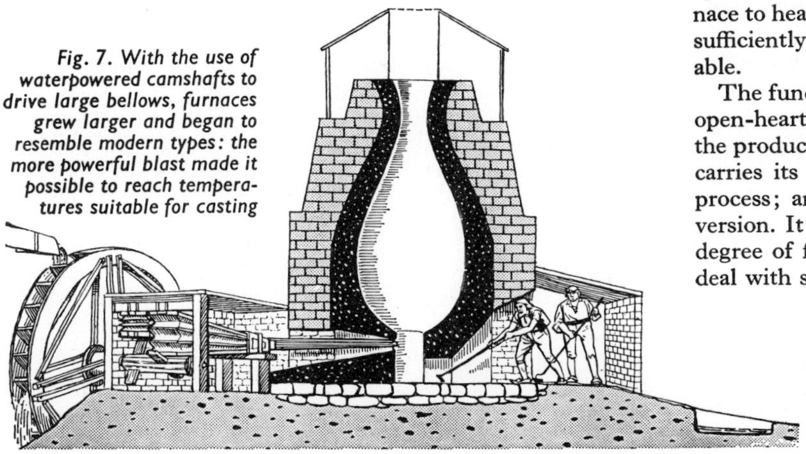

Fig. 7. With the use of waterpowered camshafts to drive large bellows, furnaces grew larger and began to resemble modern types: the more powerful blast made it possible to reach temperatures suitable for casting

[Courtesy: Macdonald and Evans Ltd]

Forging and working

By the middle of the 16th century, the size of bloom produced per heat had risen to several hundredweights and to deal with such masses something more than a sledge hammer had become necessary. Fig. 8 shows that in Agricola's day a well equipped smithy included a tilt-hammer in which cams on a wooden shaft, driven by water power, raised the hammer which then fell under its own weight.

In a larger forge hammer of a type which was in widespread use from the beginning of the 16th to the 18th century, the helve or shaft, about 8–9 ft long and 30–40 inches in diameter, was made of stout wood, clamped at intervals with iron hoops. The cast-iron head, through which the shaft passed, was fastened with wedges into a cast-iron collar or hurst. Above the hammer shaft ran a strong, but elastic spring-board of timber, the 'rabbet', which was firmly bolted to the post and to the frame through which it passed. The axle-tree turned by the water-wheel, was connected to a ponderous circular frame, the 'arm-case', in the rim of which were inserted blocks of wood, shod with iron on their surfaces. Each of these cams in turn lifted the hammer and jerked it against the rabbet. The blow exerted was, therefore, not merely that of the falling beam and head, but was increased by the resilience of the rabbet. Such hammers could deliver up to 160 blows per minute.

In 1838 John Nasmyth conceived the idea of a direct-acting steam hammer. Steam had already been applied to turn the camshaft of the tilt-hammer but the new design was to increase enormously the size of the forging which could be produced and, incidentally, did much to ensure the success of the Bessemer process. Hammers of 25 tons weight were in operation in Sheffield as early as 1865.

The energy exerted by such hammers is mainly expended superficially and, as ingots increased in size, these methods failed to work the central portions. As early as 1856 Bessemer took out a patent for a press which would exercise a kneading action to the very core. He does not appear to have followed up the idea, however, possibly because Whitworth had already taken out a patent in 1856 for a hydraulic press capable of exerting pressure up to 2000 tons for the forging of guns and armour-plate.

In comparison with hammering, the rolling of metal is a comparative newcomer. Small hand-operated rolls for the reduction of ingots of gold or lead had come into use relatively early; they were certainly used in the production of sheets for coinage blanks as early as the 16th century.

Probably the earliest record of any mill for this purpose is contained in one of Leonardo da Vinci's sketches about AD 1500. Slitting rolls were used in the iron industry in the 17th century for splitting bars into long, thin strips for nail making. In 1783 Henry Cort introduced grooved rolls for the production of bars, rods and sections This increased output enormously.

Cort's mill was provided with a fly-wheel to even out the torque and continued in use without any essential modification until in 1857 the three-high mill stand was introduced. This eliminated the waste of time involved in the catcher passing the bar back to the inlet side, the metal now being worked in both directions. In 1866 the first reversing mill at Crewe facilitated even larger ingots.

A search was also going on for finishing methods to speed up the work on lighter sections before the metal had become too cold, and in 1861 Charles White patented a mill with the stands of rolls so close together that a bar was in three pairs of rolls at the same time. By 1872 20 mills of this kind were in use in South Wales alone. Almost simultaneously, George Bedson had patented a continuous mill for the production of wire-rod, each pair of rolls being driven at increased speed.

On such equipment Richard Johnson and Nephew of Manchester were soon producing 100 tons of ¼ in. rod per week, in coils over a quarter of a mile long.

Conclusion

In the story which I have tried to tell, the interdependence of the metallurgist and engineer—at first one and the same man—is outstandingly evident. Without the other neither could have given humanity reliable metallic raw materials.

Each development in power sources has opened the gate to new advances in metallurgical production techniques, and these in their turn have rendered mechanical engineering advances possible. Thus, steam power made it possible to increase the size of forgings and reduced cost of strong components on which its own further progress depended.

The millwright of the 18th century has developed, by gradual steps, into the engineer of to-day. He has created substitutes for much human labour; devised plant for the quicker and cheaper manipulation of the metal; and, in return, the iron and steel industry has provided him with new materials, at a price and in the form which have made it possible to contemplate further steps forward—a leap-frogging process which is by no means at an end.

The steam hammer, tilting furnaces, forging presses, heavy cogging mills and the continuous rolling mill, to mention a few only, were the essential contribution of the engineer to iron and steel making. It was not only the mechanical engineer, however, who played his part. The invention of the arc-melting furnace and the growing employment of electric drives were essential steps, while increased machinery loads have made increasing demands on the civil engineer in the design of foundations.

Fig. 8. Even in the 16th century no large smithy was complete without a water-powered forge-hammer: note the overhead controls.

REFERENCES FOR FURTHER READING

1. NEEDHAM, JOSEPH, *The Development of Iron and Steel Technology in Chinc.* Newcoman Society, London, 1959.
2. AGRICOLA, GEORGIUS, *De Re Metallica.* 1556. Basel (Translated from the Latain by Herbert Clark Hoover and Lou Henry Hoover. Dover Publications, New York, 1950.)
3. SINGER, CHARLES, HOLMYARD, E, J. HALL, A. R. and WILLIAMS, TREVOR I., *A history of Technalogy.* 1954—58. Clarendon Press, Oxford.
4. SCHUBER, J. R. T., *History of the British Iron and Steel Industry.* Routladge and Kegan Paul, London.
5. ATITCHISON, LESLIE, *A Hiitory of Metals.* 1960 Macdonald and Evans, London.

THE GREAT MASTERS

The Ingenious Mr Bessemer

by F. R. McKears*

In the course of a long and eventful life, Bessemer attained a position of great prominence and considerable wealth. En route from his obscure village boyhood, he took out more than 150 patents applicable to a wide variety of industries and founded the steel industry as we know it today. This has tended to overshadow his other achievements.

Even the most indulgent father sometimes finds it necessary to draw a line and to bid his offspring toe it. The line Anthony Bessemer drew was intended to keep his youngest son, Henry, out of the typefoundry which provided the family's livelihood. It was no use, of course: juvenile enterprise was more than equal to the occasion.

Small boys being what they are, this might have been expected. What could hardly have been anticipated was young Henry's performance: in circumventing parental prohibition, he also penetrated a carefully preserved parental secret.

His father, Anthony Bessemer, was born in London but at the age of 11 was taken by his parents to Holland, where he served an apprenticeship in mechanical engineering. While still under articles, he assisted in the building of the first steam engine ever erected in that country. At 21, he moved to Paris and, by the time he was 26, had been elected a member of the Academy of Sciences in recognition of his work on microscope design. Still on the right side of 30, he became a senior official in the Paris Mint and had much to do with the production of the French coinage and various kinds of medallions. His position was assured: he was a respected man of considerable prominence.

Then came the French Revolution. Abandoning everything, Anthony Bessemer and his young wife, escaped the mob by the skin of their teeth and landed in England with little money and no other assets. He soon restored the family fortunes by making gold chains of novel design—they looked massive but were, in fact, extremely light. London jewellers bought them as fast as he could make them and, with the capital thus earned, he bought a small country property in Charlton, Hertfordshire, to which he retired. It was here that his three children were born, Henry, the youngest, arriving on 19th January 1813.

A village boyhood

Retirement soon palled on Anthony Bessemer's active mind and, reviving an old interest in printing type, he began to cut letter punches for the famous Caslon typefoundry. Henry Caslon became a close family friend and, indeed, stood godfather to Henry. Later establishing his own foundry, Bessemer senior produced type which gained an extraordinary reputation for durability: its working life was twice that of the type made by other founders.

Life at Charlton was good. The foundry flourished, the family prospered. There was a carriage in the coach-house and there were horses in the stable. Servants came from the village to do the housework and to tend the garden. On Sundays, Papa would sport his top hat and Mama would deck herself in her church-going regalia to lead Anthony junior, Annie and young Henry into the Bessemer pew. The family was contemporary middle-class England in epitome: enterprisingly industrious, virtuously prosperous, solidly respectable. All this tended to set the Bessemers somewhat apart from the village folk, quite a few of whom depended directly or indirectly on the family, the house or the foundry, for their living.

There is no record that Henry was particularly oppressed by this sense of being set apart: on the contrary, such evidence as exists suggests he rather enjoyed it. It led, on the one hand, to a firm family unity which did much to shape his character and, on the other hand, to a degree of solitude for which he was temperamentally suited.

One of the great unifying influences in the family life was the flower garden. Anthony Bessemer, during his years in Holland, had become an enthusiastic tulip fancier—tulips were the fashionable flowers in those days—and his Charlton garden was always a delight to the eye. From his father, Henry learned a love of flowers which remained with him all his life. Every member of the family was rather more than competent at drawing and painting the garden and its individual flowers. Henry had at least his fair share of the family artistic ability, but it was his sister, Annie, who was the real expert. In later life, this outstanding ability of hers was to lead to one of the major turning points in Henry's career.

As a boy, Henry had a special understanding with one of the family dogs. Together, they would wander the Hertfordshire lanes and fields, poking their noses into everything in the immemorial manner of a boy with a dog. One of their regular ports of call was the flour mill at the end of the village, where Henry was quite capable of sitting for hours, watching the fall of the water on the overshot wheel or the operation of the great wooden spurwheel with its massive oak shaft and polished hornbeam cogs.

* *Editor, 'Torch'.*

On his walks, he would frequently pick up a bit of roadside clay and fashion it into a model of something. Leaves, flowers, birds, animals—sometimes real, sometimes imaginary—these were his subjects. Modelling, like painting, came naturally to him, and he was forever badgering his father to let him make a mould from one of his models and then cast it in type metal.

This was where the parental foot was put down. Anthony Bessemer was prepared to explain the foundry operations, but Henry was not permitted to go into the works unless accompanied by his father, and was absolutely forbidden to enter the melting house. This, of course, was as good as an invitation, and on one of his clandestine visits Henry discovered that small amounts of tin and copper were being added to the normal alloy.

This, he considered, had more to do with the durability of his father's type than had the shape of the serifs, to which it was usually attributed. In the face of this precocious talent, Bessemer senior admitted defeat, allowed Henry to leave school at the age of 15 and took him into the foundry business. Thus, the only schools the boy ever attended were the village school in Charlton and a private school in Hitchin.

It was from his father that Henry received the only engineering training he ever had. He bought him a small lathe, made sure he became a competent technical draftsman, and kept him hard at the workbench for a couple of years. Before long, Henry produced his first invention: a machine for making toy bricks of white pipeclay.

The lonely city

In 1830, the family moved to Camden Town. London astonished Henry; he had never imagined anything like it. He spent many an evening during his first weeks there tramping the streets, not so much exploring as trying to get on terms with this extraordinary city. But how different were the London streets from the country lanes he loved! Henry was quick to feel the loneliness in a big city and admitted in his later writings that one of the motive forces in his eventful career was a determination to make this sprawling, soulless place sit up and take notice of him.

For the moment he had to carry on quietly enough. He perfected a method of making metal casts of natural objects and produced copper-coated whitemetal castings of such things as a rosebud with leaves, a particularly convoluted piece of Scotch kale and a group of three prawns lying on a large vine leaf. These he exhibited at the Topliss Museum of Arts and Manufactures, which then occupied the present site of the National Gallery in Trafalgar Square.

This activity, though it certainly attracted attention, failed to produce much money. His tinkering had, however, shown him how to produce similar castings from the plaster medallions which were sold by itinerant street vendors at a penny a piece. Taking his ideas a stage further, he experimented with various alloys until he had formulated one which could be used to cast dies suitable for stamping ornamental scrolls, medallions and similar low-relief subjects in cardboard.

A firm of art publishers gave him his first order, for 500 copies of a Raphael cartoon stamped in relief on buff cardboard.

Each impression cost him 3*d*, including materials: he found a ready sale for them in wholesale quantities at half-a-crown a time. Thousands of these, in a multitude of designs, were produced for bookbinders, cardboard box makers and others, and at least one handsome quarto edition of the Bible was published, its cover decorated with a leather panel bearing a Raphael reproduction.

The young Bessemer painted by Lehmann

So Henry Bessemer produced his first commercial work. His parents, of course, were delighted. So were their Camden Town neighbours, the Allen family. A close friendship had developed between the two families, and Henry had been observed casting sheep's eyes at Ann Allen, a pretty girl who was so taken with this good-looking, well built, 18 year old six-footer that, when he proposed to her, she hesitated only as long as was ladylike. But marriage was not yet possible: Henry could not properly afford to set up house.

The Government Stamp Office

In the course of his diemaking and stamping activities, Henry had stumbled across another, distinctly different, method of making from an embossed paper stamp, dies capable of turning out thousands of facsimile impressions. No great skill was needed and the materials were cheap. Horrified to realise how easy it would be for an unscrupulous man to defraud the Government Stamp Office by supplying forged stamps for affixing to the many documents which required them, he kept his discovery to himself. Then he learned that the Stamp Office itself reckoned that it was losing about £100,000 by frauds of various kinds. These, though differing in detail, generally depended on the removal of a genuine stamp from an out-dated document so that it could be used again.

Bessemer bent his mind to the problem of designing a stamp which could neither be removed from a document nor easily forged on a stamping press. His conclusion was that it would be necessary for the stamp to perforate the paper. Arming himself with a suitable press, he gave up his other interests until he had perfected a perforating die. This, he was sure, would make his fortune. If he could save the Government such a large sum annually, they would surely reward him suitably?

He set off, one bright morning, for Somerset House, full of optimism and carrying with him specimens of his new stamps. He had also provided himself with a few of his own forgeries, just to prove his point. Sir Charles Presley, President of the Stamp Office, was, to put it mildly, astonished by the forgeries and delighted with the young man's proposals. A few days later, Bessemer was offered the position of Superintendent of Stamps, at a salary of £600–£800 a year—handsome enough, in those days. Overjoyed, he rushed off to Camden Town to tell his parents all about it—and to tell Ann that they could now be married.

This clear-eyed young lady, however, couldn't see what all the fuss was about. Why, she wanted to know, couldn't the Stamp Office just date the existing stamps as they issued them?

Somewhat deflated, Bessemer went back to tell the Stamp Office what his fiancée had said. In the event, the dies used for embossing the existing stamps were modified by means of three moveable inserts so as to include the date of issue. Sir Charles Presley was delighted all over again: no new machines, no new methods, no more swindles—and, as he explained to Bessemer, no Superintendent of Stamps, either.

So, at the age of 20, Henry lost a job he had never started. For a long time he tried to get some sort of payment in recompense for his efforts and expenses, but nothing was forthcoming. During the next 45 years the grievance rankled. Until he was 65, when belated recognition was at last made, his only profit from his excursion into stamps was the bitter taste of shabby treatment from officialdom.

From plumbago to velvet

His battles with the Stamp Office could not be allowed to monopolise his time: he had to live, and he wanted to get married. Casting about, he bethought himself of a curious tale he had been told about the manufacture of lead pencils.

The only British source of plumbago was a mine near Borrowdale, in Cumberland. So valuable had the mineral become that the mine entrance had been virtually fortified against the gangs of robbers who, for more than a century, had systematically bedevilled the lives of the mine owners. The miners were required to change into working clothes on arrival and, after their work, to change back into ordinary clothes under inspection, lest some of the mineral should be concealed about them. A military guard escorted the wagons taking the plumbago to Keswick, where it was packed and sent to London for auction. Prices averaged 40*s* a pound.

The cost of handsawing the plumbago nodules into strips suitable for pencils had the effect of doubling this already high price for, during the operation, considerably more than half the material was reduced to useless powder. A pencil was a precious object in those days.

Bessemer sought a mechanical cutting method, and found it in fine saws made from discarded watch springs. Mounting them in a reciprocating saw frame with their cutting edges uppermost, he lowered the plumbago on to the saw by means of a fine screw, operated from above. He fitted a spring friction clutch to the frame, so that it stopped if the blade ran into 'black diamond'—extremely hard carbon particles embedded in the material.

The machine worked well and reduced the amount of plumbago wasted to a small fraction of the quantity lost in handsawing. He tried to get a firm of pencil manufacturers to let him saw their plumbago for them at nominal cost, sharing with them the value of the materials saved. They would not entertain the idea: they could not suffer their precious 'prepared plumbago' to leave their premises.

In disgust, Bessemer turned his attention to the plumbago sawdust, which he could buy for about half-a-crown a pound. He developed a means of compressing it, while hot, into blocks which proved perfectly suitable for pencil-making. As a result of this invention, it became standard practice to make pencils of powdered and compressed plumbago, the various degrees of hardness being produced by different degrees of heat and pressure. The thing was worth a fortune but Bessemer sold the invention outright for £200. He still had some way to go before becoming a businessman.

On the other hand, he probably needed the money. He had just married his Ann, without waiting for the one great invention which would make his fortune and, at the same time, form the main interest in his life. But he was quite sure it would turn up one day.

He designed a new typecasting machine which, since it appeared to function perfectly, he sold to an Edinburgh firm of typefounders. Unfortunately, after a particular valve had been operated some 7000 times, it became clogged with type-metal, and no way of overcoming the difficulty could be found. The machine was allowed to drop into obscurity.

He bought, for £65, a rose engraving machine with which he engraved plates for the printing of patent medicine labels, coupons and suchlike, where protection against fraudulent imitation was necessary. He also developed new alloys for the engraved rollers used in paper embossing and printing.

He perfected a means of stamping Utrecht velvet—a harsh, stubborn, worsted material quite unlike the silk and cotton velvets which, with their beautiful figured patterns, were much in vogue with the upholsterers of the day. For a time Henry thought all was going to be well: his Utrecht velvet, stamped with reproduction Genoa patterns, had been ordered for furnishing a suite at Windsor Castle. With so august a testimonial, how could it fail? At first he could get as much as 6/- a yard for putting the fabric through the rolls of his press but, as more material was produced, so the price for embossing came down, and eventually he sold his machines and engraved rollers to a Banbury firm of Utrecht velvet weavers.

Type composition

One interest led to another. Together, they provided the young couple with a livelihood, but nothing so far seemed likely to offer the chance which Henry knew would come along one day. At about this time, however, he met James Young, who wanted someone to design a machine for composing type. A skilled hand compositor could set perhaps 1700 characters an hour: a machine should increase output and reduce costs.

The job was not easy, but Bessemer finally built a machine with a keyboard similar to that of a piano. As the operator pressed a key, so the appropriate character was released from a case mounted above the keyboard and was carried by its own weight down grooves in an inclined plane until it reached the point where, joined by other characters arriving in the proper order, it was assembled into a line. Here a boy sat, inserting spaces as required and sliding complete lines on to a galley beside him, thus freeing the assembly slide for the next line.

This, under the conditions then obtaining in the printing industry, ought to have provided the breakthrough Bessemer was looking for. The machine got off to a good start: it was installed in the offices of the *Family Herald*, and attracted the attention of Mr Walter, proprietor of *The Times*. Walter suggested that a young lady, employed in the office and without any kind of typesetting skill, should try to set 5000 characters an hour—about three times the output of a skilled man—for ten consecutive hours on each of six consecutive days. She managed this quite easily and was presented with a £5 note for her efforts. Uproar followed, since skilled compositors were not prepared to see untrained girls take their living from them—and another Bessemer invention slid into limbo.

Bronze powder and automation

Now Henry's sister, Annie, came along with the simple request which was to change his entire life. Her young brother was supposed to know all about decorative lettering, so would he please letter the title page of her album of flower paintings? Of course Henry would. No ordinary black ink for Annie, either: he would gild the letters. Off he went to buy some 'gold' powder, and was mightily astonished to be charged 7/- an ounce for it. After all, it was only brass . . .

But the manufacture of the powder was a monopoly and involved a long, laborious and secret hand process: there was obviously a future for anyone who could design a machine to do the job. The story of Bessemer's two years' work on the project is too long to be told in full here, but it is an entertaining one which illustrates his emergence as a man of business ability.

After several false starts, he finally developed a mechanical method of producing from bar a powder which consisted of tiny, microscopically thin flakes. Having tried samples out on the trade—and established that he could sell it at a thumping profit—Bessemer told his friend James Young about it. Young put up the capital and Bessemer built his bronze powder factory at Baxter House, St Pancras. He did not patent the process but went to great lengths to ensure that the secret did not leak out. He designed each of the five necessary machines himself, doing all the detail drawings personally and having the components made by a number of different firms—further confusing the trail by detailing parts of different machines on one drawing—so that no-one could form any conclusion as to what the proposed machines would actually do.

He installed each machine in a separate room and made them all 'self-acting', as he termed it: they started work automatically when material reached them and rang a bell when they had finished a batch so that they could be stopped from a remote point. Taking a hint from the Cumberland plumbago mine, he restricted access to the works: until his factory had become established beyond doubt as a commercial success, he was the only person to see the assembled machines or to enter the rooms in which the process was worked. Even the man who looked after the steam engine was on the other side of a brick wall: all he ever knew of what went on was that his engine drove a shaft which disappeared into a stuffing box.

Bessemer made a lot of money out of his bronze powder. His success, of course, was at the expense of the original German suppliers, who sent over an emissary with instructions to see what he could do by way of industrial espionage. Bessemer, incognito, and hugely amused, met the man and, while saying loudly that he could tell him nothing, carefully contrived to let fall a few strategic details of one of his more troublesome false starts—just enough to ensure the maximum waste of Teutonic ingenuity and an interesting time on the journey up the garden path.

With commercial success assured, Bessemer invited his wife's three brothers, Richard, John and William Allen, to run the factory for him. He never disclosed precise details of his process, and for the next 40 years none but he, his brothers-in-law and two assistants ever saw the inside of the premises.

He was now in his thirties and the father of three children, Henry, Alfred and Elizabeth. His very comfortable income at last allowed him to satisfy his hankering for the country life he had loved so well at Charlton but he could not bring himself to go quite so far afield. Eventually, he bought an estate at Highgate—then, still a country village—and settled down to much the same kind of family life as he had known in childhood. He called his new home 'Charlton', and there he did all he could to foster that same family affection and unity by which his own father had set so much store.

Not that this meant any kind of retirement: on the contrary, he entered on so many new schemes that he soon had to take in a partner, Robert Longsdon, who later married his sister. In the late 1840's and early fifties, he developed a new kind of press for extracting the syrup from sugar cane, a steam fan for ventilating mines, a method of treatment of pit coal, and a centrifugal pump capable of raising 20 tons of water an hour for land drainage.

Those were the days of the development of the railways and he simply had to have a finger in that pie. He designed a continuous braking system which applied hydrostatic brakes to all the coaches of a train, instead of merely to the engine wheels. Above each of the carriage wheels was mounted an iron box within which was a rectangular rubber cell carrying a wooden brake shoe. The assembly was connected to a continuous pipe running the whole length of the train to the locomotive, where the engine driver could turn the water pressure on or off instantly. With the pressure on, the flexible rubber expanded and brought the wooden shoe into contact with the wheel. If the water pressure was relieved the rubber contracted, raising the shoe from the wheel.

He also designed the forerunner of the modern vestibule vehicle and, by showing that the tractive effort required to overcome the wind resistance offered by a single suitcase measuring 3 ft × 1 ft, carried, as was the practice, on top of a railway coach travelling at 30 mile/h, was equivalent to that required to pull a weight of 2880 lb inside the carriage, he was at least partly responsible for the introduction of luggage vans.

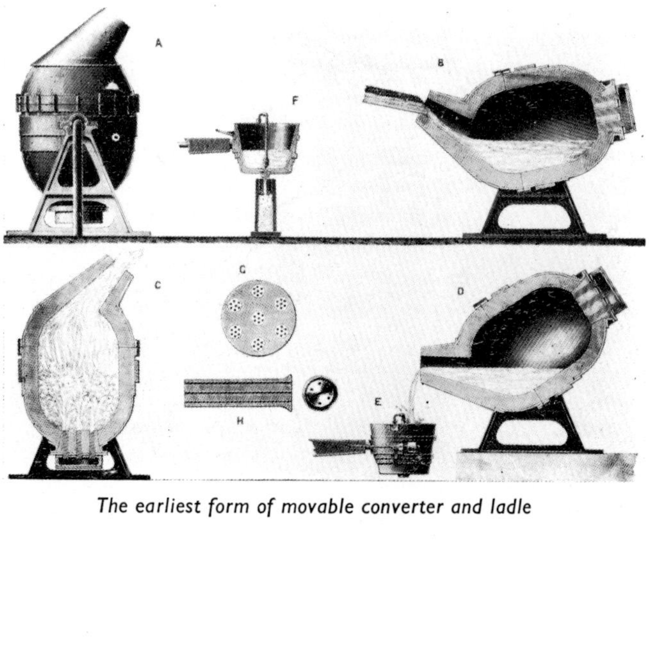

The earliest form of movable converter and ladle

The list of his activities during his thirties is a long one: optical glass manufacture, sheet glass rolling, plate glass polishing, processes for the paint, oil and varnish industries—all these, and many others, kept him busy and kept his name before the public. Most of them succeeded in bringing him profit, for he had developed an acute commercial sense and wasted little time on ideas which seemed unlikely to earn their keep. Not for nothing had he become known as 'the ingenious Mr Bessemer'.

The converter

In 1853, Henry Bessemer was 40 years old. His life had brought him both fame and substance, yet only now was he on the brink of that one great discovery which he had known all his life would be his one day. It seems almost churlish to say that his invention of the Bessemer converter, which established the world's modern steel industry, was made by accident, but it is certainly true that he was trying to do something else when it turned up, and that there was no-one more surprised than he at the result of his experiment.

He had become excited about the condition of British ordnance when the Crimean war broke out. No rifled cannon existed: the smooth-bore guns could fire only round shot. Bessemer designed an elongated projectile, capable of carrying a bigger charge of explosive, and gave it rotation by means of grooves cut in its side. Such a shell would, of course, emerge from a smooth-bore gun barrel and keep to its trajectory, travelling nose first all the way. Now he found himself embroiled in another set-to with officialdom: according to the War Office, the invention was not of a nature to be used, or experimented upon, by the War Department. So he took his projectiles to the French Army which, in the person of Commandant Minié, was impressed to the point of enthusiasm. But, they wondered, was it altogether wise to fire projectiles weighing 24 lb and 30 lb from a 12-pounder gun which was made of none-too-trustworthy cast iron?

Bessemer returned to England, determined to improve the malleability of cast iron. The story of his experiments is an important chapter of industrial history which cannot fairly be compressed into a few words. It must suffice here to say that it took him about two years to design his converter which, accepting a charge of molten pig iron, allowed air to be blown through it to remove the carbon and thus to produce soft, malleable iron. The addition to the melt of small, controlled quantities of carbon produced steel.

Steel itself was no novelty, but for many centuries it had been made in such small amounts that its uses were inevitably restricted. Now, for the first time, it could be used freely for all kinds of purposes.

Bessemer introduced his process at a meeting of the British Association in Cheltenham and his paper was reported in *The Times* of the following day, 14th August 1856. He refused to sell his patent outright, even though one firm offered him £50,000 for it. In less than a month, he received £27,000 for individual licences—but disaster lurked round the corner. Ironmaster after ironmaster reported total failure: not a single licensee could make the process work. Bessemer, the cry went up, was a fraud, a charlatan.

The inventor then returned all the licence money and spent two years—and some thousands of pounds—trying to find out what had gone wrong. Eventually he discovered that the cause of the trouble was the phosphorus content of the iron: by chance he had used low-phosphorus iron in his experiments whereas the ironfounders had used high-phosphorus metal. Announcing his findings, he tried to

Ingot production from Bessemer steel: the converter on the left is being charged

induce them to have another go. But they had had enough and refused to touch the process.

Determined not to be beaten, Bessemer, in partnership with Robert Longsdon, built his own steelworks in Sheffield, and proceeded to turn out large quantities of good quality steel at £10 a ton—as compared with the £65 or £70 a ton charged for steel produced by the earlier processes.

Continental countries were quick to adopt the Bessemer process—sometimes under licence and sometimes simply pirating it—but British ironmasters kept aloof because British iron ore tends to be high in phosphorus content. Not until Sidney Gilchrist Thomas showed that such iron could be converted in a vessel, lined with a basic firebrick instead of the acid type which had been used previously, did the British industry capitulate and finally accept the Bessemer converter.

A prophet in his own country . . .

When the 1856 patent expired, Bessemer calculated that in 14 years he had received £1,057,748 in royalties. Further, the Sheffield works had been enormously profitable, distributing in dividends during the same period 57 times the subscribed capital. When the works were sold, they fetched 24 times the subscribed capital, so that in all the partners received 81 times their original investment in 14 years.

Now the name of Bessemer was truly world-famous. He had his detractors, of course, as every famous man has. Those who annoyed him most were the men in high official positions whose concern for their own dignity was exceeded only by their inertia. It also took a long, long time to convince the Army that its guns ought now to be made of steel, that cast iron was no longer good enough.

It was an uphill struggle to get Bessemer steel accepted for structural purposes by official architects and engineers. Finding himself irritated beyond endurance by delays at the Admiralty, Bessemer wrote to *The Times* in 1864, pointing out that the Bessemer steel shells with which Russia was arming herself would easily penetrate the iron plates of the warships then being built in Britain; that France was building steel warships; that it was high time Britain woke up, and that this was a grave question which demanded a speedy answer: however it took the British Government 12 years to arrive at it.

And it was virtually only in his own country that he was thus treated. Foreign countries piled honours upon him; no fewer than 13 American towns were named after him—but in Britain recognition was slow. Come, however, it eventually did. The Telford Gold Medal, the Prince Albert Gold Medal, the Howard Prize of the Institution of Civil Engineers, the Presidency of the Iron and Steel Institute, a Fellowship of the Royal Society, the Freedom of the City of London—that city in which once, long ago, he had felt so lonely. These were the highlights, but there were many others.

The accolade

At the age of 65, in 1878, Henry Bessemer had another tilt at Authority. His Stamp Office adventure still rankled and, as he had done so often before, he wrote to *The Times* about it. This time the newspaper took up the cudgels on his behalf and, thus encouraged, Bessemer made representations to the then Prime Minister, Lord Beaconsfield. The outcome was his knighthood, an honour which particularly pleased him because he could share it with his wife, who had made her crucial contribution 45 years previously.

By the time he received his knighthood in 1879, he had retired from business life—but not from inventing. For nearly 20 years he lived in a big house in Denmark Hill, inventing, among other things, machines for polishing diamonds and for grinding mirror glass, an asphalt paving technique, and a ship with a saloon which remained steady, no matter how violently the vessel rolled. This ship, the s.s. *Bessemer*, came into being solely because Bessemer was a poor sailor and suffered agonies on the cross-Channel journey. He designed a suspended saloon, mounted on large working beams as shown in the illustration, each of which carried a hydraulic cylinder at either end. The saloon was supported on an axis running its entire length. A set of hydraulic force pumps, driven by a separate steam engine, supplied water to the hydraulic cylinders under the control of a 'steersman', whose job it was to keep an eye on a spirit level which indicated the movement of the saloon floor from the horizontal. The controls required had been reduced to a single handle: all the steersman had to do was to move this as required.

Before the ship was built, Bessemer had tried his system out on a small model built in a meadow at the back of his Denmark Hill house and established that, at its maximum oscillation of ten complete rolls per minute through 30 deg., the saloon floor could easily be kept within 2 in. of the horizontal. He expected that the greater inertia of the full-scale saloon improved this result.

The saloon was not the only novel feature of the ship: it was a 'double-ended' vessel, ie, it could travel equally well in either direction, and could thus leave harbour without having to turn round. At least, that was the idea: in the event, on 8th May 1875, the vessel crashed into the jetty at Calais and was wrecked before being properly tested and no more was ever heard of the idea.

In between these activities, his experiments on telescope design, his work on a solar furnace and the more domestic tasks of designing small household gadgets and superintending the running of his large garden, his retirement was, to say the least, not inactive. But in June 1897 his wife died. They had lived affectionately and devotedly together for 63 years and he never really recovered from her loss.

A few months later, on 15th March, 1898, Sir Henry Bessemer, FRS, a charming, enterprising, alert old gentleman who wore his 85 years lightly, himself died, leaving behind him a world which had changed beyond recognition during his lifetime, a world which he greatly helped to shape.

What we owe him

During his extraordinarily prolific career, Bessemer made contributions, some of them spectacular in their day, to many industries. None of his other inventions, however, have survived in quite the same way as the converter, which remains today essentially unaltered. Some have been forgotten, superseded by newer methods; some survive only in the sense that they served as springboards for the ideas of later inventors; others again—his telescopic connection between railway coaches and his early method of copper plating white metal castings, for example—were allowed to lapse at the time, only to be reinvented when the need for them arose.

Bessemer was by no means unique among inventors in occasionally designing something for which there was, as yet, neither technical nor social requirement. Perhaps his most illuminating performance of this kind was his 1856 patent covering the use of pure oxygen instead of air in his steel-making process. In this, he was at least 80 years ahead of his time: there was no prospect whatever, in his day, of oxygen being available in anything like adequate quantities. Yet there was no guesswork in his reckoning: his patent was based on his recognition of oxygen as the active agent in the process—and on a hardheaded determination to protect this knowledge.

His acquaintance with chemistry was something of a *tour de force* in itself: it must be remembered that at the time of his birth there had not even been a proper system of chemical nomenclature, let alone an understanding of the principles of chemical combination.

His perpetual readiness to study a new subject from first principles, coupled with an endearing alacrity to 'have a go' and a dogged resolution of purpose was, predictably, accompanied by a scornful impatience with the incompetent, the mediocre and the pretentious. Nothing would satisfy him but an honest job which embodied the best of which he was capable.

In considering his life and work as a whole, it is almost impossible to overrate the importance of his early years at Charlton. From his father, he inherited and acquired an inventive turn of mind, an eye for beauty, a love of nature and

An ingenious invention of Bessemer's which never came to anything: section through a vessel with a stabilised saloon

a respect for small detail which influenced him throughout his life. The sense of being somewhat different from others bolstered his self-confidence when he stood, as pioneers often must, alone among his critics.

Yet true loneliness had no part in Bessemer's life. Always behind him he had the support of his loving wife and family—and if, in his middle years, he sometimes appeared the perfect exemplar of the Victorian *paterfamilias*, he was never, to them, a forbidding autocrat: a twinkle was never far from his understanding eyes.

His career would be difficult to duplicate today: our modern world confines its engineers to relatively small sectors of the vast field in which Bessemer exercised his inventive talents and yet, ours is a world of steel which owes much of its explosive rate of technical and social advance to the metal which Bessemer was the first to make cheaply. Had he never done anything else, this achievement alone would have ensured him an honoured place in history.

Acknowledgement

All illustrations not otherwise acknowledged are reproduced by courtesy of the Mansell Collection.

REFERENCES FOR FURTHER READING

1. 'Sir Henry Bessemer FRS, an Autobiography', 1905, *Engineering*.
2. JEANS, W. T., *The Creators of the Age of Steel*, 1885, Chapman and Hall.
3. WYMER, NORMAN, *Sir Henry Bessemer*, Oxford University Press.
4. MURRAY, DAVID, *Steel Curtain*, 1959, Pall Mall Press.

Though at the very core of mechanical engineering, technical development of lubricants and of all but the simplest plain bearings have had to wait for the spread of power-driven machinery during the Industrial Revolution. Individual examples, however, were far ahead of their time and most modern techniques have had precursors throughout recorded history.

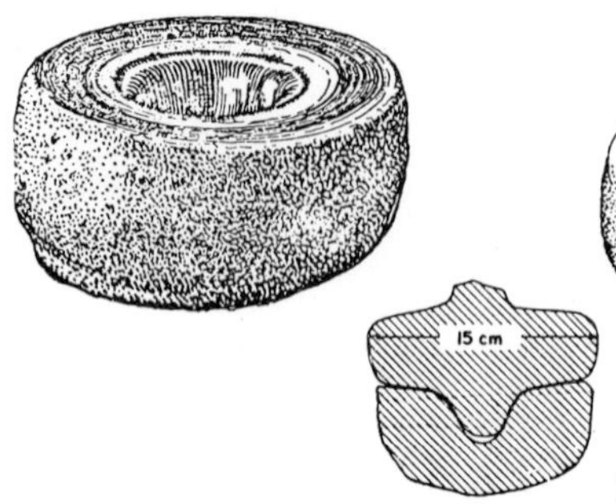

[Courtesy: The Clarendon Press]

Fig. 1. These polished stone bearings are believed to have been used to support a potter's wheel. Dating from before 2000 BC, they are some of the earliest made by man

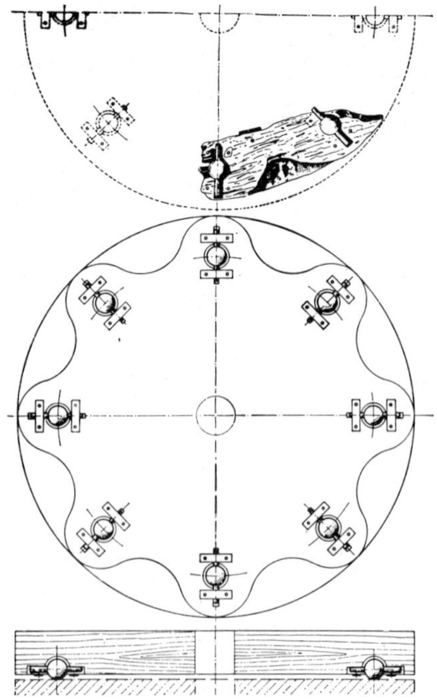

[Courtesy: Libreria Della Stato, Rome]

Fig. 2. Remains of a 3 ft diameter bearing found in Italy. A close approach to the rolling contact idea

Bearings and Lubrication

by H. Naylor, PhD, BSc (Associate Member)

Although there are many interesting things relating to bearings and lubrication to be found in the early records, they cannot reasonably be held to provide the foundations of present day bearing technology. Some laws and ideas that form part of current practice were first discovered centuries ago but they have been re-discovered in more recent times, rather than developed from their earlier origins. Lubrication is, then, a modern science and almost all of what we now consider significant has been discovered in the last hundred years.

Attempts by early man to employ rotary motion led immediately to the invention of bearings. Semi-rotary fire-making bow-drills from Palaeolithic times, using crude cup-shaped thrust bearings, probably represent the earliest examples.

Ancient civilisations

The use of the wheel in any form requires some kind of supporting bearing. The earliest extant record of the use of wheels on vehicles comes to us in the form of a conventionalised pictograph inscribed on tablets found in Sumer, dating from about 3500 BC. The first potter's wheels are also thought to date from this period, the earliest surviving one being that found at Ur. A number of examples of polished bearings, believed to have been used to support potter's wheels and originating before 2000 BC, have been found at various sites in the Near East. Typical examples are illustrated in Fig. 1.

By 1500 BC wheels for use on chariots, where lightness was an important factor in vehicle performance, had reached a high level of development and it is reasonably certain that animal fats were used to lubricate them.

Although, from these early beginnings, the use of mechanical devices was continuous, as can be seen from the records of the Egyptian and other ancient civilisations, the design of bearings and the art of lubrication remained in a rudimentary state until the eighteenth century. It is curious to note that, as long as man or animals provided the driving power, little interest was taken in improving the efficiency of machinery. It was the Industrial Revolution, leading to the large scale adoption of power-driven machinery, that eventually provided the incentive for improvement.

However, the record through the ages is not entirely without interest and many examples are to be found in

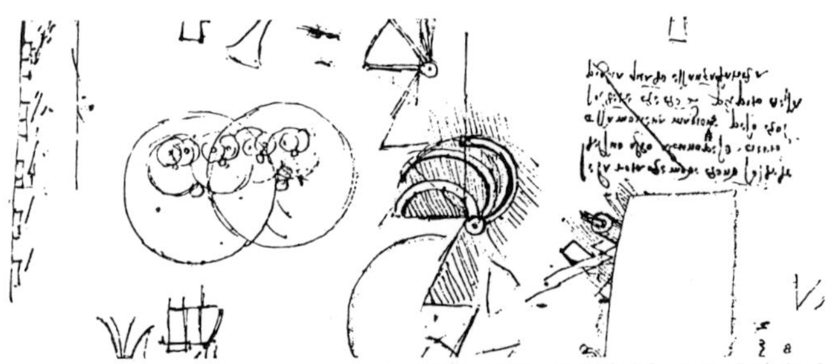

[Courtesy: British Museum, London]

Fig. 3. The sketch books of Leonardo da Vinci contain hundreds of diagrams of mechanical devices, including this series of rollers arranged to form a low friction bearing. The principle, known also to Ramelli in the 16th century, is used in some present day balancing machines

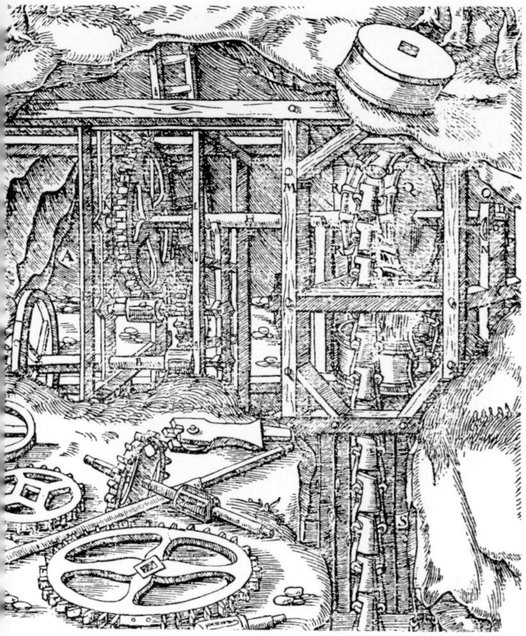

Fig. 4. The numerous crude bearings shown by Agricola include (at P) a roller support for the shaft of a chain pump

which later developments were anticipated by the isolated use of novel devices. For example as long ago as 3000 BC the Egyptians made use of the lubricating properties of viscous fluids by employing layers of viscous mortar on which to 'float' large blocks of stone into position. Until recently it was thought that they had used wooden rollers for the transport of such loads but a study of the available evidence now suggests that logs were placed with their axes in the direction of motion and merely formed a bed along which the load could be pulled. If this conclusion is correct, then the first example of a true rolling contact bearing is that used by Diades, a Greek, in his battering ram, circa 330 BC. His linear bearing employed free-rolling elements, separated by a form of cage, and contained between raceways.

In 1928 the now famous remains of a large bearing were discovered in Lake Nemi in Italy. Shown in Fig. 2, it is believed to have been constructed about AD 40 and to have consisted of a circular wooden platform, some three feet in diameter, on which a number of bronze balls were supported in trunnion bearings, together with a stationary race. Although not a true rolling contact bearing, it was a close approach in concept to the flat thrust bearing of today.

These examples were apparently quite isolated ones. Contemporary practice was unable to make effective use of such advanced ideas and they remained undeveloped.

The Renaissance and after

The Middle Ages appear to have yielded virtually nothing or value in the science and technology of bearings. It is in the writings of Leonardo da Vinci that recorded thought on the problems of friction and wear is first encountered. He made an experimental study of friction and discussed the differences between sliding and rolling friction. He propounded two of the laws of dry sliding friction, *viz*, that it is proportional to load and that it is independent of the nominal contact area. He obtained a value of the coefficient of friction for all the material combinations he tested of 0·25. Among his sketches is one, Fig. 3, showing the use of a series of rollers to provide a low friction support for a spindle. But this part of da Vinci's work had no influence on subsequent developments since it remained unpublished until 1894.

The sixteenth and seventeenth centuries saw the beginnings of serious technological progress and the spread of printing and engraving processes provide us, for the first time, with a clear record of the events of the period.

An excellent example is provided by *De Re Metallica*, published in 1556 by Georg Agricola, which contains illustrations and descriptions of mechanical devices and equipment used in mining and metallurgical operations. The illustrations indicate the extensive use of simple machine elements including gears and, of course, bearings, but the latter are in general little more than holes in the wooden or iron framework. However, one diagram, reproduced in Fig. 4, shows the use of a bearing incorporating (at *P*) a roller " . . of pure steel . . " to support the shaft and reduce friction.

Theory of friction

In 1687, Book II of Sir Izaac Newton's *Principia* was published. Section IX considers, in a context quite unrelated to bearings, the circular motion of fluids and examines the hypothesis:

> The resistance arising from the want of lubricity in the parts of a fluid is, coeteris paribus, proportional to the velocity with which the parts of the fluid are separated from each other.

The significance to lubrication theory of this proposition, now well known as Newton's Law of viscous flow, was not to be appreciated for two hundred years.

Meanwhile, in 1699 Amontons, the French physist, reported the first experimental work on friction since that carried out by da Vinci (which, at that time, was still unpublished). He rediscovered the two laws of sliding friction and, like da Vinci, concluded that the coefficient of friction was the same for all surfaces, arriving at a value of $\frac{1}{3}$, compared with da Vinci's $\frac{1}{4}$. In 1750 Euler tested Amontons' conclusions and agreed with the value of $\frac{1}{3}$.

Experimental work on unlubricated friction is difficult even today and it is not surprising to find that not all the experimenters of the period agreed with Amontons. For example, the Rev. Samuel Vince in 1785 attempted to establish the laws of solid friction and, while claiming that his experiments "were made with the utmost care and attention . . . that I do not scruple to pronounce them to be conclusive" nevertheless arrived at the result that friction "increased less than proportionally with the weight and increased with the area of contact."

The results of a comprehensive investigation of friction were published by Coulomb, also in 1785. He confirmed the two laws of Amontons and added a third—that friction is independent of the sliding velocity. These three laws still provide us with a useful approximation for a wide range of operating conditions. He obtained information on the frictional properties of numerous materials, including agate, quartz, glass and steel, and was the first to speculate on the nature of sliding friction. He also carried out experiments on lubricated surfaces and on rolling friction.

Coulomb's work is unusual in that it was undertaken to meet the specific needs of the engineers of the day and found immediate application in the, by then, rapidly expanding industries. He was awarded the prize offered in 1799 by the Academie des Sciences of France for the best work on friction.

After Coulomb, no major contributions to the theory of solid friction were made until well into the present century although investigations provided more accurate data over wider ranges of materials and conditions. Jenkins and Ewing, for example, published in 1877 the results of a comprehensive study of friction under dry and lubricated conditions for surfaces such as steel, brass and wood. Data of this kind quickly found their way into the early textbooks on applied mechanics and millwrighting and were put to practical use.

Design of bearings

At the beginning of the eighteenth century, bearing design had advanced little from the days of Agricola but some illustrations exist from this period showing bearings with split bushes and separate retaining caps. Good examples are to be found in Plumier's book on turning, published in 1701, and Fig. 5, taken from this, shows details of a lathe incorporating bearings of this type.

Generally however, the simple sleeve bearing could still meet most of the demands placed on it and there was little incentive to improve it.

Only in the case of carriage-wheel bearings had there been any real need to elaborate and, of the bearings in general use, these consequently represent the most highly developed types. A description of the advanced practice of the day comes to us from a discourse to the Royal Society in 1684 by Robert Hooke. On the problem of reducing friction in carriage wheels he states:

> . . . the less rubbing there be of the axle, the better it is for this effect; upon which account, steel axles, and bell-metal sockets, are much better than wood, clamped, or shod with iron; and gudgeons of hardened steel, running in bell-metal sockets, yet much better, if there be provision made to keep out dust and dirt, and constantly to supply and feed them with oil, to keep them from eating one another; but the best way of all is to make the gudgeons run on large truckles, which wholly prevents gnawing, rubbing and pitting.

This discourse also contains a reference to the rolling friction between a wheel and the road surface in which it is pointed out that the non-elastic yielding of the floor is the great impediment to rolling motion.

Newcomen had built his first steam engine in 1712 but its operating speed was very low and consequently it placed no great demands on its bearings. Modified by Watt and others, the slow speed beam engine was to survive almost until the end of the 19th century, primarily because of the undemanding nature of its construction.

However, as a result of the development of these large prime movers and their conversion to rotating motion, power-driven machinery began to make its appearance on a rapidly increasing scale. Operating speeds and loads began to rise and the deficiencies and limitations of the crude bearings still in general use began to be apparent. By the early 19th Century the cost of friction began to be appreciated, especially by the railway engineers, and provided a stimulus that led eventually to the important discoveries of the 1880s.

Some examples of the more elaborate carriage-wheel bearing designs of the early 19th century are shown in Fig. 6.

The earliest paper on lubrication to be presented to the Institution was published by W. Bridges Adams in 1853 and entitled *On Railway Axle Lubrication.* It is an interesting assessment of the problems involved and is no doubt typical of the thinking of engineers at that time.

> The word lubrication literally signifies slipperiness, but this does not express the precise action. Oil, or grease, or soap, interposed between two metallic bodies moving one upon the other, is composed of a series of small globules, which keep the bodies separated, and serve as rollers. The surfaces of metallic bodies, however apparently smooth, are composed of salient and re-entrant angles of larger or smaller size, according as the material is hard and polished, or soft and rough. Therefore the more imperfect the structure of an axle and bearing, the more viscid must be the lubricating material to keep them from contact. If the cushion of lubricating material be insufficient in extent, contact ensues between the salient angles of the metal, and heating takes place to such an extent as to boil away the lubricating material and drive it off.

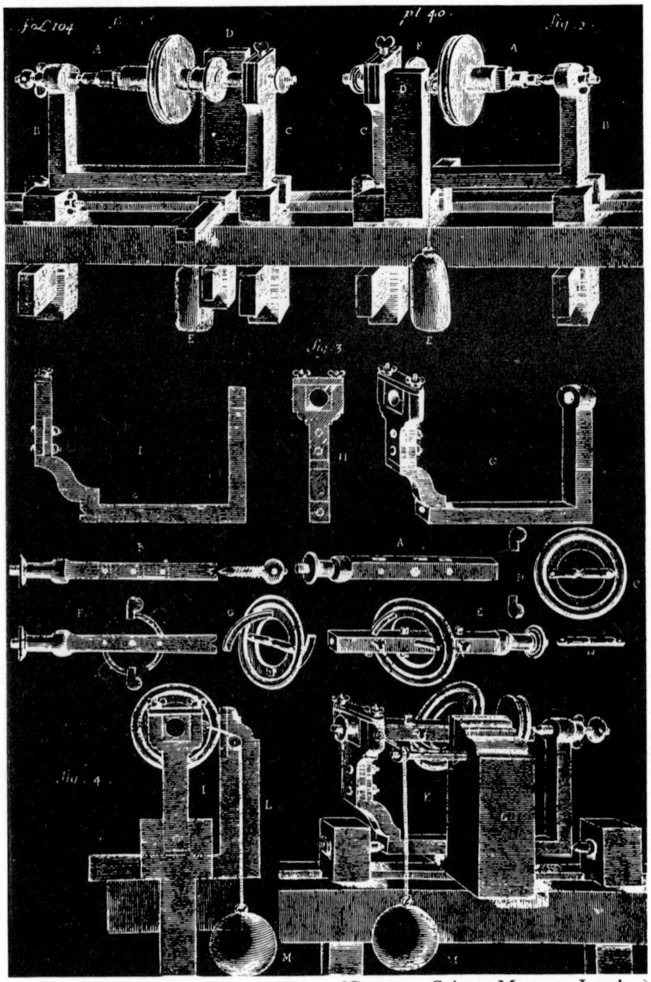

[Courtesy: Science Museum, London]

Fig. 5. By 1700 bearing design had begun to develop towards the types used today. The lathe bearings shown in this engraving from Charles Plumier's 'L'Art de Tournier en Perfection', have split replaceable bushes and removable caps

Adams went on to compare the well developed axle bearings of highway carriages with the crude design of railway axleboxes, pointing out that the former would operate without attention for some 5000 miles over bad roads whereas the latter required greasing every 100 miles or less. He proposed that the boxes should be completely enclosed and provided with a seal to keep out the dirt and that the lubricant should be applied to the underside of the journal rather than to the loaded side. His design is shown in Fig. 7.

By this time the bearings built into stationary equipment had evolved from those described by Plumier into the general form that has survived into the present. Separate brasses, retained by a removable cap, was common practice and, from about 1840, a white metal lining was frequently used.

The patent literature of this period reveals great activity in bearing design. Many ideas are described for raising the lubricant from a reservoir formed in the base of the pedestal to the bearing surfaces—for example the collar and channel arrangement shown in Fig. 8. Some of these methods are still in use today.

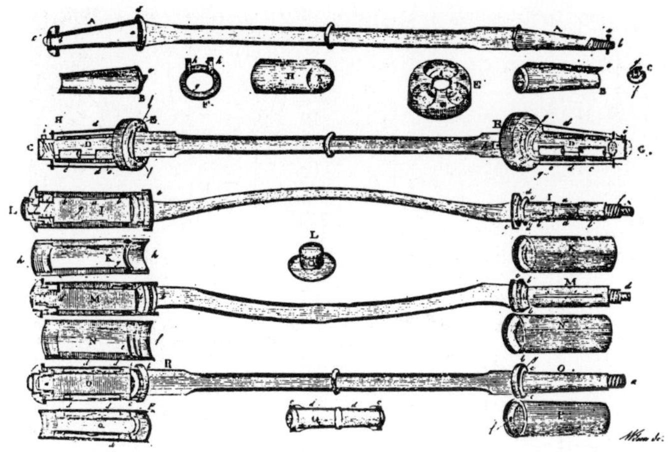

[Courtesy: Science Museum, London]

Fig. 6. Until quite recently, the most advanced bearing designs have often been associated with wheeled vehicles. These carriage wheel bearings, described in Felton's 'Treatise on Carriages', were in use about AD 1800

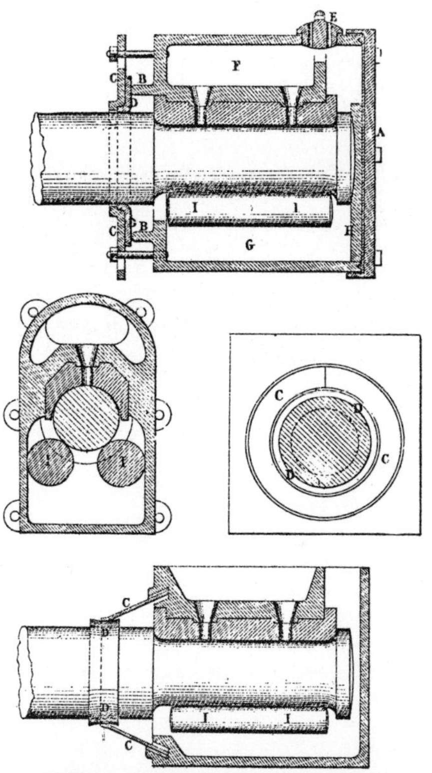

Fig. 7. This design of axle box was proposed by Adams in 1853 to overcome the problems of friction and wear experienced by the railway engineers of that time ◀

Although a satisfactory plain bearing design had now emerged, it was still not possible to determine its load capacity and frictional torque except by experience; and the available experience was becoming more and more confusing. In some cases, for example, friction was seen to increase with speed whereas in others it fell spectacularly as the speed was increased. The effect of load appeared to be equally ambiguous.

Hydrodynamic lubrication

It was with an appreciation of these problems that Thurston in the USA set out to make a series of experiments on a lubricated journal bearing for a range of speeds, loads and temperatures. Because the implications were still not understood, the results, published in 1878, appeared to be inconsistent and were accepted with some reserve. However, Thurston demonstrated for the first time the existence of a minimum in the curve of friction against speed for a lubricated journal bearing and clearly defined a region in which friction increased with speed.

In 1883 Petroff in Russia published the results of extensive investigations of lubricated journal bearing performance, during which he tested animal, vegetable and mineral oils covering a very wide range of viscosity. He concluded that under certain conditions the journal and bearings were separated by a film of lubricant.

He also attempted to obtain a quantitative expression for the bearing friction and, using Newton's equation of viscous shear, he deduced his now well known equation for a shaft running concentrically in a bearing. Although published in 1883, Petroff's work remained virtually unknown outside Russia for many years.

Prompted by the existing state of confusion, the Institution set up a Committee on Friction in 1878 which commissioned Beauchamp Tower to conduct an experimental investigation of journal-bearing lubrication. As a result, a series of reports were published of which the first two were particularly important. An impressive feature of the first, published in 1883, is the accuracy with which Tower observed and reported facts, the significance of which it was impossible for him to understand fully at the time. For example he writes:—

> Early in the experiments it was found that, immediately after the motion of the shaft was reversed, the friction was greater than it was when the shaft had been running in the same direction some time. This increase was greatest with a new brass, and diminished as the brass became worn, so as to fit the journal more perfectly. . . . The phenomenon must be due to the surface fibres of the metal, which have been for some time stroked in one direction, meeting point to point and interlocking when the motion is reversed.

A correct explanation was soon to be provided. Later in the same report there is a reference to high oil pressures.

> The oil flowing over the top of the cap made a mess, and an attempt was made to plug up the hole, first with a cork and then with a wooden

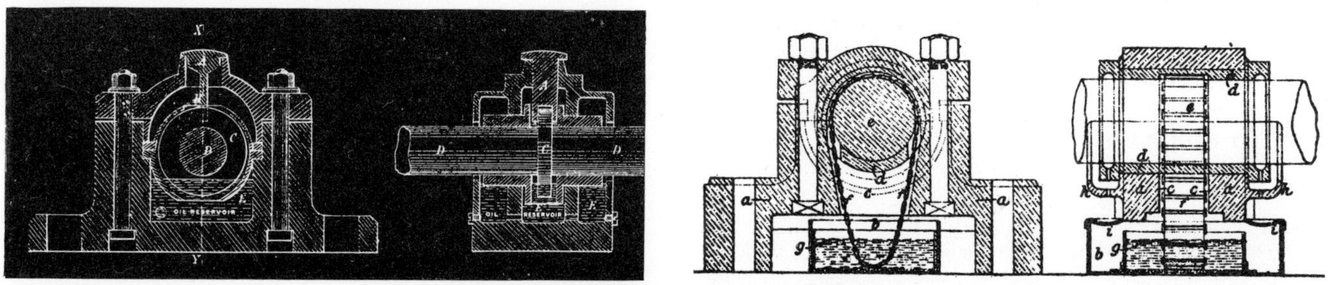

Fig. 8. Improvements in lubrication methods no doubt contributed to the rapid development in the 19th century. These drawings from patent specifications show self-lubricating designs

plug. When the machine was started the plug was slowly forced out by the oil in a way which showed that it was acted on by a considerable pressure.

In the second report, published in 1885, Tower followed up this discovery by measuring the pressure distribution throughout the bearing. Two further reports contained largely inconclusive results and it seems doubtful whether Tower was involved in their authorship.

Not everyone appreciated the value of his work. A critic, writing to an engineering journal in 1884, complained:

> I had anticipated that Mr Tower's researches would have had some practical value; but as far as I can see they might just as well never have been undertaken. . . . we gain absolutely nothing of any value from the discovery, if such it be, that the use of an oil bath diminishes friction enormously. We cannot use oil baths, and the fact is therefore of no importance.

In 1883, Osborne Reynolds published a paper in which he differentiated clearly for the first time between viscous and turbulent fluid flow in pipes and determined the condition, now known as the critical Reynolds Number, governing the transition. At a meeting of the British Association in Montreal in 1884 he discussed the hydrodynamic equations of viscous flow without, however, attempting their integration. Two years later, in 1886, he presented to the Royal Society his masterly paper *On the Theory of Lubrication and its Application to Mr Beauchamp Tower's Experiments, including an Experimental Determination of the Viscosity of Olive Oil*. The paper contains solutions of Reynolds' equation for thin fluid films contained between surfaces in relative tangential motion and in normal approach. These show that, if the motion of the surfaces is tangential, then a necessary condition for the generation of pressure between them is that the lubricant film must converge in the direction of motion; Reynolds explained how this condition was fulfilled automatically in a cylindrical journal bearing.

Thus, in less than ten years, hydrodynamic lubrication had been discovered and a firm theoretical basis provided.

Although he was able to account in a satisfactory way for Tower's observations, owing to certain mathematical difficulties, Reynolds obtained solutions only for eccentricity ratios up to 0·5. It was the German mathematician Sommerfeld who, in 1904, published an explicit solution of the equation for an infinitely long 360° bearing for all values of eccentricity ratio and so completed the basic theory of the plain journal bearing.

Rolling-contact

Although, as we have seen, the first true rolling-contact bearing on record made its appearance about 330 BC, the chain of practical development leading to present-day designs did not start until the time of Agricola and Ramelli about the middle of the sixteenth century. The writings of Leonardo da Vinci contain numerous designs involving rolling contacts but since they remained unpublished his ideas could not influence development.

The bearing devised by the Italian artist, Benvenuto Cellini to support his statue of Jupiter is usually regarded as noteworthy in that it represents one of the earliest recorded instances of the use of balls to reduce friction. In his autobiography Cellini gives the date as 1534 and the translation runs:

> Having with the utmost dilligence finished the beautiful statue of Jupiter I placed it upon a wooden socle, and within that socle I fixed little globes of wood which were more than half hidden in their sockets. . . .

It seems clear that the balls were in sliding contact with the base containing them and, like the Lake Nemi bearing, the device did not therefore form a true rolling-contact bearing, but rather a set of castors.

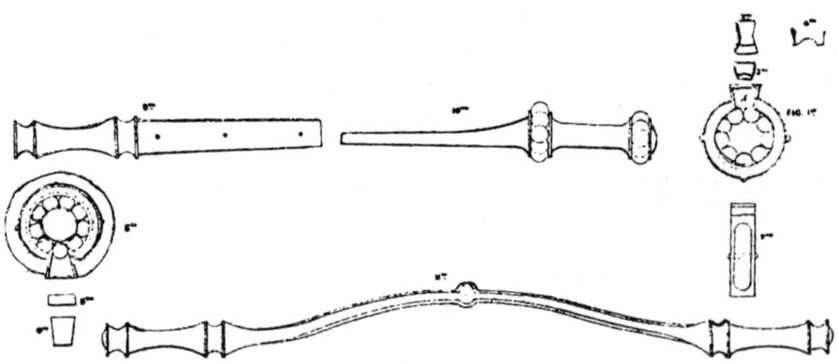

[Courtesy: Science Museum, London]

Fig. 9. Another example of advanced design in axle bearings: this patent application of 1794 describes the first radially loaded ball bearing on record

From about 1700 the roller bearing entered into a phase of continuous development whilst the radially-loaded ball bearing did not make its appearance until about the end of the century. This difference was no doubt due mainly to the relative ease with which the elements of a roller bearing could be made with the still primitive equipment then available. By 1787 the evolution of the roller bearing was almost complete: in that year John Garnett obtained a patent for a bearing incorporating caged rollers, located by spring retaining rings and having a spherically-seated outer race to ensure axial alignment.

The earliest known working example of a true rolling-contact ball bearing, dates from this same period. It is a large thrust bearing with races, two feet ten inches OD, provided with 2¼ inch diameter, cast iron balls and is very similar in design to current types. Originally installed in a post mill at Sprowston, Norwich, in 1780, it is now in the Bridewell Museum there.

The first patent for a radial ball bearing was that obtained by Philip Vaughan in 1794. The specification, the diagram from which is reproduced in Fig. 9, describes a complete axle with uncaged balls, rolling directly in stationary raceways formed in the shaft. The revolving outer races were provided with a filling slot, closed by a wedge-shaped insert.

Improvements in materials and machine tools, together with an increasing demand, ensured a steady rate of development throughout the 19th century. Ball bearings for bicycle wheels came into use shortly after 1860 and many of the patent applications in the following 20 years concerned improvements to various wheeled-vehicle bearings.

By this time cages of various kinds were in general use and inner and outer races, separated from shaft and housing respectively, were a common feature. Towards the end of the century machines were developed for grinding bearing balls accurately and cheaply, and the way was open for the rapid expansion of applications that was to follow.

Although, unlike the case of the plain journal bearing, the general principle on which the rolling-contact bearing operated was clear, the problem of designing for a specified load capacity was formidable and it was not until the present century that a satisfactory method became available to the designer.

The foundation for a theoretical approach to the subject was provided by Heinrich Hertz in 1881. Using the methods of classical elasticity, he studied the problem of curved solid bodies in stationary point-contact systems and produced methods for the calculation of contact area, deformation and contact pressure in terms of the applied load and the geometry and mechanical properties of the system. The work was not immediately applicable to the design of complete rolling-contact bearings but it provided a starting point for many other investigations, notably for those of Professor Stribeck in Germany and Professor Goodman in England.

Stribeck's work was commissioned by the Deutsche Waffen und Munitionsfabriken, Berlin and began about 1897. He measured the deformation of steel balls compressed together along the line of their centres, and of balls loaded against plane surfaces. He then used the results to confirm the theory of Hertz. These experiments were followed by others on balls in grooved tracks and on complete bearings with various track configurations. From the results of these experiments, in which load capacity, frictional characteristics and operating temperatures were determined, Stribeck was able to propose semi-empirical methods for the design of ball-bearings of all types. Goodman's work on rolling contact bearings was started about the same time and followed, independently, a somewhat similar course. It provided empirical formulae for load-capacity which have been of great practical value.

These three investigators supplied the theoretical basis on which a highly successful bearing industry was founded, and many of their methods are still used today.

Lubricants

Bitumen, seeping to the Earth's surface, is known to have been made use of since 3000 BC. In the Middle East it was employed as a building material and in the embalming of the dead. The Egyptians used viscous liquids in the moving of large building blocks—probably the first application of the hydrostatic principle to lubrication—as early as 3000 BC and it has been established by chemcial analysis of remains found in tombs dating from 1400 BC that a form of grease made from animal fats and lime was used as a lubricant on chariot wheels. From this time onwards, vegetable and animal oils are known to have been used extensively throughout Europe and the East.

Distillation of crude oil was first attempted *c.* AD 100. By 1300 fractionation columns had been developed but were not used on any large scale. By the 16th century mineral oil was being sold in small quantities throughout Europe but was still used mainly to provide bitumen, burning oil for lighting and for medicinal purposes. The term petroleum appears for the first time in a book by Agricola and in *De re Metallica* he describes the methods of refining then in use.

The situation did not change materially until the beginning of the 19th century when, again under the spur of industrialisation, serious efforts were made for the first time to develop materials for use specifically as lubricants. As in the case of bearing design, the demand for improvements was created, in the first instance, by the users of wheeled vehicles.

Until the drilling of the first successful oil well by Drake in 1859, mineral oil was not available in large quantities and, consequently, lubricants were based on vegetable and animal oils and fats. Typically, an early patent application by Hardacre in 1812 described a composition to prevent "to a very extensive degree the effects of friction". It consisted of plumbago and "pork lard or beef suet, tallow, oil, goose grease or greasy substance . . . mixed together in certain proportions".

There followed a period of intensive experimentation and the discovery of the value of many basically important lubricating materials, such as alkaline soaps, lead compounds, graphite and sulphur. Hardly anything was left untried and a specification of 1848 covered compounds made of

> laurel leaves, ivy leaves, oxymuriate of mercury, subnitrate of bismuth, sulphate of copper, sulphate of zinc, tartarised antimony, pulverised Roman alum, Greenland or Southern whale oil, Gallipole or olive oil, sperm oil, palm oil, the vegetable substance called caoutchouc, brown common soap, minium, white peper, cerriss, common salt, powdered alum.

The new approach to the formulation of lubricants brought with it a need for mechanical testing equipment and a number of patent applications involving machines for assessing lubricating quality were made in the second half of the century. For example, a machine was described by Ingram and Stapfer in 1866 in which a given quantity of oil was subjected to friction between certain surfaces at a uniform rotation and a constant external temperature. The equipment measured the number of revolutions required to generate a given temperature by friction and was automatically switched off by a mercury thermometer which operated an electromagnet.

The first paper on lubricants to be published by the Institution was presented by John Lea in 1853. In it Lea describes a lubricant consisting of whale oil with additions of lead oxides and India rubber, the purpose of the latter being to improve the temperature/viscosity characteristics of the oil.

Following the successful operations of Drake and the extremely rapid development of oil fields in America and elsewhere, mineral oils had substantially replaced animal and vegetables oils as lubricants by the end of the century and the basic modern idea of lubricants consisting of a mineral oil base with additives to improve specific properties had become well established.

Conclusion

The major developments in engineering have come about in response to external pressures—technical, economic and social—and because the complementary arts and sciences have provided the necessary background to make them possible. The history of bearings and lubrication serves as an excellent illustration.

Early history saw inspired attempts to introduce devices centuries ahead of their time but all were ignored or neglected because they lacked a favourable environment, both economically and technologically. For example, the cost of a truly spherical ball must have been prohibitive in previous centuries.

Not until the power-driven machines of the Industrial Revolution made their appearance and the economic penalties of inadequate performance became more evident in the shape of fuel consumed were the processes of invention and enquiry finally set in motion to reduce friction losses. But of course the very concept of energy loss depended on the establishment of theoretical thermodynamics. And the same factors that had forced the introduction of mechanical power—together with just sufficient physical understanding—now ensured the successful solution of the accompanying practical problems.

Theoretical studies of bearings and their lubrication made vast progress in the last twenty years of the 19th century and the results of this work have provided the foundations for modern theory and practice. Another period of basic re-thinking appears to be upon us at present.

This illustration from Diderot's Encyclopedia shows, among other things, a pole lathe; probably the oldest kind of machine tool; the treadle-operated cord was wound round the workpiece while the springy pole provided power for the return stroke. In this 18th century model a screwed mandrel gave longitudinal traverse. The evolution of machine tools is traced in an article opposite

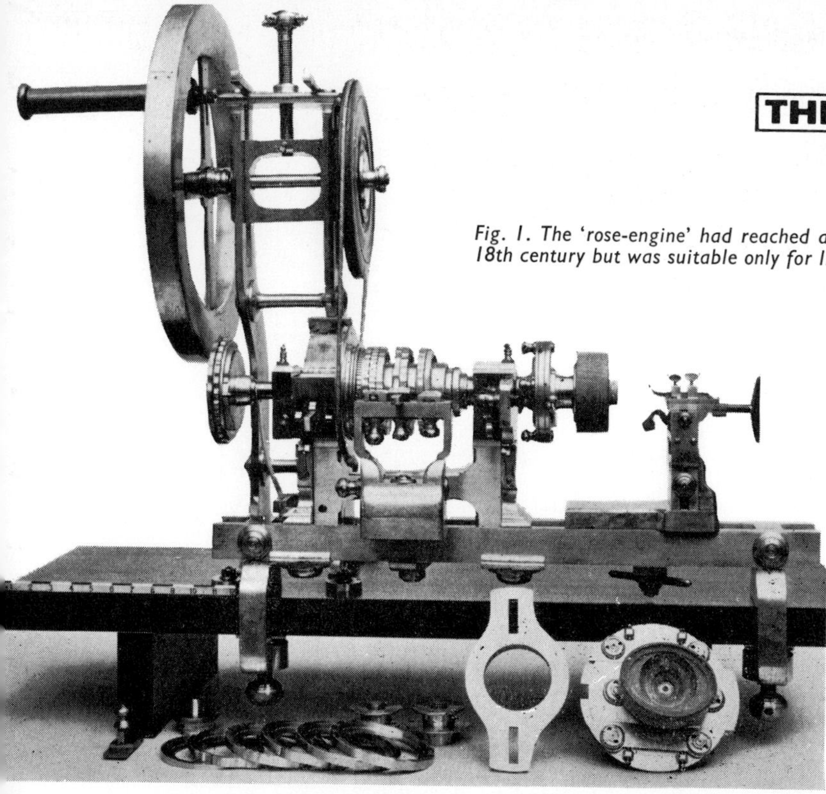

Fig. 1. The 'rose-engine' had reached a high pitch of excellence by the 18th century but was suitable only for light ornamental work

[Courtesy: Science Museum, London]

After the age of the millwrights with its wooden structures, progress in mechanical engineering depended on faster and more accurate production methods of which casting and cutting were the two most important. The 19th century was the heyday of the general purpose machine tool but it also saw the forerunners of modern multiple and automatic equipment. Towards the end of the century the names of the great pioneers such as Bramah, Maudslay and Nasmyth were increasingly supplanted in the story of machine tool development by joint stock companies.

Metal-cutting Machine Tools

by W. Steeds, O.B.E., B.Sc. (Member)

One of man's first steps towards civilisation was to acquire the ability to make and use tools. At first these were very primitive and comprised only stones used as hammers, and sticks used for digging and as levers; but a great advance was made when man learned how to work flints into edged tools. More sophisticated tools and implements were later made from animal and fish bones while great improvements followed the discovery of ways in which metals could be worked.

These steps extended over many thousands of years. Possibly the first implement that had any suggestion of a machine in it was the bow drill which is known to be of very considerable antiquity. The potter's wheel may have suggested the lathe, the origin of which goes back beyond the beginning of recorded history.

Apart from the lathe and the drill, there were apparently no machine tools, as late as the 12th century AD[1]. Since then progress has been accelerating and the modern world is so dependent on machine tools that it is difficult to imagine what it would be like without them. Although metal-cutting machine tools are gradually being replaced by forming tools, we still depend on them for the majority of engineering parts.

The variety of machine tools now available is so vast that it is impossible in a short article to deal with all of them; for present purposes a machine tool will be defined as a machine capable of removing metal in the form of chips or swarf so as to bring an object to a required shape, size and finish. This excludes machines in which the shape is produced by plastic flow; wood-working machines and press-tools. It has also been found necessary to exclude gear-cutting machines.*

Early lathes

Undoubtedly the lathe is the most important machine tool ever produced but it remained remarkably primitive up to the middle of the 18th century. A then common form was the pole lathe shown in which the workpiece opposite was driven by a cord passed round it and attached at the top to a springy pole, mounted near the ceiling as a horizontal cantilever; the bottom end of the cord was attached to a treadle. This gave the work an oscillatory motion of which only one half could be utilised.

A great improvement was the introduction of an endless cord which passed round a pulley on the lathe spindle and also round a large flywheel mounted separately at the rear of the lathe and rotated by one or two labourers. This method continued in use up to 1830–40. The treadle and crank mechanism had come into common use during the 18th century.

In these early lathes the tool was generally supported by a simple rest such as is still used in wood-working lathes and the operator had to exercise considerable skill in order to obtain any accuracy. Screw-cutting was not usually possible but in the mandrel lathe opposite which was known as early as the 15th century, the work was held on the end of a

* *but cf article on power transmission, p. 118, Volvme 1.*

spindle which had a screw thread working in a fixed nut so that, as it rotated, it also traversed axially; a fixed tool could therefore cut a thread on the work.[2]

This developed into a form in which the mandrel had several short threads of different pitches and half-nuts, carried by levers, and could be engaged so that threads of several pitches could be cut without change of spindle.

A great advance in the cutting of screws was made early in the 18th century by Thiout (1692–1767) who constructed a lathe in which the spindle was fixed axially and provided with a threaded portion engaging a nut which was connected by a link to one end of a lever pivoted at the tail-stock end. The other end of this lever was coupled by a second link to a carriage, free to slide on a guide parallel to the work. The tool was held in this carriage and thus cut a thread of the opposite hand on the workpiece.

The lathes depicted by Diderot in his encyclopedia which, presumably, represented practice in the early 18th century, were still very primitive. He does, however, show several in which the toolholder is carried on a slide. He also illustrates a slide rest of almost modern construction; but this seems to have been used only on a lathe for turning ornamental shapes. The earliest known illustration of a toolholder with a screw adjustment of the cut occurs in a German book published in 1480!

Boring machines

The ornamental lathe or rose-engine, an example of which is shown in Fig. 1, had reached a high pitch of excellence in construction by the middle of the 18th century; many examples survive and exhibit great skill and craftsmanship. But these lathes were not of much use to contemporary engineers who wanted to machine the comparatively large components of steam engines and mill machinery.

The only records of machine-tools for such heavy work relate to the boring machines used for the machining of steam-engine cylinders. These machines undoubtedly developed from the much earlier gun-boring machines, of which there were two kinds. One is shown in Fig. 2: the gun was mounted vertically and the boring tool was rotated by means of a horse-gin. In the other type the gun was mounted horizontally and in some of them the gun seems to have been rotated.

The earliest cylinder boring machines used in England were horizontal and in them the cylinder was supported on a carriage which could be hauled along a track, parallel to the axis of the bore. The cutter head was driven by coupling its shaft to a waterwheel through gearing and its only supports were a number of pads which bore on the cylinder surface. The results were naturally not very satisfactory.

A great advance was made by John Wilkinson (1728–1808) who, in 1775, made a machine in which the boring bar was supported at both ends and the cutter head traversed along it, the cylinder being fixed.

Although the vertical form of gun boring machine is mentioned in Spanish literature as early as 1603 and was in use in England in the early part of the 18th century,* it does not seem to have been used for cylinder boring until the middle of the 19th. One was designed by Peter Ewart for the Soho Foundry about 1795 but it seems never to have been installed. It is surprising that the advantages of the vertical arrangement were not better recognised; in 1840 James Nasmyth was still pressing them.

* *cf. Fig. 5, page 133, Volume 1.*

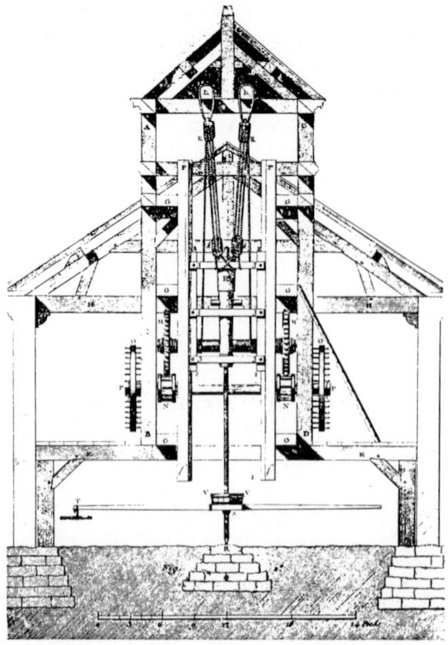

Fig. 2. During the early industrial revolution the only heavy machine tools were horse-gin driven gun boring machines, one type of which is shown here

The last quarter of the 18th century saw great improvements in the lathe; these were the work of several famous engineers. In England the first of these was Joseph Bramah, who, in 1794, designed a slide rest which is generally thought to have been constructed by Henry Maudslay, then in Bramah's employ.

Maudslay is universally regarded as the father of the modern machine tool and one of the greatest engineers of all time. During the last decade of the century he built several lathes of advanced design and remarkable workmanship. In France, Jacques de Vaucanson (1703–82), built a very fine lathe which still exists. In America David Wilkinson,[3] whose father and four of whose five brothers were all engineers of importance, also designed and patented a slide rest and built lathes prior to 1800. There is no reason to suppose that these improvements, though approximately simultaneous, were not the original ideas of their makers. The lathes produced by these pioneers constitute the transition from a machine for producing ornamental effects and light mechanisms to one for producing the relatively heavy and large components of engines and mill-work.

Lead-screws

One of the most important improvements was all metal construction but it is interesting to notice that lathes with wooden beds were still in use in the fifth decade of the 19th century. Another important advance was the use of lead-screws of considerable accuracy, robust enough to enable relatively heavy cuts to be taken. The principle of using a lead-screw in conjunction with change gears was known to Leonardo da Vinci and was used by Henry Hindley (an instrument maker and inventor of the globoidal worm gear) as early as 1741.

Nevertheless, the screws of Maudslay and Vaucanson are the earliest surviving examples of lead-screws in the modern sense. The accuracy of lead-screws is of fundamental importance in machine-tools and their production engaged the attention of many engineers during the 19th century. Maudslay devised a machine for this purpose and in 1826 Bryan Donkin (1768–1855) devised a screw-adjusting machine in which corrections were applied by giving slight rotations to the nut by means of a cam-plate and a lever. This principle was again adopted in the construction, between 1900 and 1904, of a standard screw-cutting lathe by Sir W. G. Armstrong Whitworth & Co., installed in the National Physical Laboratory. The lead-screw of this lathe had an intrinsic accuracy of 0.0002 in. over its length of 53 in. and, with the correction gear, gave results accurate to 0·0001 in. at any point.

During the first half of the 19th century great improvements were made in lathe design and construction. Self-acting feeds became usual and multi-ratio gearing was introduced to increase the range of feeds. Half-nuts were fitted which could be easily engaged and dis-engaged from the lead screw. Beds were made heavier, spindle bearings improved and accuracy increased.

Some of the landmarks in this progress are worth mentioning. One, now in the Science Museum, was built in 1817 by Richard Roberts who, after being employed as a boatman on the Ellesmere canal and as a pattern-maker at the Bradley Ironworks, Bilston, worked for Maudslay and in 1816 set up for himself in Manchester. In August 1821 Joshua Field visited his works and saw 'wheel-cutting by milling cutters.'

Roberts is generally credited with the introduction of templates and male and female gauges for pieces that had to be produced in quantity. He became a manufacturer of power looms and other textile machines and about 1828 joined Thomas Sharp, a business man, to found the firm of Sharp Roberts & Co., which subsequently became famous as locomotive builders. Roberts was a prolific inventor and has strong claims to rank with Maudslay and Whitworth among the most important engineers of the 19th century.

Joseph Clements (1779–1844) also contributed much to the improvement of machine tools. After being employed by both Bramah and Maudslay & Field, he set up on his own in London and produced some remarkable machine tools. One of his lathes, now in the Science Museum, exhibits outstanding workmanship. A gap-bed lathe built by him had a continuously variable driving mechanism consisting of two cone pulleys connected by a crossed belt and provided with a control gear operated by the cross-slide so as to produce a constant cutting speed when facing discs. Some eighty years later this feature was incorporated in the headstocks of lathes built by John Lang & Co.

In a lathe attributed to the year 1837, separate ways were provided for the saddle and the tailstock; the saddle also had a rise and fall arrangement for quick adjustment of the tool height, a feature which became common on American lathes some twenty-five years later.

Other landmarks are James Fox's 22ft bed lathe of 1830, Whitworth's sliding bed lathe of 1839, Nasmyth's double ended or 'Ambidexter' lathe and Beattie's double-headstock lathe, both of which were produced about 1850.

[Courtesy: Science Museum, London]

Fig. 3. Roberts' planing machine, which he claimed to be the first of its kind, is certainly the earliest to have survived

Planers

After the lathe the most widely used type of machine tool is certainly one for producing flat surfaces but there are no definite records of such machines before the second decade of the 19th century. Inconclusive evidence indicates that planing machines may have been produced during the 18th century. A primitive planing machine was invented by Nicholas Forq, a French clockmaker, in 1751.

There are statements that Matthew Murray invented a planing machine for machining the slide or D-valves of engines before 1800. But the earliest surviving planing machine is that shown in Fig. 3, which Richard Roberts said he had built in 1817 and described as the first of its kind. The machine was hand operated but it employed the double column or standard and the cross-beam which remain the commonest form today.

Clements built a planing machine of advanced design about 1825 in which the table moved on wheels, about two feet in diameter, and was driven through a bevel-gear cluster and dog-clutch reversing mechanism, coupled by a belt to a pinion shaft whose two pinions engaged racks on the underside of the table. Two tool heads were carried on a cross-beam, which itself was carried in plummer blocks that could be adjusted up and down the columns by screws. The whole cross-beam could thus be rotated about a horizontal axis so that one tool could cut on the forward stroke and the other on the return stroke—undoubtedly the first example of double-cutting, an idea which was to exercise the ingenuity of many engineers during the 19th century.

The dog-clutch for reversing the motion of the worktable was operated through levers by strikers bolted to the side of the table and a lost motion device in the bevel gear cluster enabled the momentum of a flywheel to be used to assist in slowing down and then accelerating the table on reversal. This principle was used by Wm. Sellers, the designer of the spiral pinion drive for planers, some 80 years later.

The two tool heads could be tilted to enable dovetail slides to be machined and the table could be fitted with centres to support cylindrical work. A long screw, carrying a dividing plate, could be fitted to the table and be engaged with a nut so that the table could be indexed for the purpose of cutting racks.

In 1835 Sir Joseph Whitworth who had once worked for Clements, patented a planing machine in which the worktable was fixed and the standards, together with the cross-beam, moved over it on wheels—a principle that was used again about 1900. Whitworth's machine used two screws for the drive and reversed by means of a bevel-gear cluster and dog-clutch.

Roberts in 1838 employed a single belt drive. The belt was shifted across an intermediate loose pulley and a few years later, the use of two shifting belts became standard practice. Non-shifting belts in conjunction with friction clutches were used towards the end of the 19th century and compressed air was also used to operate the clutches. Magnetic clutches were used with some of the early electric motor drives. Belts and clutches were finally made obsolete by the introduction, about 1910, of the Ward-Leonard electric drive.

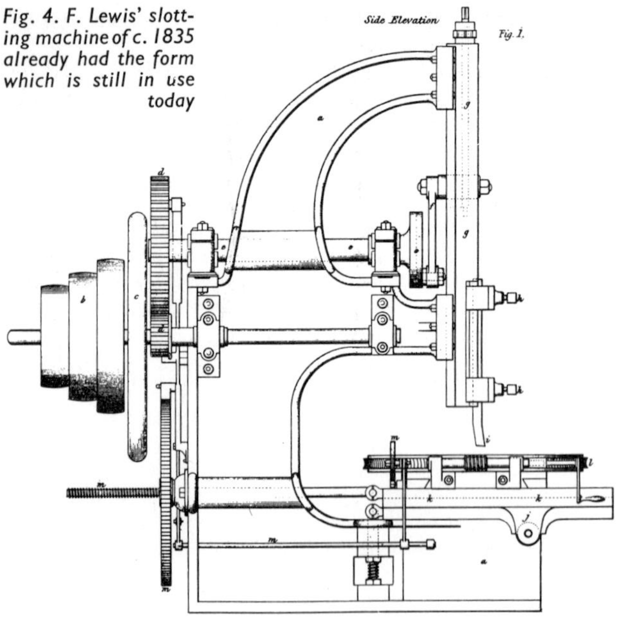

Fig. 4. F. Lewis' slotting machine of c. 1835 already had the form which is still in use today

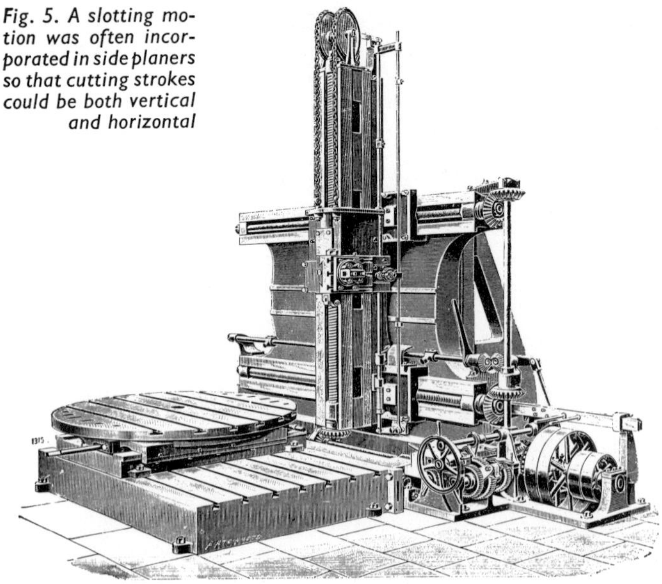

Fig. 5. A slotting motion was often incorporated in side planers so that cutting strokes could be both vertical and horizontal

[Courtesy: 'Engineering']

Shaping and slotting

For small work, the shaping machine is handier than the planer but it seems to have developed later. Nasmyth's, usually regarded as the first of its kind, is said to date from 1836. Shaping machines quickly developed into the two types still current; that in which the ram reciprocates on the top of a column or pedestal while the worktable is moved on a saddle attached to the face of the pedestal; and that in which the ram is carried in a saddle which can slide along ways machined on the top of the base casting.

The drive to the ram was originally by means of a crank and connecting rod and this method remained common for the second type of machine, except where a Whitworth quick-return motion was used. The crank and slotted link mechanism, housed in the pedestal, became the commonest drive for the pedestal type machine but rams driven either by racks or hydraulically have been built and, in one make, the cut was taken on the inward stroke of the ram.

The slotting machine is closely related to the shaping machine and probably preceded it since, as Nasmyth points out, the principle involved was that used by Maudslay in the wood mortising machine which he made for Brunel in the first decade of the 19th century. However, the slotting machine actually appeared in a rather specialised form—the key-groove cutting machine or key-seater—the earliest example of which is credited by Nasmyth to Richard Roberts.[4] This machine may have been made as early as 1821 and certainly well before 1836.

Nasmyth improved it by placing the ram actuating mechanism below the worktable so that, apart from weight, there was no limit to the diameter of the work.

Slotting machines quickly assumed the form shown in Fig. 4 which they retain to this day but other types, which are no longer used, were also developed. One notable improvement was the quick-return motion invented by Whitworth.

A slotting motion was often incorporated into side-planing machines such as the one in Fig. 5, so that the cutting stroke could be vertical as well as horizontal; the tool slide was usually moved by a screw and nut and this drive was also used in large machines of the conventional column type.

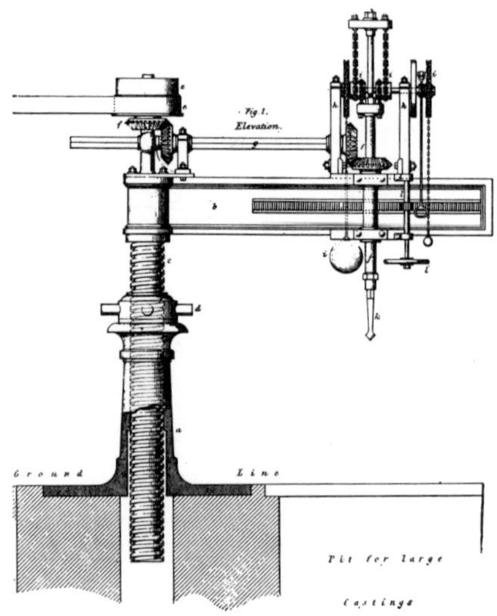

Fig. 6 One of the earliest illustrations of a radial drill shows a square threaded column which could be raised by means of a tommy bar. Rack and pinion moved the spindle along the arm

A machine for producing blind slots was made about 1907 in which a cutter which had teeth resembling those of a side and face milling cutter was secured to, or made integral with, a stem and given an oscillatory cutting motion, accompanied by a downwards feed. Thus it produced a hole whose depth was greater than the radius of the cutter.

One of the principle applications of the slotting machine is in the production of non-circular holes and this is also the particular province of broaches. These have been used for probably about 200 years but in the early days were driven through the hole by hammering and later on by presses. The broaching machine which pulled the broach through the hole did not appear until about 1880 and then only in the simple form of a key-groove cutter. The Lanchester Motor Co. were pioneers of splined shafts but in 1907 they were still using broaches which had to be pushed through the work. By 1910 broaching machines with rack and pinion, screw and nut and hydraulically operated rams were on the market. Vertical machines did not appear until later.

Drilling

The drilling of holes is an ever recurrent necessity; it has been performed by hand tools for many centuries and the bow drill was certainly known from antiquity. But the earliest drilling machine proper to survive is the beautifully made one of Vaucanson; this must have been produced for some special purpose, as it is not suitable for the general run of drilling operations. In form it resembles a travelling column horizontal boring machine, the column being about a foot high and the bed about 18 in long. The column is moved along the bed, and the spindle head along the column, by screws and nuts. Large machines of this form were very popular about 100 years later for drilling boiler shells.

One of the earliest illustrations, reproduced in Fig. 6, shows a radial drill with a square thread column. A large nut with tommy bar handles was used to raise the whole column and arm assembly. The spindle head was moved along the arm by rack and pinion and the drive was through bevel gears and a square shaft.[4]

All the other forms of drilling machines which are now common have been developed since the middle of the 19th century. Improvements have been in respect of greater rigidity, wider speed and feed ranges, easier control and increased capacity. Multiple spindle machines, which are now widely used, were first built in the 1850's; a notable one was that for drilling the plates for the Hungerford bridge which had 80 spindles and a feed imparted to the worktable by hydraulic arms.

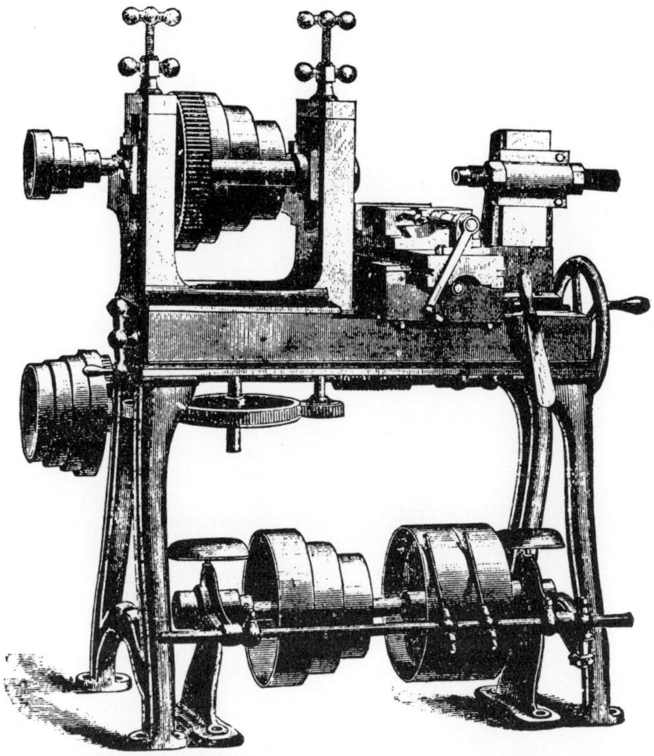

Fig. 7. Eli Whitney is credited with the first practical milling machine which he developed for small arms production. This is another early American machine, obviously descended from the lathe

Multi-spindle machines in which the spacing of the drills could be varied and universal radial drills in which the arm and the spindle could be rotated about mutually perpendicular axes came soon after. Drilling performances were greatly improved by the extended use of twist drills, made possible by the development of the universal milling machine.

Another machine of the same period had four spindles which could be indexed round a central column, but had only one work holding fixture. The type of machine in which several spindles and work fixtures are used and the fixtures are indexed from station to station, all the spindles working simultaneously, did not appear until some fifty years later.

Milling

The principal alternative to planing and shaping machines is the milling machine and the use of rotating cutters dates back at least to the 17th century when they were used by English and French clockmakers. There is an illustration in Diderot's encyclopedia which shows a rotating cutter machining the outside of a rifle barrel. Richard Roberts said that in 1815 Maudslay and Field used a machine with a rotary cutter for truing up cast-iron blocks but its use was discontinued because of the intolerable noise.

The first milling machine that was a practical success is universally credited to Eli Whitney who made it about 1818. An early American machine is shown in Fig. 7.

James Nasmyth made a milling machine for Maudslay in 1829 and later supplied many similar machines to mechanical engineers. In 1835 the American firm of Gay & Silver made a machine with a rise and fall spindle head and an overarm to support the outer end of the spindle.

The future of the milling machine was bound up with the development of suitable cutters. It was not until the middle of the century that these could take more than very light cuts. The credit for this improvement has been given to the Phoenix Iron Works of Connecticut which produced cutters with teeth of greater pitch and better clearances in 1850 but J. R. Bodmer (a versatile Swiss engineer who came to England and started an engineering works at Bolton in 1834) is also credited with the production of greatly improved cutters.

Another notable advance was the invention of the machine-relieved form cutter by Joseph R. Brown in 1864 which made form milling, particularly of gear teeth, an economic proposition. His father, David Brown, had started a small works in Providence R.I. in 1833 and J. R. Brown built the first U.S. dividing engine and later founded Brown and Sharpe.

Milling machines are particularly suitable for small arms production and it was for this purpose that Whitney had made his original machine. Their use quickly spread; in 1855 no less than 74 were sent from America to the Royal Small Arms Factory at Enfield which was being modernized. Nevertheless, they were crude in design compared with the lathes of that period. But after 1880 progress was rapid and by the end of the century machines were being produced which incorporated nearly all the principle features of today's designs except automatic cycles and push-button control. A multi-station automatic indexing milling machine was in use in the late 1890's.

Automatics

Reverting to the lathe, the middle of the 19th century saw the emphasis being placed on production capacity, the accuracy then attainable being all that was required at the time. The lead was taken by America with the invention of the turret lathe by Stephen Fitch in 1845. These early machines were little more than centre lathes in which the tailstock had been replaced by a saddle carrying a turret; but gradually a design evolved which was specifically adapted to the use of the turret. Almost from the start there were two different arrangements; in one the turret indexed about an axis parallel to the spindle as shown in Fig. 8; in the other about a vertical axis. The latter became the most common. By 1890 the turret lathe, and its variant the capstan lathe, had become standard machines.

The turret lathe opened the way for the spindle automatic lathe and automatic screw machine patented by Christopher M. Spencer in 1873. But he seems to have been forestalled by Jacob Schweizer, a German who, in 1872, built the single spindle sliding-head machine in Fig. 9, forerunner of today's 'Swiss' automatic. The firm of Brown and Sharpe was prominent in developing automatic machines and adopted a new position for the turret axis which it still employs.

The next step towards the production engineer's ideal machine in which all the tools work all the time was the development of the multi-spindle automatic. The first of these (Fig. 10) was invented about 1893 by R. Hakewessel and E. C. Henn.

The development of both turret and automatic lathes was aided by the production of good self-opening die-heads for the machining of screw threads. The conventional die-head, with four dies moving in radial slots, opened and closed by a cam-ring, may be traced back at least to 1819 when Simon Goodrich saw it at a works in Guildford. Screwing machines or bolt-cutters with die-heads were fairly common after 1830 but up to the 1870's turret lathes mostly relied on chasing attachments, presumably because die-heads were not sufficiently accurate and reliable.

The chasing attachment usually consisted of a shaft carried at the back of the lathe, at whose ends arms were fixed. One arm carried a half-nut which could engage a short lead-screw situated behind the headstock and the other carried a tool slide. The device was brought into use by rocking the shaft about its axis. Chasing attachments continued in use until the 1930's but by then the die-head had become the principal method.

Although vertical boring-machines were used from very early days the vertical lathe did not come in until the latter half of the 19th century. By 1870 it was well established in its two principal present day forms, one with a cross-rail integral with a column casting and the other with two columns and an elevating cross-rail. In the early days the tables were several feet in diameter and later ones were as much as 40 ft but machines with tables of only 20–30 in diameter were found to be very useful.

Very large machines were made for the stators and rotors of electrical generators at the turn of the century. For the machining the large circular bases of naval gun mountings Sir W. G. Armstrong Whitworth & Co. produced a machine employing an annular revolving table 30 ft in diameter with two cross beams arranged at right-angles and supported by columns at the centre and the outside. This was generally described as a 'circular planing machine.'

Grinding

Grinding by abrasive wheels has been known as a machining method for hundreds of years and in the 17th and 18th centuries workshops equipped with a dozen or so large grinding wheels driven by waterpower seem to have been common. This however was all 'off-hand' grinding. The use of abrasive blocks and stones for finishing the bores of guns also dates back several centuries but grinding machines that were machine tools in the modern sense did not appear until about 1830.[5] Since then progress has been steady if not very rapid. Prior to 1841 Nasmyth & Gaskell had produced a surface grinder employing two segmental wheels of about 7 ft diameter which rotated about a common horizontal axis, the work being on external tables. The edge-wheel type of surface grinder appeared first in 1853 and the planotype edge-wheel machine in 1870 or thereabouts.

The earliest grinding machines were little more than lathes equipped with an abrasive wheel instead of a cutting tool and the Americans Ambrose Webster and Charles S. Moseley are credited with producing the first precision machines in the 1860's. Moseley's machine employed a travelling wheel-head, an arrangement that later became common in very large machines.

Browne & Sharpe also played an important part in the improvement of grinding machines; some of this was due to Charles H. Norton who worked for them and who later founded his own firm. The great improvements made in wheel manufacture by F. B. Norton (no relative) were also a factor in

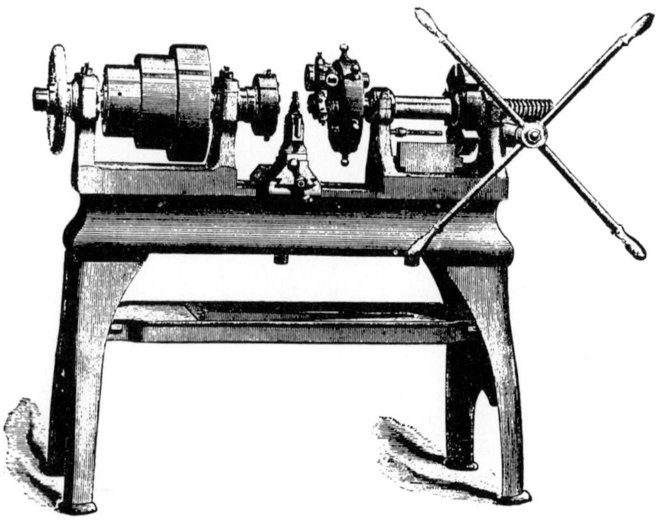

Fig. 8. Demand for faster production led to the development of the turret lathe in America. This early model (c. 1860) has the turret axis parallel to the spindle

[Courtesy: Alfred Herbert Ltd]

Fig. 9. The first automatic metal working lathe was built by Jacob Schweizer in 1872. A single spindle sliding-head machine, it was the forerunner of the present 'Swiss' type

[Courtesy: National Acme Co., Ohio]

Fig. 10. Invented in America about 1893, this multi-spindle automatic came closest to the production engineer's ideal; all the tools working all the time

the progress made in precision grinding. The universal cylindrical grinder made by Brown and Sharpe in 1863 acquired a very wide reputation for accuracy, as did a large machine for grinding rolls made by Morton Poole in 1870.[6] This is of some interest because it used two wheels on opposite sides of the work, carried on a saddle which was supported on knife-edge pivoted pendulum links. It may have led to the subsequent invention of centreless grinding.

The lack of suitable spindles for small wheels, and the difficulties of driving such wheels at the high speeds necessary to obtain satisfactory results retarded the development of internal grinders but in the 1880's Brown & Sharpe produced a successful design.

The grinding of holes in large components was not really feasible until the planetary wheel type of machine appeared. A patent for the double-eccentric mechanism for adjusting the size of the hole was taken out by R. Peacock in 1887 and machines embodying this principle were later built by Beyer Peacock. In Germany Mayer and Schmidt produced a planetary wheel machine about 1900, but with a different size adjustment: and in 1903 Hulse & Co. produced a machine in which the wheel rotated about a fixed vertical axis and the work was given a translatory motion in a circular path by giving the worktable two component motions along mutually perpendicular slides. A somewhat similar principle was also used at this time by the Pratt & Whitney Co. in a machine for grinding spheres and cup shapes but in this case it was the wheel that received the motion, the work merely rotating about an axis.

Tool and cutter grinders developed alongside cylindrical and surface grinders, but chiefly in America, and by 1870 several makes were on the market.

Into the 20th century

The period 1890 to 1910 saw numerous small changes and two important ones in the construction of machine tools. They were made heavier and more rigid because of the demand for higher cutting performance with the recently introduced high-speed tool steels; and the cone-pulley was displaced by the single-pulley drive and, later, by the self-contained electric motor. Single-pulley drive was dependent on the introduction of multi-ratio gearboxes which could be easily manipulated, were reasonably silent and could handle the increased torques and speeds required. Development of these was probably stimulated by the motor-car, then emerging from its infancy.

Attempts at individual electric motor drive were made as early as 1890 but real success was not achieved until about 1903 and belt-drive did not become obsolete until well after the first world war. At first, individual motor drive was used only on large or very large machine tools and then in conjunction with belts or chain-gearing. But toothed gearing soon took the place of belts and as electric motors and their control gear became less bulky and more reliable, their use extended to the smaller machines.

The early electric drives were mostly by d.c. variable speed motors providing, electrically, speed variations up to about 4 to 1, which had to be supplemented by gearing. Later the variable speed motor was displaced by the constant speed motor in conjunction with gearboxes and d.c. was displaced by a.c.

By 1910 almost all the types of machine tool now in use had been brought to a well developed state. Progress since then has been in the improvement and refinement of detail, increase in accuracy and ease of operation, and changes brought about by the advent of carbide cutting materials. This latter event has tended to make the actual cutting time frequently the shortest event of the whole machining cycle and has stimulated the use of automatic loading and unloading devices. However, these were fitted on automatic lathes in the 1890's.

Research into the action of cutting tools began to receive attention in the late 1880's and by 1910 some insight into the problem had been obtained; in this connection the names of F. W. Taylor, John T. Nicolson, Dempster Smith and the Verein Deutscher Ingenieure will be remembered.

Progress in machine tool design and construction has, of course, been closely related to progress in other fields of engineering. Improvements in foundry techniques have enabled more complex and better basic structures to be used. Better steels for both tool bits and workpieces have made the machining process intrinsically easier.

Advances in ball and roller bearings have facilitated spindle design and have mitigated the effects of bearing wear. The great advances made in the accuracy and availability of workshop measuring equipment have both demanded, and led to, improvements in machine tools themselves as well as of the products.

On the other hand the ability to produce mechanisms of greater precision or, alternatively, of the same precision at lower cost, which has resulted from the improvements in machine tools, has led to advances in most fields of mechanical engineering.

REFERENCES FOR FURTHER READING

1. KLEMM, FRIEDRICH, *A History of Western Technology*, translated by Dorothea Waley Singer, London, 1959.
2. SINGER, HOLMYARD, HALL and WILLIAMS, *History of Technology*, Oxford.
3. ROE, J. W., *English and American Tool Builders*. 1926 New York.
4. BUCHANAN, ROBERTSON, *Practical Essays on Mill Work and Other Machinery* 1841 3rd edition, London.
5. WOODBURY, R. S., *History of the Grinding Machine*.
6. ROSE, JOSHUA, *Machine Shop Practice*, c. 1880 London.

THE GREAT MASTERS

William George Armstrong —Baron

by A. T. Bowden, Ph.D., B.Sc. (Member) *

Portrait by Ramsay in 1830

Son of a cultured and well-to-do corn merchant, Armstrong set out in life to become a solicitor and had actually begun to practice when, in the course of some legal work, he became involved in the management of an engineering firm. Having once discovered his true calling as an inventor and entrepreneur, he never looked back. He became world-famous as an armament manufacturer, was raised to the peerage despite some differences of opinion with the War Office, and died the principal employer and benefactor of his native town.

Lord Armstrong ranks as one of the great engineers from the North, taking his place with Watt, Robert and George Stephenson, Parsons and Gibb.

However one may define the attributes of an engineer, there must above all be the inborn urge to do, to create; the intuitive appreciation of form and function and the will—the burning desire—to dedicate oneself through learning and experience to the successful accomplishment of those things one's heart is set upon. These are common to all artists and the engineer is essentially an artist.

In Armstrong one can see the engineer working as a leaven throughout the learning and practice of the law, to which he initially applied himself.

In the story which follows, I have tried to trace the processes which brought this great engineer to the top and at the same time to consider those personal qualities which allow us to acknowledge him, not only as a great engineer, but as a great man: slowly but surely the mould formed into which the whole of his energies were to be directed.

There is no royal road to success in engineering and it is an extraordinary fact that, before he became a great engineer, Armstrong was already an accomplished lawyer.

Childhood and education

William George Armstrong was born at 9 Pleasant Row, Shieldfield, Newcastle upon Tyne, on 26th November 1810. He was the only son of Alderman William Armstrong, a corn merchant, who had migrated from a village near Carlisle, and established a business in Cowgate, Newcastle. William Armstrong was a man of some culture and learning, was much liked and had a considerable circle of friends. One of the closest of these was a solicitor in Newcastle, Armorer Donkin, a man of considerable character and personality, who had built up a sound legal business and amassed a considerable fortune in the process.

Donkin was a bachelor and, when Armstrong's children were born, he took a great interest in them. First a daughter, Ann, was born in 1800, then William George in 1810. The son was rather delicate but from an early age he showed an enquiring and ingenious turn of mind.

William George was educated first at private schools in Newcastle, and later, when his health improved, at Bishop Auckland Grammar School, where he lived as a boarder with the master for two years. While there, he visited the engineering works of William Ramshaw who was so struck with the intelligent interest that Armstrong showed in the works' activities that he invited him to his house where he met Ramshaw's daughter, Margaret, whom he married subsequently.

As Armstrong grew up, Armorer Donkin became more and more attached to him and they spent a great deal of time together during the school holidays at Donkin's house at Rothbury. It was through this connection that Armstrong took a liking to the Rothbury district and became a most accomplished fisherman. Nothing pleased him more than a good day's fishing on the River Coquet.

On leaving school, Armstrong entered Donkin's office as an Articled Clerk. In the meantime his sister Ann had married W. H. Watson, a special pleader at the Temple in London.

After a few years with Donkin, Armstrong joined his brother-in-law in London for further experience. He eventually returned to Newcastle, joined up again with Donkin and was admitted into partnership in 1835.

* *Director, C. A. Parsons and Co. Ltd.*

From a portrait in the Institution Meeting Hall

Hydraulic crane

During a fishing outing to the Craven district of Yorkshire, in 1836, Armstrong's attention was arrested by a mountain streamlet which was used at the bottom of its fall to turn a water wheel. It occurred to Armstrong that, if the brook were conveyed from the summit in a pipe, the whole fall, instead of only the tail end, would be available for power. This was the beginning of his long interest in hydraulic power and from then onwards his mind was always working on that subject.

Possibly his professional work proved uncongenial, for he began to take a scientific interest in electricity and to develop his interest in hydraulics. His demonstrations and lectures on these subjects at the Literary and Philosophical Society in Newcastle and other centres brought his name to the notice of learned societies. In 1846 he was elected a Fellow of the Royal Society. Thus, when he developed his hydraulic inventions, he was already a fairly distinguished man of science, and his views and ideas carried considerable weight.

In 1838 Armstrong formed a friendship with Henry Watson of the firm of Watson's High Bridge Works, Newcastle, and acquired the habit of going to the works almost daily.

There he studied with the greatest interest the manufacture of turret clocks, theodolites, telescopes, paper-making machinery and other equipment for which Watson had already become famous. He continued this practice for about six years, discussing and assisting in, the design and manufacture of some of the work. In this way, to all intents and purposes, he served his time as an engineer and acquired a solid foundation of technical and mechanical knowledge. In 1839 he had his first experimental hydraulic engine built at Watson's works, which operated on the town's water supply with an output of some 5 hp. He then proceeded to design and make a working model of a hydraulic crane, in which the lifting operation was performed by the single stroke of a piston. However, as there appeared to be little support for his ideas, he gave up the subject and it was not until five years later that an event occurred which offered him new facilities and opportunities for exploiting his invention.

In 1845, the Whittle Dene Water Company was formed to supply the town of Newcastle with water and the acting solicitors were Donkin, Stable and Armstrong. At the first meeting of the shareholders, in July 1845, Armstrong was appointed Secretary.

The formation of this company gave him the opportunity to return to his hydraulic crane invention and in 1845 he communicated to the Finance Committee of the Newcastle Town Council his proposal to convert one of the cranes on the Newcastle Quay to hydraulic operation. This proposal was accepted with the provision that, if it proved successful, four further cranes could be similarly converted. With the financial help of his friends, he proceeded to convert the first crane, the work being carried out at Watson's Works. It proved a success and Armstrong and his friends decided to form the Newcastle Cranage Company to convert the other four cranes, the work again being undertaken by Watson's.

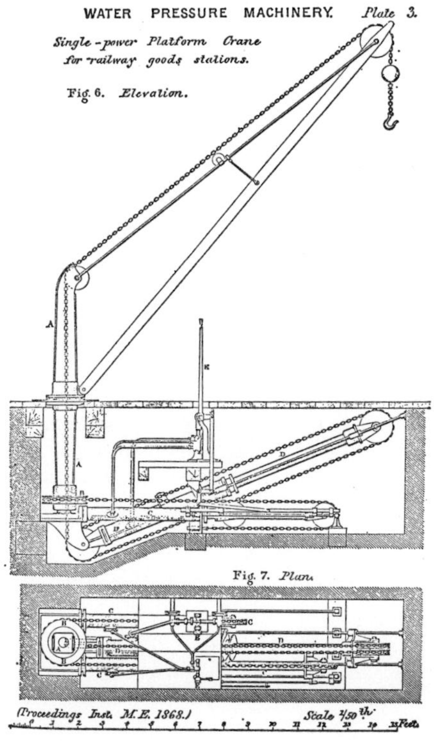

A hydraulic railway crane illustrating a paper read by Armstrong to the Institution in 1868 ◀

W. G. Armstrong and Co.

Early in 1847 the partners decided to make their own hydraulic machines and for this purpose formed W. G. Armstrong and Co. Armstrong, then 37 years of age, resigned from the solicitor's office and the Water Company, in order to give his whole attention to the project.

In the summer of that year, the Company bought two

The battleship 'Victoria' was one of many men-of-war launched from Elswick Shipyard ▼

The Tyne swing bridge in Newcastle, built in 1876, is still in operation

The Elswick works in 1847

fields of five and a half acres in the then pretty country village of Elswick on the outskirts of Newcastle as a site for the factory. The earliest orders came mainly from docks, railways and mines for cranes, warehouse hoists, dock gates and their associated hydraulic machinery, but they included one from the *Newcastle Chronicle* for a hydraulic engine to drive their printing press.

A steady volume of orders for general engineering as well as hydraulic machinery poured into the factory. Contracts for building bridges were also undertaken. Two of the earliest were the Inverness Bridge, completed in 1855, and the Soane Bridge, ordered by the East India Railway Co. and completed in 1857.

In 1854, trouble arose in the Near East between Russia and Turkey and Britain found herself involved in the Crimean War. At an early stage of the campaign, Armstrong was asked by the War Office to design and make submarine mines for blowing up the Russian ships, sunk to block the entrance to Sevastopol harbour. Although he did so, they were not used. Nevertheless, Armstrong had been induced to think about military matters and had made useful contacts at the War Office.

In 1854 newspaper reports of the battle of Inkerman recorded the use of two 18-pounder guns which had silenced an entire Russian battery. But they weighed two and three-quarter tons each and, when required to be moved forward, could not be manhandled over the rough ground. Armstrong at once began an investigation into the subject, with a view to the construction of field guns which would be more powerful and easier to handle. He had an interview with the Duke of Newcastle, Secretary of State for War, and put before him his detailed proposals to enlarge the standard military rifle to the calibre of a field gun, and to use elongated lead projectiles instead of cast-iron balls.

The Armstrong gun

As an idea, the rifled gun was not new. Experiments with rifling had been carried out in Russia and in Prussia, as long ago as the seventeenth century, and in England the Ordnance Select Committee had already considered many scores of proposals for elongated projectiles and rifled guns. Everyone concerned with artillery knew well enough the kind of gun that was needed. But, up to 1854, the metallurgical and mechanical problems involved had not been solved. It was to these problems that Armstrong, with the Duke's backing, now turned the combination of intuition, training and perseverance which made him such an outstanding engineer.

He was able to deliver his first gun in 1855. This was constructed of a steel core with polygroove rifling and a wound iron wire jacket—without doubt one of the most outstanding engineering conceptions of all time.* The projectile was soft coated and fitted intimately into the grooves as it travelled along the gun which weighed a quarter of a ton and fired a 3 lb shot.

The initial trials were successful, but the Ordnance Committee did not consider the gun big enough and sent it back to Elswick to be re-bored for use as a 5-pounder. In December 1856, two years after he had first seen the Duke of Newcastle, Armstrong's 5-pounder was tried with equal success and he was then commissioned to submit a still larger gun. This, an 18-pounder, was successful in every way. In November 1858 a Committee appointed to pronounce on the best form of rifled guns for field and garrison service reported in favour of his artillery.

Armstrong was already known as a successful engineer, but his new inventions brought him increasing fame, the more so when he refused the financial reward that was offered him and presented his patents to the nation. Always a man of commanding personality, he became a prominent figure at the War Office and, in February 1859, he entered the service of the Government as Engineer of Rifled Ordnance. At the same time the Queen conferred upon him a Knighthood for his services.

The Secretary of State for War gave a contract for rifled guns to the Elswick Ordnance Company, formed simply by setting up a new department at Elswick with public funds. Armstrong, now a Civil Servant, had no financial interest in the new firm, which was to manufacture exclusively for the Government.

For a time, all went well with the Elswick Ordnance Company and orders poured in for armament to be used in China, New Zealand and Japan. There was, however, bitter hostility to the Armstrong guns. Faults were found in a hundred and one different quarters: the Service officers of the day were intensely conservative. The theoretical superiority of the breech-loader over the muzzle loader was soon admitted, but not its satisfactory operation in the field. There were also allegations that costs at Elswick were unduly high and that money was being thrown away there which could be saved if manufacture were kept in the hands of the Royal Arsenal at Woolwich.

In 1862 the Elswick contracts were terminated and the Government abandoned breech-loading, returning to muzzle-loading with which the Services felt more at ease. For many

* *cf. p. 131, Volume 1.*

years thereafter, the Elswick works had no part in the design and manufacture of British guns.

These events led to Armstrong's resignation from his post as Engineer of Rifled Ordnance and to the amalgamation of the Ordnance Company and the Engine Works. The new company set out to capture foreign orders for their armaments and, by the late eighteen-sixties, Austria, Italy, Spain, Egypt, Denmark, Holland, Chile and Peru, were all customers of Armstrongs.

Warship construction

Early experimental proof of guns had been carried out in a variety of places, mainly on land, but during the search for a suitable proving ground, the Company had carried out some proofs at sea and found that a relatively small barge would suffice for proving a 12-ton gun. This led to an interest in ship-building. The first armed vessel Armstrongs produced was designed at Elswick in 1868 and built at the yard of C. Mitchell and Co. at Low Walker on the Tyne. This was the shallow draught coast defence gun boat *Staunch* of 160 tons displacement and 95 h.p., carrying the service 9-inch muzzle-loading gun. In 1822 the two firms amalgamated and two years later a shipyard devoted exclusively to warship construction was opened at Elswick, the Walker Yard henceforth concentrating on merchant shipping.

The opening of the Elswick Shipyard had been made possible by the replacement of a low-span, eighteenth century stone bridge by the Tyne Swing Bridge, constructed in the Engine Works and installed in 1876. Operated hydraulically, this bridge (weighing some 1450 tons) swings on a centre pivot to allow a clear passage for ships up and down the river. In its day, it was an outstanding development and a magnificent engineering project.

In 1878, at a time of strained relations with Russia, breech-loading guns were again introduced. A new 6-inch breech-loader had been developed which Armstrong himself considered the greatest improvement since the introduction of rifled ordnance. The firm got large orders for these. As hydraulic equipment in warships became more complex, the Admiralty once again turned more and more to Elswick "for the simple reason," as Armstrong said, "that they could hardly do otherwise". A steel works, producing acid open hearth steel, was added to the Elswick works in 1884.

At this time, Armstrong, turned seventy, had gradually retired from active participation in the business, leaving the management of the works in the hands of the experts who had joined the Company during the years of his leadership. Of the five original partners in 1847, only Armstrong himself lived to see it converted into a public concern in 1882. He was raised to the peerage in 1887.

In 1897, Sir W. G. Armstrong and Co. amalgamated with the firm of Sir Joseph Whitworth and Co. to become Sir W. G. Armstrong Whitworth and Co. Ltd. Later, in 1927, the latter joined with Vickers Ltd to become Vickers Armstrong Ltd, one of today's great heavy engineering companies.

When Joseph Wilson Swan invented the incandescent electric lamp in 1879, Armstrong was particularly interested. Newcastle was the first town in England to apply electricity to public lighting installations. Armstrong, with characteristic quickness to appreciate the significance of the new method of illumination, adopted it for his Rothbury house, using a water turbine to supply power. Cragside was, therefore, the first large house in the country to have electric light. Characteristically, Armstrong associated himself with Swan, Spence Watson and J. T. Merz, to form Swan's Electric Light Co. Ltd, and opened a factory at South Benwell, Newcastle, that is believed to have been the first in Europe to manufacture a complete electric lamp.

A practical mind

To revert to more personal matters, Lord Armstrong, as the pioneer of hydraulic machinery and engineering, takes a foremost place among the men of genius who contributed to the material progress of the nineteenth century. There have been many thinkers more strictly scientific, for he was not deeply versed in any special branch of learning. His strength lay rather in the activity of his intellect, the quickness of his perception, and the penetrating skill with which he reduced theory to practice.

His discoveries and inventions proceeded upon invariable lines: close observation, ingenious deduction, followed by

[Courtesy: Vickers Armstrong Ltd]

A signed photograph of Armstrong and two friends

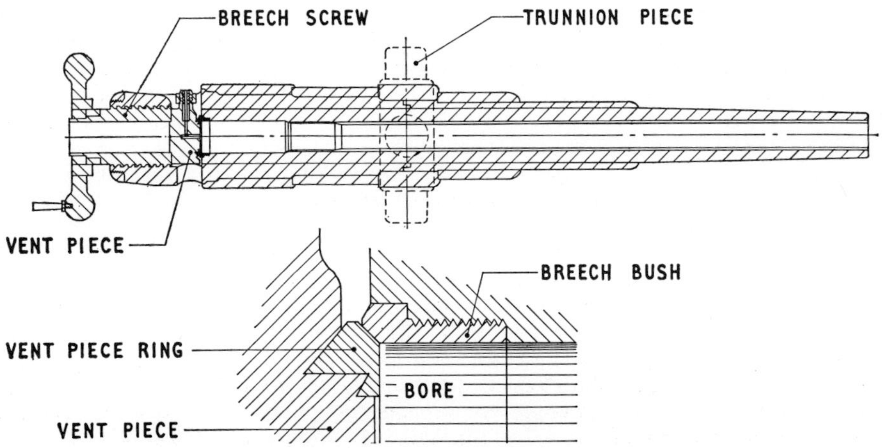

A wound wire jacket was a feature of Armstrong guns which became world famous

assiduous experiment. His industry in research upon any subject which interested him amounted to complete absorption for the time being. During his early studies of artillery he would be up day after day at sunrise to fire his gun on the moors or on the seashore. He thought of nothing else, but concentrated the whole force of his mind upon his investigations to the exclusion of everything else. His talents were rigidly limited to practical issues.

It would be an overstatement to describe him as lacking in imagination, but he had no taste for transcendental or speculative enquiries. Solid as were his theories, his work, whether he was building an engine or a house, was more thorough still. It wanted no detail which industry or perseverance could give it.

In private life Armstrong was the most charming of companions. In the prime of life he was an excellent walker, and was fond of fishing and shooting. He was an excellent host, and in congenial society a striking conversationalist. No man could have lived a simpler life or borne his honours with less ostentation.

Strangers sometimes found his manners rather cold and reserved. He had a keen eye for pretentions together with a distaste for them which he did not always take the trouble to conceal. Yet nobody had a more generous appreciation of real ability and worth.

Lord Armstrong was twice President of the Institution of Mechanical Engineers, and once of the Civil Engineers. He read papers to the British Association: in 1845 upon his hydro-electric machine; in 1854 on the application of water pressure to machinery; and in 1882 on the treatment of steel for ordnance and other purposes. He was President of the British Association, in 1863, when their meeting was held at Newcastle. He also read three papers to this Institution on bridges and hydraulic machinery.

He was a clear and interesting lecturer, though he had no gift for extempore speaking. He took a considerable share in the public life of Northumberland, acting as High Sheriff in 1873 and serving for some time as County Councillor for the Rothbury Division, but he declined to be nominated for the Mayoralty of Newcastle.

Public spirit

During his rise to fame, Armstrong was perhaps the greatest benefactor that Newcastle has ever had. He took a direct interest in the housing of his workpeople, the education of their children, the establishment of hospitals, parks, natural history museums and many other worthy causes.

It was in 1863 that the then Sir William revisited Rothbury, where he had spent many happy holidays at the Donkin home, and was again so impressed with the beauty of the scenery, that he resolved to purchase the Debdon Valley. He built his house at Cragside and the surrounding hills became his park. To the end of his life he lavished money upon his estate, creating and re-creating landscape like an eighteenth century nobleman.

At this time the health authorities were becoming aware of the necessity to provide more open spaces for the public. The Corporation of Newcastle, already blessed with the Town Moor, opened several public parks after 1873. To crown this municipal effort, Armstrong presented the town with the area of land known as Armstrong Park in 1880 and, three years later, added Jesmond Dene and the Banqueting Hall with its furniture and pictures. By this time Newcastle had more open spaces for public recreation than any other large town in the Kingdom.

During the rapid development of the Armstrong Works, the housing of the workpeople who had moved into the district became a major problem. In 1851 the population of Elswick was only 3539; in 1861 it had reached 14 345 and by 1891 it had leapt to 51 608. The building of houses proceeded rapidly; street upon street sprang up on the hillsides beside the Elswick Works and Armstrong himself was responsible for the building of many of them.

He took an interest in the hospital services of the town and made possible extensions and improvements to the Infirmary at Forth Banks. In 1888, he opened the Fleming Memorial Hospital for sick children and in 1896, as a memorial to Lady Armstrong, presented it with an out-patients department.

Finally, there was his interest in education. He opened a school for the children of his employees which, in 1889, was by far the largest in the city. When his father died in 1857, he carried out one of his wishes and presented to the Literary and Philosphical Society in Newcastle his collection of 1284 scientific works. Later he gave them a new lecture theatre and helped to provide accommodation in the Hancock Museum for the collection of the Natural History Society. He was active in the formation of the Durham College of Physical Science in 1871 and, in 1904, this college adopted the title of Armstrong College to commemorate him.

Armstrong's last charitable act was to purchase from Lord Crew's Trustees in 1894, the historic Bamburgh Castle in Northumberland. His intention was to restore and endow it as a hospital or a convalescent home 'suitable for impoverished persons of cultivated habits and acquirements'. Unfortunately he did not live to see the work completed.

Lord Armstrong died in 1900, and a statue was erected to his memory, between the Hancock Museum and what is now the University of Newcastle upon Tyne, both of which owe so much to his public-spirited munificence.

REFERENCES FOR FURTHER READING

1. JONES, E. R. *Heroes of Industry*. 1886 Low, London.
2. COCHRANE, A., *The Early History of Elswick*. 1909 Mawson Swan and Morgan, Newcastle.
3. SCOTT, J. D., *A History of Vickers*. 1962 Weidenfeld and Nicholson, London.

Bamburgh Castle, Northumberland, which Armstrong intended to restore and convert into a convalescent home.

Part of a 16th century printing shop: for the evolution of printing machinery see overleaf

THE EVOLUTION OF ENGINEERING

Printing and Papermaking Machinery

by A. D. Cummings, MSc, PhD

Unlike most technologies printing went through two revolutions: the introduction in Europe of movable type in the 15th century, as well as the application of steam to presses in the 19th century. The former may well have been a precondition of the latter, both by helping to spread knowledge and by pioneering precision toolmaking. With the introduction of continuous papermaking, newspaper presses and mechanical typesetting—all made possible by mechanical engineers—printing became the first medium of mass communication.

Fig. 1. The work of the European papermaker in the 19th century (according to Diderot's Encyclopaedia) differed little from that of his Chinese precursor. The man on the left is shaking the pulp; the coucher in the middle is stacking sheets with alternate layers of felt; and on the right the stacks are palleted and pressed

Printing is a process by which one man's thoughts can be communicated to many; as such it has had an enormous effect in world history. It was also the first mechanical invention involving exact measurement. Within the last 150 years power-driven machines have made newspapers and cheap books available to everyone who can read. For this engineering triumph, as for many of the material benefits we enjoy today, the world is heavily in debt to a handful of inventors. Although in recent years broadcasting has to some extent replaced newspapers, we still rely on printing for confirmation and for permanent record.

The art of the printer remained static for a long time. It seems that the printer himself is not inventive for most improvements have come from outside the trade. The engineer, on the other hand, has responded to the challenge of a difficult technical problem.

Hand-made paper

The early civilisations of Egypt and Mesopotamia developed separate systems of picture writing, representing first ideas and later sounds that became written script before 3000 BC. In Egypt they were fortunate in having a material, papyrus, ready to hand to form a writing surface. The plant does not grow in Mesopotamia where they had to fall back on tablets of soft clay that hardened in the sun after marking with a chisel-shaped instrument.

Papyrus is an aquatic plant with a fibrous part that can be cut into strips, laid on a board side by side and crossed with another layer of strips at right angles. When this lamination is pressed it forms a coherent sheet that only needs smoothing with a flat tool to form a writing surface, suitable for a reed pen using a black, water-based ink. The sheets were joined edge to edge and the completed 'book' formed a scroll. This material had a resonably good life as long as it remained in a dry climate but it did not travel well outside the Mediterranean area; in humid climates it disintegrated.

In other places parchment (prepared from the skins of calf, sheep or goat) was the only writing material. Obviously this was more expensive, though it could be cleaned and used over and over again.

Paper was invented in China about AD 100 for writing and drawing purposes. Various vegetable fibres from plants (e.g., linen and hemp) and parts of shrubs (e.g., mulberry) were extracted by softening in wood ash and then beating in water with mortar and pestle to fray them out. Sheets were formed by dipping a draining sieve or 'mould' into the diluted pulp, lifting and shaking from side to side to felt the fibres, leaving the wet but coherent mat to dry on warm ceramic tiles. A sheet suitable for writing on with pen or brush could then be peeled off. Cotton rags from discarded clothing were later used by both Chinese and European papermakers and remained the basic material for a very long time.

Considering how simple this process is and how useful the product (new uses are still being found, the world consumption of paper is going up and up), the art of papermaking spread westwards incredibly slowly. It reached Europe by

Fig. 2. Seals were the earliest form of printing. Babylonian rollers for impressing clay tables anticipated several features of modern printing

way of Arabia, North Africa and Spain about AD 1200. The method used in medieval papermaking, which is shown in Fig. 1, remained unaltered until 1800 and followed the Chinese method exactly, except that a brass wire sieve or mould with a wooden edge or 'deckle' was used instead of a reed sieve; and the sheets were not dried on tiles. The mould had parallel wires and frequently a design in wire was attached to the upper surface. This impressed its pattern on the fibres of the paper to form a 'water-mark', a form of marking still used, but now impressed with a roller.

Rag-pickers collected rags in large towns and sold them to papermakers; they were a shady lot, accused of stealing clothes from drying grounds and of spreading disease but they were essential to the industry. The rags were boiled, then 'rotted' and beaten in water to form the pulp. This was kept until wanted to fill up the vats of diluted 'stuff' or pulp containing about one per cent fibre.

Two men worked at a vat; the 'vatman' plunged the mould into the stock and, after draining and shaking, the 'coucher' transferred the wet paper to a piece of woolen felt. They passed moulds to and fro until the coucher had built a pile about 18 in. high of alternate felts and papers (called a 'post'). This was pressed to consolidate the fibre and squeeze out water. Later a third man, the 'layer', separated sheets of paper from felts, made a pile of sheets and pressed them again. The sheets were finally dried in heated lofts.

Today paper is made in many sizes, controlled originally by the size of the mould. A convenient size was about 20 in. by 25 in., such as a man could comfortably handle, but larger sheets were sometines made and then up to nine men had to raise the mould.

[Courtesy: Clarendon Press, Oxford]

Fig. 3. These hand type moulds of 19th century origin were similar to those used by Gutenberg 400 years earlier. They were about 4 in. square

The paper made in this way was good and had a very long life, as the present state of many 16th century, and even 15th century, books clearly shows. Most modern paper is made from wood pulp, often chemically bleached, and generally its life will be shorter.

A small amount of paper is still handmade for special purposes.

The beginnings of printing

In antiquity all books were hand copied. In the middle ages this was a regular part of monastery work, one monk read aloud while others produced anything up to a score of copies. The books were given in exchange for other texts or were sold to travelling monks. Later, universities took up this work and paid copyists were employed. The work was slow and even monastic scribes had to eat, so that books were rare and costly possessions; often they were kept chained or locked up to prevent theft.

The art of making a repeated pattern on a surface by means of a stamp, as in printing, comes from several sources, most of them a long way back in time. As early as 2000 BC important Babylonians signed documents by impressing a private seal, often in the form of a small roller as shown in Fig. 2, on a soft clay tablet. Several centuries before the Christian era coins and medals were impressed by striking blanks with a reversed inscription in a punch made of hard metal, and metalwave was decorated with repeating patterns in a similar way. Furthermore, Pliny mentions the process of stamping cloth with a pattern wetted with a mordant and boiling in dye to produce a permanent design; this method probably originated earlier in India. Thus the idea of impressing patterns on surfaces was not new; fabric printing comes very near to paper printing.

The earliest examples of printing on paper in Europe were playing cards, proclamations, indulgencies and 'block books' in the 13th century. These latter were intended for a wide audience and consisted of a few pages of religious story with a crude picture and a few lines of text to a page, printed from woodcuts. The pages were probably hand-pressed.

The next step was to find something faster and more flexible. Printing with metal type was not invented in Europe until the 15th century and little is known about early experiments with separate wood and metal type. Needham in *Science and Technology in China* writes of a work printed there with movable wooden type in the 11th century AD. Movable metal type was first used in Korea in the 14th century.

In Europe, separate metal type that could be composed, broken up and used again was first used by Gutenberg, a German goldsmith of Mainz, about 1440. There is some doubt how much of the printing process is due to him, for nothing was recorded until 30 years after his death and three separate inventions are involved; movable type; an oil-based ink that would take on a metal surface; and a press to bring paper and type into close contact.

Gutenburg would be acquainted with making hard metal punches for medals and jewellery; he would also know about repetition casting of small objects in soft metal. Printing requires a large number of characters and obviously a separate mould for each was out of the question. Gutenberg invented an adjustable mould, similar to the one shown in Fig. 3, with a slot to take one of a number of matrices. The character was first formed on the end of a steel punch, as shown in Fig. 4, and hardened. The punch was next impressed by a sharp tap on the face of a copper bar, the matrix, measuring about 1½ in. by ¼ in. by ⅛ in. This was slipped into the closed mould where it formed the base of a die When molten type metal was poured in it solidified in the form of a short shank with the letter in relief on the end.

Compared with the achievement of interchangeable type, the press was easy; early printers merely took over the household press that in various forms had long been used for grapes, olives, cheese and linen. But it had to be a precision press, the platen had to hit one exact spot every time and the paper had to come away again without smudging. Moxon (the first to write, in 1683, a full account of the mechanical side of printing)[1] gives clear directions for the construction of the press. While it would not be beyond the skill of a good carpenter, it would need a knowledgeable man to direct him. It was, perhaps, the first precision machine ever built.

Once printing was invented it spread rapidly over all Europe; from Mainz it reached Italy in 1465, Switzerland in 1468, France in 1470, Spain in 1472 and England in 1476. This illustrates two things—that the apparatus was simple enough for one man who had learnt his job thoroughly to take his skill elsewhere and set up for himself; and that the market for books was considerable.

Caxton learnt the art of printing in Cologne and brought it to England. By 1500 he and his successor Wynkyn de Worde had published about 500 different books. The quality of the work was amazingly high, as is shown by Fig. 5, part of a page of a Gutenberg Bible. This is in the Gothic script that imitated the style of the hand-copyists, but by 1470 the clearer Roman type we use today had been adopted which was both easier to read and to make.

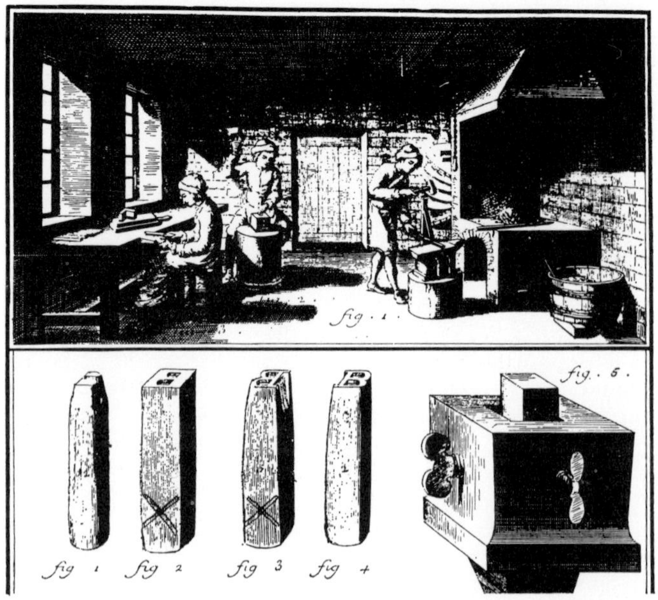

[Courtesy: University Press, Oxford]

Fig. 4. A punchmaker's shop and the stages in making a punch for the letter 'B'; this was used to form the copper die for the type metal

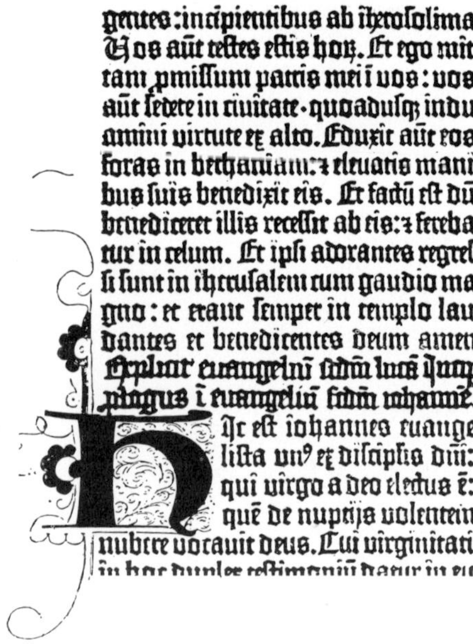

gentes: incipientibus ab iherosolima.
Uos aūt testes estis hor. Et ego mit-
tam pmissum patris mei i vos: vos
aūt sedete in civitate · quoadusq; indu-
amini virtute ex alto. Eduxit aūt eos
foras in bethaniam: et elevatis mani-
bus suis benedixit eis. Et factū est dū
benediceret illis recessit ab eis: et fereba-
tur in celum. Et ipsi adorantes regres-
si sunt in iherusalem cum gaudio ma-
gno: et erant semper in templo lau-
dantes et benedicentes deum amen.
Explicit euangeliū scdm lucā. Incipit
plogus i euangeliū scdm iohanne.
Hic est iohannes euange-
lista un9 ex discipulis dni:
qui virgo a deo electus ē:
quē de nuptijs volentem
nubere vocavit deus. Cui virginitatis

Fig. 5. Part of a page from the first important book printed from movable type: Gutenberg's 42-line Bible. The print is slightly reduced in size

Handprinting development

As will be apparent from Fig. 6, printing shops changed very little for centuries. An early printing shop would include a man to make the steel punches (dies) and one or more to make type. Compositors set up the author's words, picking out type from a sloping case with their right hand and setting them in a composing 'stick' held in the left. The lines of type were then made into pages, wedged in a frame and carried to the pressman. A trial 'pull' was taken and inspected for errors; if all was well, printing could begin.[2]

Other men would hang the printed pages up to dry (the paper was dampened before pressing to make it flexible and take a good impression) and still others would arrange the sheets in order and send them to the binder. The inky type was washed in alkali and redistributed into cases.

The press itself consisted essentially of a flat plate which was brought down by a steeply pitched screw to squeeze a sheet of paper into close contact with the inked type lying face up on the table. The 'forme' or frame of type would be put in position, inked and covered with a sheet of paper held in a frame hinged to the type-bed. An early improvement was to put the latter on metal rails with a mechanism to locate and slide it smoothly under the press.

Pressmen worked in pairs, one inked the type while the other placed a sheet of paper in the frame and brought it down on the type. He now turned a handle to bring the forme under the platen with his left hand and pulled the press-handle with his right hand once, moved the second half of the forme under the press, pulled again and ran the forme out again. Pulling the press-handle was heavy work and after a while the men changed jobs. Between them they could print 250 sheets an hour on one side and this was the pace of printing from Gutenberg to the application of steam power early last century.

Printing has had its ups and downs in quality. In the earliest days, while printers were still looking back to the slow but artistically satisfying work of the copyists, some reached standards that have seldom been surpassed. In the 17th century the volume of publication greatly increased but standards of production declined. Fewer illustrated books appeared, the look of the page was poorer because of irregular type-setting and careless registering on the page. The 18th century saw a revival of good work. The Birmingham printer John Baskerville, who had made money as a manufacturer and had taken up printing as a hobby, set new standards by using blacker ink and paper smoothed by pressing between hot copper plates.

The Industrial Revolution not only brought inventions in manufacture but set ideas moving in other fields. One change of great eventual consequence was an increase in literacy. The commercial middle class, especially their womenfolk, acquired a taste for reading, and even poor country people shortened winter evenings by hearing their sons and daughters read tales and romances.[3]

Continuous papermaking

When the 19th century opened, steam power had been in use for almost 100 years but had only recently been used to provide rotary motion. Watt's patent had just expired, men were looking for new ways of using the steam engine. Until then the well-tried hand methods in use in the papermaking and printing industries had sufficed, but the new reading public had a big appetite for books, newspapers, pamphlets and hand methods would no longer do. Ideas about liberty of, amongst other things, the Press, were spreading from France and America, not only to this country but to all Europe. In spite of taxes on paper, newspapers and advertisements, newspapers were selling. *The Times* began to come out daily instead of weekly with a circulation of 10 000, though most other papers had very much smaller circulations. It was the newspaper owners who stimulated mechanical developments.

A Frenchman, Louis Robert of Essones, invented and patented in 1798 a machine to make paper in continuous lengths. He sold this to his employer and the latter's English brother-in-law brought a crude model and some rough drawings to this country and persuaded two stationer brothers named Fourdrinier to finance the development of the machine by a brilliant young engineer, Bryan Donkin.

After about two years' work a partly successful machine, able to make a continuous web of paper, was evolved. It looked like a domestic wash-tub with a wringing machine on the side. Paddles stirred up the pulp and splashed it on to an endless wire sieve passing over rollers and delivering a web of wet paper to press-rollers which squeezed water out and consolidated the paper. This was then cut into sheets and hung up to dry as formerly. Two of these machines started work in Hertfordshire, the first in 1804. By 1810 the process was launched, but the Fourdrinier brothers, who had borne the financial burden, were ruined, mainly through trouble in establishing their patent.

Steam heated drying cylinders, patented by T. B. Crompton in 1821, greatly increased the speed of manufacture. Donkin went on to make more and better machines and by the time of the Great Exhibition in 1851, at which he won a gold medal, he had produced 191, one of which is shown in Fig. 7. Modern machines, still called Fourdriniers, retain similar features; a continuous wire mesh (now of stainless steel) press-rolls and drying rolls. The changes in papermaking since then have been in the use of new raw materials and increased speed of output. Rags have given way to wood-pulp from trees grown as a crop for this purpose. Mechanical wood-pulp, made by grinding up tree-trunks, was patented by Keller of Saxony in 1845 and chemical pulp, made by cooking wood chips in calcium sulphite, was patented by Tilghmann of Philadelphia in 1867. Both these and treated waste paper are used today. The finished product now comes off the machine at 20 mile/h and up to 20 ft wide.

Flat-bed printing

The mechanisation of printing took place by stages during the first half of the 19th century. The gradual evolution from flat type on the hand-press to continuous rotary printing with the type on a cylinder is still not complete today. In the newspaper field, where speed is of the first importance, it was achieved by 1865, but in book and magazine work the need for speed was less and for quality greater.[2]

In the first decade of the 19th century *The Times* was being printed on a number of Stanhope presses, the invention of a clever, eccentric nobleman, the third Earl of Stanhope. This improved hand-press used a combined screw and lever movement and was constructed of cast iron. It reduced physical labour but did not increase output.

A German printer, Friedrich Koenig, came here in search of British engineering skill and capital for his idea of a steam-driven press. He made two fortunate contacts, John Walter, owner of *The Times*, and a man named Nicholson, editor, translator and founder of a scientific journal. The latter was interested in printing and had patented a process using a rotating cylinder in 1790. This contained the germ of many later rotary machines.

Koenig's first effort was a platen press, rather like a hand-press operated by steam power. Nicholson introduced him to the rotating cylinder for paper and Koenig then went ahead to produce the type of machine shown in Fig. 8, but driven by steam. The flat type-bed travelled to and fro beneath a cylinder of about 2 ft diameter with three separate printing sectors, printing three sheets for each complete revolution, the sheets being fed in and taken off by hand.

[Courtesy: Burndy Library, USA]

Fig. 6. Printing techniques did not change much between Gutenberg and the industrial revolution. On the left a 16th century print shop is shown; on the right two presses in an 18th century establishment; note bracing against roof beams and hinged paper frame on type forme

The Times installed two of these and, on 28th November, 1814, the reader was told that he "held in his hands the first issue of a newspaper printed by steam power". These were the days of machine-wreckers—hand operatives who feared the loss of their livelihood—and trouble was anticipated. The new machines were erected secretly in a separate building and on the night of 13th November, the printers were told to wait for news from the Continent and not start work at the usual time. At 6 a.m. next day Walter went in to them and said that the entire edition of the paper had been printed by steam power without their help, handing out copies to prove it. He went on to say that if they attempted violence he had a force ready to suppress it but that if they accepted the new method peaceably he would find work for them all. The men submitted.

In 1813 Bryan Donkin, who had succeeded with his continuous papermaking, also attempted a continuous printing machine, but this time he failed. His machine had four flat faces of type set up in the form of a square prism. This rotated about a horizontal axis against another square prism with curved faces round which the paper was wrapped. Flexible drives held the two in contact, so that the faces carrying the type and paper were always pressed together. It is said to have produced good work for the Cambridge University Press but it was not a success.

The Koenig press served *The Times* well until 1827, with modifications brought in by the paper's own engineers, Applegath and Cowper. In that year they installed a new four-cylinder machine of their own design, still with a reciprocating flat bed of type, that increased output from 1000 to 5000 impressions per hour. Paper was fed in by four men from conveniently placed tables; the sheets passed downwards and round the cylinders while the bed of type went to and fro beneath them, inking rollers passing over the type at three points. The great disadvantage of this type of press (which is still widely used in book wook, where the page is smaller) is the heavy weight of type moving backwards and forwards with a dead stop at each end. The big page of the newspaper now proved a stumbling block to progress.[4]

Fig. 7. A papermaking machine about 1850: pulp from the tub on the left is lifted on to a vibrating wire mesh belt, moved over suction boxes and pressed between rollers on continuous felt; finally it goes over steam heated drying rolls

Fig. 8. Applegarth's and Cowper's single cylinder flat type-forme machine (about 1828) developed from Koenig's. It could be either hand- or steam-powered. In principle today's flat bed presses are not very different

Rotary presses

Developments were going along parallel lines in the United States where the firm of Richard Hoe (still going as Richard Hoe and Crabtree) rivalled *The Times*. Both had rotary machines ready in 1846 which are shown in Figs. 9 and 10. The Hoe cylinder worked horizontally with columns of type in flat formes wedged in its surface; grouped round this type cylinder were four impression cylinders. Each of these was attended by men who fed in sheets and by others who took them off.

The Times' machine was similar except that the 64 in. diameter type cylinder ran vertically, with eight impression cylinders round it. This also worked with flat type-formes, each a column wide, the circumference being actually a large number of flat faces. This monster required the full-time attention of 25 men and boys to keep it fed with paper. It was claimed that this vertical machine was safer from the danger of throwing off masses of type than the horizontal machine but the 'Hoe Horizontal' displaced it from *The Times*' own printing shop in 1858.

Wood-pulp paper could now be made in big rolls and reel-fed printing had for some time been possible for small type areas such as on labels and bank notes, but newspapers still carried a tax (a penny a copy) requiring a stamp on each paper and this seems to have tied them to sheet printing. However, that tax was repealed in 1855, the one on paper itself in 1861, and reel-printing was in.

In 1865 Bullock at *The Philadelphia Register* first printed a newspaper from a reel of paper though, for some reason, he cut the paper into sheets immediately before it passed the printing rollers. His machine is seen in Fig. 11 and is not unlike the one on which part of the CME is printed! Once again *The Times* was not far behind, the new Walter press, first used in 1868 and the first rotary machine of the modern type, printing from stereo plates attached to cylinders, using a reel of paper and cutting it after printing at up to 12 000 impressions an hour. All these machines were steam-driven until 1891 when electric motors were first introduced in Birmingham.

Stereotyping, invented in 1727 and now coming in for newspaper work because of the need to bend the forme, has a long history. It was always uneconomic to print an edition of a book and break up the type if another edition might be called for later. On the other hand, it was expensive to keep type set up indefinitely on the uncertain chance of another printing. Since the 18th century it had been customary in the book trade to take plaster casts of pages of type and store these so that casts in lead alloy type-metal could be taken from them.

For newspapers there was no need to store but, as circulations increased, it became necessary to print the same edition several times on separate machines and for this stereo copies were labour saving.

Small stereos were easy, but the big semi-cylindrical plates to fit a newspaper machine were held up until the invention of an ingenious matrix material called 'flong' (French—*flan*). It is a heavy tough paper board, patented by John Dickinson in 1809, but not applied to this use until much later. After moistening, it is pressed against the flat bed of type between rollers, dried and fitted into a semi-cylindrical moulding box. The type-metal has a low melting point so that is sets quickly without distorting the flong matrix which can be used up to 20 times.

The mechanical problem in newspaper production is running off a large number of copies in a short time. As the number of copies per hour rose, a problem developed of carrying a band of paper under tension between pairs of rollers at high speed. One of these, the plate cylinder, has a very rough surface; the other, the impression cylinder, must have a soft surface—originally a blanket-like material was used—into which the paper can be pressed. As speeds increased, the drag on this material wore it out quickly, necessitating frequent renewal, until a hard-surfaced 'blanket' with a resilient interior was devloped.[5]

Except for very large editions, the book trade, with its lower capital resources and less exacting time schedules, can still do its work on flat-bed, rotating cylinder presses and the advertisements in this Journal are done on them.

It has been mentioned that the ever-growing reading public took in books, periodicals, political pamphlets and weeklies. Some suggest that the Victorian Sunday stimulating reading; there was not much else to do. The Society for the Diffusion of Useful Knowledge published *The Penny Magazine* from 1833 to 1847, a miscellany of everything except fiction and politics, which reached 200 000 copies weekly; it had eight pages and five woodcuts, well printed but depressing to a modern eye.

Nevertheless, it was one of the pioneer illustrated weeklies. The other was the *Illustrated London News* (1842) which is happily still with us.

Illustration techniques

The earliest 'block books' had woodcut pictures as integral parts. The wood was cut away to leave lines of the drawing standing up; after very crude beginnings the art was brought to a high degree of perfection by Dürer in the 15th century.

During the 16th to 18th centuries the woodcut was supplemented by copper plate engraving. The engraver cut grooves forming the mirror image of the illustration into the surface of a polished plate which was inked and wiped, leaving ink only in the indentations to be transferred to paper in a press like a mangle.

In the etching process (and others related to it) the copper plate is covered with varnish, the artist draws his design on this surface with a needle, breaking the varnish so that, when the plate is washed with acid, the metal is eaten away. leaving hollows which hold ink for printing, as in engraving.

This is the ancestor of a number of increasingly important chemical machining processes, notably printed circuit production.

Modern processes nearly all depend on photography, even when the original is some form of line drawing. Line blocks were introduced by Firmin Gillot of Paris in 1850. In the half-tone method, the invention of Meisenbach in 1881 (but based on an idea of Fox Talbot's in 1852), the picture to be copied is photographed through a galss screen containing a grating of fine lines dividing the area into tiny squares of clear glass.

This image is transferred to a zinc or copper plate coated with light-sensitive, acid-resisting coating. This is exposed and developed like a photograph, washed and etched with acid. On any square that has been exposed to light, the acid resist will disappear and during etching, the acid will eat into the square, thus reducing the amount of inked area on the picture.

[Courtesy: The Clarendon Press, Oxford]

Fig. 11. Bullock's rotary for the 'Philadelphia Inquirer' was the first to print on a continuous web of paper

Fig. 9. Applegath's vertical cylinder eight-feeder: the type was attached to the central cylinder, sheets were fed on the upper deck and taken off by the men below

[Courtesy: The Clarendon Press, Oxford]

Fig. 10. Hoe's ten-feeder rotary for the 'Philadelphia Ledger' which replaced Applegath's machine at 'The Times' in 1857: it did up to 20 000 impressions per hour

The final development of the modern half-tone is due to Max Levy in 1890.

Photogravure, the photographic development of the old copper-plate engraving principle, was developed by Karl Klic of Bohemia. The principle is the reverse of the line block, volatile inks being used which remain only in the lines of the drawing which have been etched into the surface of the plate.

Lithography, invented by Senefelder, a Bavarian, about 1800, depended on the attraction of ink to a design drawn in grease pencil on a smooth stone surface. When the stone was wetted with water the ink adhered only to the parts of the design which were greasy and transferred to paper accordingly. In the modern process, metal plates have been substituted for the heavy stones and, again, a photographic process is used, similar to that in blockmaking but in this case the photographic image remains in the form of a greasy coating which attracts the ink.

Mechanical typesetting

After the introduction of fast presses, typesetting became the bottleneck. The last, and as it turned out, the hardest problem in mechanisation was composition. In newspaper work particularly, even the 1500 characters an hour of manual typesetting were proving too costly and slow. During the last quarter of the 19th century about 200 attempts were made to design a machine to replace these nimble-fingered and well-organised craftsmen.

The early efforts all attempted mechanical setting of the actual type, but this proved to be of an unsuitable shape for automatic handling in chutes. One such machine, the *Pianotype*, required seven operators to work a keyboard, space the type, fill up the reservoirs and turn a handle. Another line of attack was to emboss, by means of a typewriter, some soft metal so as to make matrices, one line at a time.

The solution was to go back to the individual matrices of Gutenberg.[6] This was the work of Mergenthaler, a German watchmaker settled in America. In 1886 he installed the first *Linotype* machine, the principle of which is shown in Fig. 12. There are brass matrices for each character which are set a line at a time by one man working a keyboard. After each line is set, a slug of type metal is cast from it and the matrices are then re-distributed automatically to their storage container as shown. Selection is through code notches in the side of the brass shanks.

Another development, the *Monotype*, built by Tolbert Langston in 1887, performs a similar function for individual characters. It casts them one at a time very rapidly from a matrix containing 272 moulds. The coordinate location of this is controlled by a punched tape, produced by a keyboard operator.

Both these ingenious machines have their own field of service, the *Linotype* chiefly for newspapers, the more flexible *Monotype* for books.

Conclusion

Since the spread of literacy last century, paper and print have continually expanded. Whenever the need for greater output was sufficiently strong, engineers have responded by devising new machines.

In the past the makers of printing and papermaking machinery have been in the vanguard of mechanical engineering. Pioneering precision presses and continuous handling of delicate material such as paper, they entered an altogether different field from the rough engine and millwork of their colleagues. Typesetting machines are probably the earliest example of automatic assembly of small parts and their punched tape or code notch control was a precursor of many 20th century techniques.

Today, however, printing machinery—while still the mainstay of world communications—is in some danger of being superseded by faster electronic techniques such as television and Xerography. But, no matter what the basic principles will be, mechanical engineers will be there to produce the hardware.

REFERENCES FOR FURTHER READING

1. Moxon, J., *Mechanick Exercises.* (1683—new edition, edited by Davis and Carter 1958)
2. Singer and others. *A History of Technology.* 1957. Vol. 3, '*Printing to 1800*' by Michael Clapham. 1958. Vol. 5, '*Printing in 19th Century*', by W. Turner Berry.
3. Steinberg, S H., *500 Years of Printing.* 1961. Penguin Books.
4. Cowper, E., 'The Printing Machines of The Times Newspaper'. *Trans. Inst. Civil Eng.* Vol. 9. 1850. 409–427.
5. Isaccs, G. A., *The Story of the Newspaper Printing Press.* 1931.
6. Monotype Corporation. *Ottmar Mergenthaler 1854-1954.*

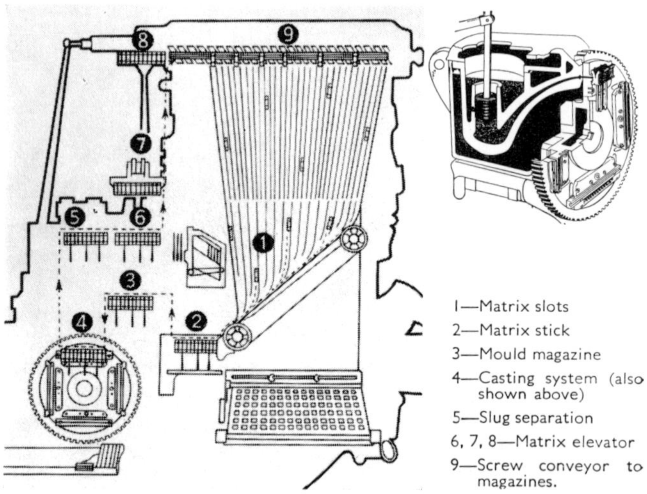

Fig. 12. In the linotype the keyboard pneumatically releases the matrices to form one side of a mould from which a line is cast in type metal. Matrices are then raised and move along a screw conveyor to fall into their appropriate magazines, released by code notches in their shanks

[Courtesy: Clarendon Press]

Fig. 1. Egyptian bead makers c 1450 BC, using a multiple-spindle bow drill on glass beads

THE EVOLUTION OF ENGINEERING

Glass Manufacture

by E. Venis, BSc(Eng) (Member)

Glass is an awkward raw material and its manufacture has exercised the ingenuity of the mechanical engineer. Some ingenious devices, including rotary transfer machines, were pioneered in the course of developing glass bulb and bottle-making machinery. The arts of plate and sheet glass making have seen little basic change until quite recently.

Fig. 2. Diderot's Dictionnaire Encyclopédique shows how crown glass was made in the 18th century. A lump is gathered, blown and rolled into the shape shown on the left; after reheating and flattening the end to which an iron rod is attached for handling, the rim is reheated and opened out (middle). Further heating and rapid spinning (right) turns the bell shape into a large disc of glass with only a small boss where the rod was attached

Glass is an artificial compound produced by the fusion of silica with an alkaline flux, plus ingredients to make it durable, brilliant, tough or heat resisting. It was probably first used as early as 4000 BC in the form of a glaze covering soap-stone beads, while the first glass objects may be regarded as imitations of precious stones and date from about 2500 BC. A little later quite elaborate figurines were shaped in clay moulds while seals and even larger articles were made by treating the glass as if it were rock crystal. Fig. 1 shows that glass was one of the earliest materials subjected to 'machining'.

Early manual work

The early engraving appears to have been done by hand, with a sapphire point secured in a handle. Later, a bow drill with a similar point was used and, later still, a copper tool, fed with sapphire dust.

Glass can be drawn out into rods and into fine threads. About 1500 BC vessels were made by winding a thread of ductile glass round a small linen bag filled with sand and tied to the narrow end of a tapering iron rod. Coloured rods fused together were cut transversely to make polychrome glass, then pressed in a mould to form mosaic glass dishes and finished using the tools of the lapidary which included the grinding wheel.

This technique led to a new process of moulding—a flattened lump of half-molten glass was made into the form of a bowl by pressing it on the outside of a convex core. These bowls were often left with a smooth finish provided by a final heating but were more commonly ground and polished, using the lapidary's wheel.

There is very little evidence as to the very early methods of heating the raw materials. At 1000°C, the temperature of a charcoal fire, glass becomes fluid; even in pre-Christian times the furnace temperature was increased by better draught and by using bellows. This permitted the revolutionary discovery, made about 4 BC, of glass blowing, which spread across the Roman world from Syria.

Molten glass was gathered on an iron tube, dipped several times into a crucible, spun to maintain symmetry, rolled on a slab to preform the shape, blown to form a bubble, then probably shaped in a mould of clay or carved stone. Skill developed quickly and it was found possible, not only to elongate the preliminary shape by swinging it, but to work it with simple tools such as tongs and pincers, obtaining any desired shape without using a mould. The finished glass article was cut from the blowing iron with shears.

Windows and mirrors

The blowing iron, the punty (the iron rod to which the partly made object is transferred), shaping tools, tongs and shears have continued in use until the present day but, although Jerome in AD 331 spoke of a window as a plate of glass produced by casting on a large flat stone, even in the 12th

Fig. 3. *Broad glass making (also from the Dictionnaire): the blower rotates the glass into a hollow cylinder (left) which is then snipped open and pressed into flat sheets*

century the windows of English churches were not always glazed, while even much later it was only the large houses which were so protected. These windows were made by the crown process which is shown in Fig. 2.

About 10 lb of glass was gathered on the end of a blowing iron by repeatedly dipping it into the pot and twisting it before rolling it into a globular or cylindrical mass on a table called a 'marver' (from the French for marble). The glass was then blown into a large globe. Transferred to the punty, it was cut from the blowpipe to leave a small opening at the end. Reheated and spun, the glass flared out into a broad flat disc up to 5ft diameter, with a bull's eye in the centre. When the bull's eye was cold, the disc was annealed in an oven.

Mirrors of a crude kind were made in Germany as early as the 12th century; superior ones in Lorraine in the 14th century, probably by the cylinder or broad process. The Venetians learnt this soon after AD 1500 and created a large industry for the European market. The broad process, as practised in France during the 18th century, used a very large ball of glass, up to 40 lb, which was gathered, marvered, blown into a globe, swung in a deep trench to elongate it and then blown into cylindrical form. The ends were opened and cut off and the resulting cylinder split with shears longitudinally whilst hot as shown in Fig. 3.

Reheated until soft, the cylinder was unrolled and flattened into a sheet with a wooden tool, the surface sometimes being ironed with a heavy wooden block before passing to an annealing kiln, to remove internal strains and stresses. This produced sheet about 3 ft by 4 ft. An improvement was made by R. L. Chance about 1830 at the Birmingham glassworks by allowing the cylinder to cool, splitting it cleanly with a diamond; then, after reheating, he flattened it on a bed of glass instead of using an iron plate.

Plate glass

The most important new development in the late 17th century was casting of plate glass in large sheets. The credit for the discovery is due to Bernard Perrot or Perrotto, the naturalised descendant of a famous family of Italian glass workers, who solved the problem of making larger and thicker sheets. The process was taken up by Louis-Lucas de Nehon of Tourlaville in Normandy in 1688, the manufacture being transferred eventually to Saint Gobain in Picardy where one of the world's largest glass manufacturers is still in existence.

A ton of glass from a very large crucible was poured on to a copper table covered with sand, the size of sheet being controlled by guides. The glass was rolled with a copper roller to produce uniformity of thickness and an even surface, suitable for mirrors which had, by this time, acquired a scientific importance in optical instruments. That the process had not changed fundamentally by the middle of this century is shown in Figs 4 and 5.

Plate glass came to England in 1773 from St Gobain but its commercial success was not achieved until the English firm substituted an iron table for the copper one which repeatedly cracked. Plate glass is not only flat and transparent but has plane polished surfaces and it is the rolling, grinding and polishing operations that distinguish it from sheet or window glass. The mirrors of the 17th century were ground by sliding a smaller plate of glass upon the one being ground. This smaller plate was cemented to a plank which, with the aid of a wheel, was pulled backwards and forwards by workmen. Later, large grinding tables of flat stone slabs were used while a large iron beam suspended from the roof by chains swung with a circular motion and served to drag iron-shod grinding blocks over the surface, using sand as the abrasive.

A Boulton and Watt engine supplied the motive power for glass grinding as early as 1789 but it was not until 1824 that Hoyau, a Belgian, introduced the principle of the rapid rotation of several grinding pads, inverted over a large circular table. The early polishing had been done with a small roller, covered with felt, using Tripoli stone and emery as abrasives. The roller was equipped with handles at each end and connected to the ceiling through a strong bow which acted as a spring, both for maintaining constant pressure and return stroke.

The 19th century saw considerable mechanisation of production generally but, because of the complex problems posed by the physical properties of glass, this industry changed very slowly. In this country the real impetus came with the repeal of the excise duties on glass in 1845 and, six years later, the end of the window tax.

Vol. IV, Glaces, Pl. XXIV.

Fig. 4. *Plate glass was cast on to a flat table and flattened by rolling*

The process of casting remained the same but sizes increased. With greater demand, production was raised: about a ton of glass was heated in a regenerative furnace, removed by a crane, skimmed, then partly inverted over a flat cast iron table, about 16 ft wide by 30 ft long, which had been covered with fine sand. The molten glass was poured in a continuous stream, just ahead of a water-cooled roller weighing about 5 tons. Moved by chains, this rolled the glass flat, the thickness of the glass being regulated by the use of two strips.

The roller was raised on wedges after passing over so that the glass could be pushed from the table into the first of a series of annealing ovens.

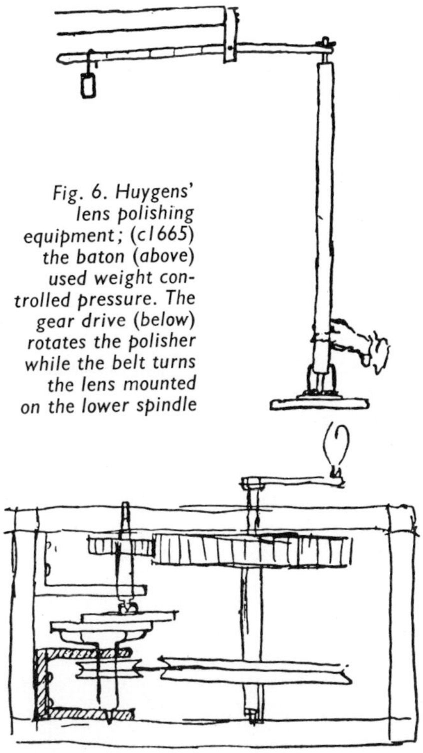

Fig. 6. Huygens' lens polishing equipment; (c1665) the baton (above) used weight controlled pressure. The gear drive (below) rotates the polisher while the belt turns the lens mounted on the lower spindle

Furnaces

The most usual type of furnace in Europe up to the 18th century was the 'beehive'. The fire was built on a grate so that the heat passed upwards through a flue to the oven where the pipe clay pots rested upon the hearth. The pots were reached by the glassblowers through openings round the circumference of the furnace. A second flue led to the annealing oven built at the top where a small door gave access to the finished articles which were slid along a tunnel where they cooled very slowly. Later a conical glass furnace became usual in England because of its improved draught.

Timber was used as fuel up to 1615 when, in an attempt to prevent deforestation, its use was forbiden for glass furnaces. For coal burning the fire was laid on an iron grate which gave higher temperatures but was so dirty that it necessitated covered pots for glass of a higher quality.

In 1856 Sir William Siemens obtained an English patent for a process of heat regeneration by using the hot waste gases to preheat incoming fuel and air. In 1861 he invented a gas producer which made it possible to use gas derived from low-grade coal. His first application was in a Midlands glass works, one of the outstanding contributions of glass technology to furnace development.

The much higher temperatures reached by these furnaces reduced the lives of the covered fireclay pots, a problem solved by the introduction of tank furnaces by F. Siemens of Dresden, in 1872.

Optical glass

For use in instruments the lens was even more fundamental than mirrors. The study of optics dates from the 4th century BC but the use of spectacles only from about AD 1280. Descartes, Newton, Huygens, and Leeuwnhoek were expert lens makers but the finest account of lens making and of machines designed to lessen its labour appeared in 1671 in a book by Père Cherubin d'Orleans. He used iron and brass for the tools, but copper for the mallets to which the lenses were cemented with pitch. Grindstones broken up and graded were used for grinding while for polishing he used Tripoli or putty powder. Glass for the lenses was provided by broken Venice mirrors as they were already polished on both sides permitting inspection.

The grinding tool, mounted on a vertical spindle, was rotated by means of a rope and a suspended weight so that the hands of the operator were free. The tool was turned accurately to the desired concave radius by means of a kind of capstan and the same machine was used for grinding and polishing, the blanks being held by hand. Hevelus, in 1647, constructed a kind of vertical lathe to the mandrel of which a wooden disc with a hemispherical upper surface could be fitted and on which the small glass blanks were mounted in pitch. A metal tool pressed down on them as they revolved so that they were ground to the curvature of the tool. Huygens, in 1665, proposed mounting the blank on a pivoted lever, pressed down on the tool by a movable weight, either tool or blank or both being rotated mechanically by a gear drive and a belt, as shown in Fig. 6.

The real problem was to produce glass of homogeneous optical quality. In 1798 Guinand, a Swiss bell founder, suggested the idea of stirring the molten glass in the crucible in order to found a homogeneous mixture. The method was kept a secret for a long time but it revolutionised glass founding. After cooling, the contents of the pot were broken down; the useful lumps of glass were remoulded into suitably sized lenses but the yield of any particular size could not be forecast. Later as much as 1000 lb of glass was annealed in one

[Courtesy: Esso Petroleum Co.]

Fig. 5. The principle of plate glass making had not changed since Diderot; clay pot of molten glass being removed from the furnace and carried by crane to the water-cooled rolling mill

piece and sawn into slabs by means of a soft iron or copper wheel, about 9 in diameter and from 1 to 2½ mm thick, revolving at about 350 revs/min, using diamond dust as the cutting medium held in the lightly nicked edge of the wheel and soda water as lubricant.

The lens was then ground to shape by fastening the blanks with wood pitch upon a circular iron table. A central vertical spindle, rotated by a belt and pully drive, carried a cast iron grinding disc of the correct shape.

Pressed glass

The cutting or, more strictly, grinding of glass on the lapidary's wheel was employed from the beginning as a method of imparting final form to a glass article but not until Roman times was it used for decoration and fine engraving. The glass article held in the hand was pressed against the wheel turned with the aid of a treadle (later a water wheel). The wheel, a disc of metal or stone, was fed with a stream of abrasive and water.

The art of glass cutting and engraving reached its peak during the 18th century when the products executed by highly skilled craftsmen-artists appealed to the rich. After 1814 a crude but cheap imitation of this was developed to satisfy the increasingly sophisticated demands of the bourgeoisie: glass pressed in moulds came into general use for domestic purposes. The use of moulds with low relief decoration, naively imitating cut glass, found much support in the USA but the first introduction of simple machinery had to wait until 1827 when a workman named Enoch Robinson in New England arranged for a gob of glass to drop into a mould which was pushed under a plunger. The action of the plunger forced the glass outwards and upwards so that it filled the mould. In the early machines, a huge side lever actuated a simple vertical plunger carried on a crosshead. This was soon superseded by a counterbalanced curved steel arm with handier reach and shorter return. In 1847 the first US patent was granted for moulds of cast iron to Joseph Magoun but it was not until 1866, when chilled iron moulds were introduced, that the product was really successful.

A power-driven press built by the O'Neill Machine Co. in the USA in 1893 (shown in Fig. 7) gathered the glass in a ladle, fed it through two rolls which cut it into gobs, and dropped it into moulds for pressing. Decades later, a tremendous production increase resulted from continuous tanks with mechanical feeding, and one press plunger being used for up to 16 stations. The moulds varied in complexity from having those with a body in one piece and a movable base to eject the pressed article, to those made in several sections which swung open, allowing automatic tongs synchronised with the machine, to transfer the finished article to the annealing oven.

Blown glass

The first development in the mechanisation of bottle blowing was the re-introduction of open moulds of clay, soapstone, wood or brass. But these moulds left the neck unfinished, while size control was left to the craftsman. Closed moulds gave uniform size while marvering allowed some preshaping but the final blowing required considerable effort to produce a smooth surface.

In 1878 Cahuc of Dorignies, in France, devised a revolving mould whilst the first use of compressed air for blowing is recorded at the Baccarat factory, also in France. Ideas of adapting the pressing machine as the first stage of bottle making were studied in great detail about the middle of the 19th century. The outstanding patent was granted to Arnall and Ashley in 1866 for the so-called plank machine. A movable mould was attached to a plank in an inverted position with a plunger fitting this mould to press the molten glass, poured through the open base end, down to fill the mould for the neck. A valve allowed air through the neck mould to blow a hollow in the glass and at the same time raised the plunger to attain the desired length of the bottle.

This machine failed, but it inspired further experiments and further patents were granted to Ashley in 1887 and 1889, the improvements including the use of a preliminary mould with rotating tables. Molten glass was gathered on an iron and a quantity dropped into the first mould, pressed by a plunger to form a neck, followed by compressed air to expand the glass to fill the neck mould; the neck was then held by hand and inverted, whereupon a second mould closed round the glass and the bottle was blown.

In 1882 Philip Arbogast of Pittsburg was granted a patent covering the pressing of the gob in a side lever machine into a neck in its finished form and then withdrawing the plunger. The bottle was removed from the press mould and inserted in a separate mould for blowing the body.

These two dissimilar experimental machines established the fundamentals for blowing bottles, namely that the first operation must finish the mouth of the neck, followed by an intermediate stage to form the preliminary shape, leaving blowing as a final operation.

At this time automation was restrained because manual assistance was needed for gob feeding, opening and closing of moulds, transfer of the pre-form and, in the case of narrow neck bottles, its inversion. Nevertheless, between 1890 and 1892, Ashley had ten three-head rotary machines at work at Castleford, each capable of producing over 2000 bottles a day.

During the last quarter of the century inventors attempted to eliminate the gatherer. An automatic feeder, patented by Homer Brooke in 1903, failed in practice but the outstanding machine of the period, the fully automatic glass bottle machine of M. J. Owens, was brought to the commercial stage.

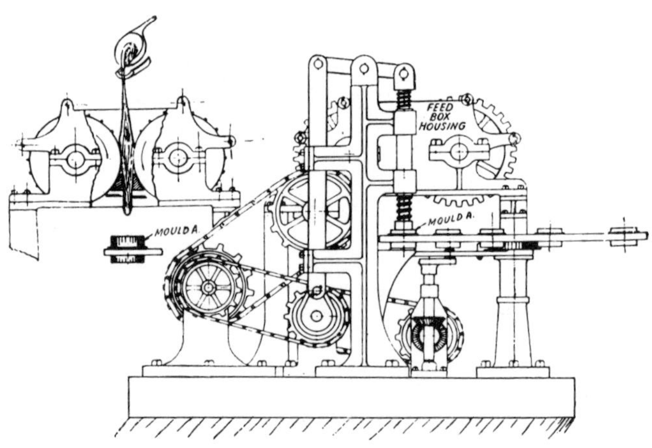

Fig. 7. The power-driven O'Neill press of 1893 gathered the glass, fed it through two rolls abd dropped it into moulds for pressing

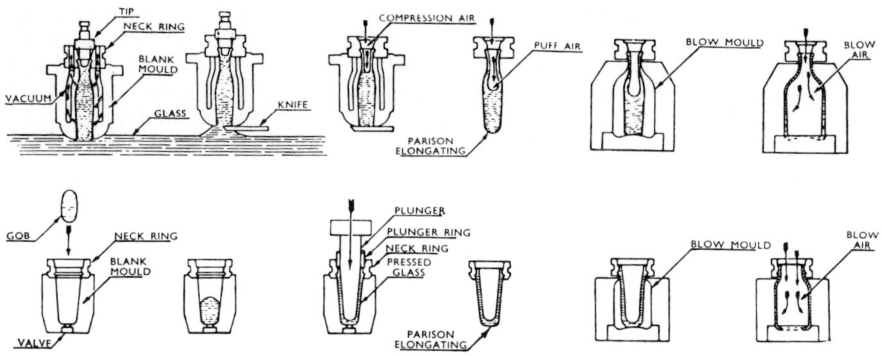

[Courtesy: Esso Petroleum Co.]

Fig. 8. (top) The vacuum and blowing process of bottlemaking on the Owens Machine. The operations shown are (l to r) dipping and filling; cut off; compression; puffing and re-heat; blow-mould; bottle blown. (below) Operations in pressing-and-blowing bottlemaking (l to r) gob delivery; gob in blank; pressing; reheating in transfer; blow mould; bottle blown

The era of automation

The 20th century saw stupendous advances in automatic machinery for producing glassware in huge numbers, its manufacture becoming one of the worlds major industries.

Problems associated with the machines for sheet glass were studied closely for some years, both in Europe and the USA. Attempts at drawing sheet glass in the form of long cylinders were made by M. A. Oppermann at Charleroi, Belgium as early as 1885, but in the USA in 1904, J. H. Lubbers was successful on a commercial scale. Molten glass was poured from a ladle into a drawing pot. A circular bell-shaped 'bait', of about 10 in diameter on the end of a blowpipe was lowered into the glass which adhered to it. The bait was withdrawn slowly while pressure-controlled air, injected through the blowpipe blew a cylinder up to 44 ft long by 3 ft diameter. The glass in the pot was cooling all this time but a cylinder of uniform wall thickness was obtained by increasing the rate of drawing steadily using tapered drums run at a constant speed for winding back the blowpipe. At the end of the draw the speed was increased so rapidly, making the wall thickness so thin, that the cylinder was broken away from the pot of glass. Subsequent operations of splitting, flattening and annealing remained unchanged.

It had been recognised for many years that a process that would avoid obtaining flat glass from cylinders would prevent much distortion of vision and be more succcessful commercially. A process patented in England as early as 1857 by William Clark of Pittsburg used a flat strip as bait. This was lowered into the molten glass which adhered to it and it was withdrawn slowly upwards. It had one over-riding snag; the weight of drawn glass overcame surface tension and the sheet necked down to a slender section. It was half a century before this was overcome. In 1902 Foucault in Belgium produced an upward movement of the molten glass within a slotted refractory float, forced down by adjustable arms. The glass is seized as it emerges from the slot and is drawn off in sheet form by a series of asbestos covered rolls, placed in pairs up the drawing block, the sheet remaining uniformly of the same thickness as the aperture. Most of the rollers are enclosed in a vertical shaft to form an annealing oven.

Colburn in America conceived his idea in 1903 after visiting a paper manufacturer. The draw is started by introducing an iron bait into the glass which is then pulled up about 3 ft to a bending roll, after which it is drawn horizontally by a series of rollers. Water chills the surface to give it stability, while gas flames directed against the bending roll keep the glass soft at that point.

In his early machines Colburn used two refractory spheres, carried at the ends of long arms and rotated rapidly in opposite directions, so as to drag the fluid glass outwards as it emerged from the bath. Later these were superseded by hooked tools, the tips of which moved in an elliptical path. Modern machines use a pair of water-cooled grooved rollers.

Continuous plate glass

The first step forward towards a continuous process of casting and rolling plate glass was made between 1884 and 1887 by Chance Bros who poured molten glass on to an inclined table from where it was fed between a pair of rolls. In 1918 Bicheroux, a German, made a significant advance, producing cast plate that required less grinding by tilting the ladle about an axis passing through the pouring edge. The distance through which the stream of glass had to fall was lessened so that overlapping and the inclusion of air bubbles were reduced. The glass was poured in front of positioned rolls so that the machine produced much smoother sheets of predetermined length. The rolled plate left the inclined slab in a horizontal direction, being drawn away on a train of cars after being sheared into sheets. Plates 12 ft wide and up to 100 ft long could be produced in this way.

In 1921 developments towards continuous casting were pioneered independently by the Ford Motor Co., the Pittsburgh Plate Glass Co., and Pilkington Bros. In the true continuous rolling process, the molten glass flows over aweir or up through a refractory slot and is passed between water-cooled rolls before it has chilled, thereby ensuring accurate thickness and width while the stretching improves flatness. The sheet, up to 11 ft wide, then passes through a continuous roller annealing oven. It is cut to conventional sizes before grinding and polishing.

Sheet glass is unsatisfactory because defects are drawn out and magnified, while production is lower than with continuous plate glass making. Yet, because of the unprecedented demands by the American automobile industry, the Colburn sheet process was successfully modified in 1925 to produce rolled plate blanks by introducing paired, knurled, water-cooled rolls near the point of drawing so that the sheet was pushed upwards without slip at a greatly accelerated drawing rate. The rolls left marks on the surface which did not matter since it had to be ground in any case.

The first continuous grinding and polishing machine came into use in 1921 as a result of developments started in 1897 at Boussois in France. The glass sheet, fixed to a table with plaster and secured with wooden pegs, was nosed through the first section of the machine and ground by rotating iron pads, fed with a sand-water slurry. The continuously moving table then passed through the polishing section where rotating circular discs, covered with felt pads, finish polished with rouge. This machine, 787 ft long, can handle sheets up to 13 ft by 8 ft.

In the Pilkington system, grooved tables slide on guides similar to a lathe bed but coupled to form a continuous bed with the table tops accurately aligned. The rough glass is laid in plaster, even across table joints. Two parallel lines of grinding and polishing machines cater for the two sides of the plate.

Bottle-making machines

Converting semi-automatic machines to be fully automatic required glass feeding devices. Michael J. Owens' mechanical genius, allied with the financial resources of the Libbey Glass Co., was an ideal partnership, Owens being allowed almost extravagent expenditure on experimental work. His conception of sucking up molten glass into a two-part iron mould was experimented with in 1898 and led to a machine which had a suction mould for gathering, and a finishing mould for blowing, and was intended to be automatic. But he found that the glass chilled when gathered from a fixed point. In 1904 he built a machine with six arms, each carrying a gathering and a finishing mould, which he used with a furnace embodying a revolving tank. Experience led to many improvements, some of them quite radical. The intermittent feature of stopping the machine for filling up was eliminated by dipping the gathering mould into the glass while revolving. The principles of operation of later versions of Owens machines are seen in Fig. 8.

Each individual head constituted a forming unit complete with a knife which swept across the tip of the blank mould, severing the glass to enclose it completely in the blank mould. The various motions were produced by cams and followers, giving wide flexibility of timing.

The problem of feeding glass mechanically from a furnace had exercised the minds of inventors for many years. The first automatic feeder to make any impact on the glass industry was the stream feeder, invented by Yorkshireman Homer Brooke. His patent, granted in 1903, covered a cup, the bottom of which was formed by a plate. The machine indexed to bring a mould directly under the furnace spout. The shear blades opened to let the glass drop. When the mould was full the blades cut the stem and closed the cup for the next load. Many defects were inherent in this proces but it was suitable for the cheap, low-quality pressed glassware acceptable at that time.

In 1911, Karl Peiler, a graduate of MIT, was assigned the specific research problem of automatic glass feeding. Proceeding systematically, he initiated two lines of investigation, the first being the construction of an automatic punty which simulated the operation of the hand gatherer, but the results were disappointing. The second project was to push, by means of a fire-clay paddle, a mass of glass over the lip of a refractory spout and sever the gob with shears. He succeeded in delivering a partly shaped gob of glass of pre-determined weight to a forming machine. But this paddle feeder did not produce a satisfactory range of narrow neck bottles. Peiler added a bowl and a plunger, reciprocating over an orifice in the bowl, cam-timed, together with the paddle and the shear, to shape the gob. This came into use in 1915.

Revolving tables suffered from inertia troubles which limited the size of the moulds and also production. The machine invented by Ingle of the Hartford Empire Co., USA, in 1925, has stationary moulds. Each section is independent with units lined up side by side. The moulds just open and close. The gob drops down a guide funnel into a blank mould where, with the plunger forming the neck, the preform is blown. The blank mould opens, leaving the glass held in the neck-rings but upside down. The rings swing through 180° and position the pre-form so that blowing-moulds close upon it. The bottle is removed by tongs.

Light blown glassware

The increasing demands of the electric lamp industry was a most important factor in the development, not only of automatic paste mould blowing machines, but also the machine drawing of tubing and rod. In drawing tubing by hand, glass was gathered on an iron and fashioned into a heavy thick-walled pre-form. A punty was attached to the end of this 'parison' and the tubing drawn by a second operator moving away from the blower.

Edward Danner, granted a US patent in 1917, fed glass to an inclined rotating refractory cylinder and passed air through the central spindle, maintaining sufficient pressure to ensure that the glass from the lower end of the cylinder emerged as tubing. Air pressure, temperature and drawing speed were regulated to obtain the diameter and wall thickness required. The drawing machine consisted of two endless chains carrying asbestos pads above and below the tubing and positioned up to 50 yards away at the end of a runway. The glass leaving the nose of the refractory cylinder changed in direction to the horizontal of the runway. This machine is capable of a linear rate varying from 16 to 900 ft/min.

A later invention, patented by Sanchez-Vello of France, in 1931, causes the molten glass to drop vertically from a

Fig. 9. The 24 spindle Westlake machine was originally equipped with a ram for each spindle. It is an early example of rotary transfer

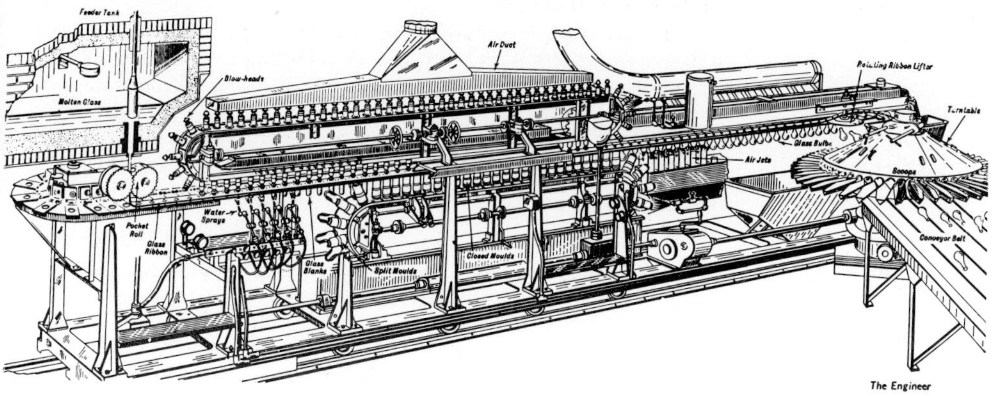

Fig. 10. The Corning ribbon machine of 1926 used gravity to pre-form the bulbs

forehearth through a circular space, surrounding a rotating refractory pipe, a change resulting in faster drawing speeds.

The blowing of seamless products such as electric lamp bulbs and domestic tumblers necessitates rotation in a mould (which must have an adherent porous lining and be water lubricated), known as a paste mould. Influenced by Owens in so far as gathering by suction was employed, the first machine (of six spindles) was designed by A. Kadow of the Westlake Machine Co., USA before 1911. It was followed by a 12-spindle machine which had its first commercial use in 1917.

A number of separate sections, each with a gathering ram, revolved round a central turret. Pausing before the opening of the furnace, the ram was thrust in to gather a gob by vacuum. As it returned from the glass surface, a knife swept across to sever the glass. The blank mould allowed the shaped gob to drop into the open jaws of the vertical blowpipe spindles which closed upon it and rotated while inverting in order to maintain even distribution of the glass as puffs of air entered, simulating the action of the blower. The moulds rose from a water bath and closed over the glass which was then blown. The moulds and spindle jaws opened to allow the glass to drop on to a conveyor where the excess glass is severed by melting-off.

The Corning Ribbon Machine was scientifically based on the measured mechanical properties of molten glass. Its first commercial use was in the USA in 1926. The machine receives a stream of glass, about 1 in diameter (controlled by a fixed refractory cylinder similar to a gob feeder plunger, from an orifice in a continuous melting tank. Two water-cooled metal rolls, one plain and the other with cup-shaped depressions at regular intervals upon its periphery, receive the glass and roll it into a flat ribbon about $\frac{1}{8}$ in thick, with bosses at appropriate intervals. The flat ribbon is received by an endless belt, each boss being placed over an opening in the belt so that it sags to form a hollow pear-shaped drop.

As the drops elongate, wet paste moulds carried in a turret surround them from below, while blowheads carried in another turret above supply air. The paste moulds rotate fast enough to avoid seams on the glass. At the precise moment blowheads and moulds leave their horizontal path, the blow moulds fall away to leave the finished glass suspended from the glass ribbon through the holes in the belt.

A moment later the bulb is lightly touched at the point near to the line of contact with the steel belt where the chilled zone makes severance easy. The bulbs then travel on through an annealing oven. This was not only the fastest automatic glass-forming machine ever produced but provided remarkably good glass distribution.

Conclusion

This brief review of the history of glass manufacture from the viewpoint of engineering outlines the evolutionary path that has led from simple tools to the highly complicated schemes yielding great efficiency at high speeds but at huge capital cost.

Many great engineering developments have arisen in response to economic and social pressures; one began a new epoch, *ie* the Siemens regenerative furnace. The higher temperature permitted replacement of soda by saltcake and, although this economy was later discontinued because of the cheapness of soda from the Solvay process, the higher temperatures attained not only resulted in rapid melting of the glass batch but allowed the founding of much harder borosilicate glasses.

The glass industry was not at once transformed by the Industrial Revolution but in spite of its craftsmanship, its secret processes, and the difficult character of the material, it eventually became a machine-based industry, the progress of which largely depended on mechanical engineering.

The important part that improved optical glass has played in the advance of science indicates the repercussions glass manufacture has had in industrial fields. One of the earliest uses of a measuring microscope for engineering purposes was in 1903 in developing the BA system of screw threads; the instrument was the forerunner of the modern toolmaking microscope.

Outstanding in the field of metrology is the Eden-Rolt millionth comparator which used optical magnification for the second stage and another idea due to Eden at the NPL was that of projecting a magnified image of a gauge on to a large screen where it could be compared with a master template.

Glass manufacture has made many contributions to engineering in other fields; bottle and bulb making pioneered rotary transfer machines whilst the techniques developed to handle this difficult material have been copied for other purposes.

As a material, glass finds wide uses, particularly in chemical engineering.

Great financial risks were taken by the pioneers at the turn of the century but there was unusual collaboration between engineers and entrepreneurs which allowed development to proceed without hiatus over two decades. Demand for cheaper products resulted in more automation, making the position of the skilled workman extremely difficult. Over a period the total number of workers did not shrink materially but the grading of their work underwent a marked change. In 1917 the output of the Owens machine was equivalent to that of 54 skilled glass bottle blowers while that of a 24-arm Westlake, producing up to 100 000 electric lamps bulbs per 24 hour day, was equivalent to that of 100 skilled glass bulb blowers.

And the Corning Ribbon Machine produces the equivalent of 8 Westlake machines!

THE GREAT MASTERS

From an engraving by W. G. Jackman

Whether or not he anticipated Trevithick with his high-pressure engine, Evans established this principle in America and became one of the founding fathers of American engineering. But his designs for fully automatic flour mills were too advanced for his age. Like many inventors, he was increasingly distracted and frustrated by quarrels about patent rights.

Oliver Evans— Pioneer of Automation

by A. D. Cummings, M.Sc., Ph.D.*

Oliver Evans' great achievement, for which he has scarcely been given sufficient credit, was that he invented the high-pressure steam engine almost independently. Trevithick, about the same time, did the same thing in Britain (and Watt called him a murderer for his pains) but he had at least worked with atmospheric engines, while Evans began without having seen a steam engine of any sort.

Low-pressure engines of the Newcomen and Watt types were fairly common in England by 1800 (Boulton and Watt had installed 500), but in the whole of America there were only six. In 18th century America, unlike England, there were few craftsmen and no real shortage of power. Cut off as they were by two thousand miles of slow ocean travel and 10 years of war, few Americans paid attention to the new power source. But some had the required vision: Jefferson said that now anything was possible, if an engine and a peck-and-a-half of coal could do as much work as a horse in a day; and Evans boldly prophesied in 1789 that

> men now living will see the western waters covered with steam boats and a child now born will travel by steam carriage from Philadelphia to Boston in a day.

Nothing could stop him from coming back again and again to the idea of steam transport. A climax in his life came when he tried out an amphibious vehicle. This failed, in spite of his drive and energy; he pulled back and spent his remaining resources on easier applications, on stationary engines and steam boats. He was then nearly fifty and the thoughts of an increasing family dependent on him, and the probability of his powers declining, made him turn to safer things.

Evans' lifetime, 1756 to 1819, was a period scarcely encouraging to inventors. They are an unaccountable breed at any place or time; apart from inventiveness, essential characteristics seem to be iron determination to succeed, colossal obstinacy in the face of advice, and a degree of personal vanity.

Eighteenth century America, poor in everything except natural resources, was a particularly inhospitable soil for inventors, yet it produced three of importance.

Evans came of mixed Welsh and Dutch stock and his portraits show the short, thick-set figure, upstanding hair and pugnacious cheerfulness common in Wales. He was the fifth of a Delaware farmer's twelve children. His maternal grandfather was a miller and, after the War of Independence (in which all the Evanses of military age were enrolled in the militia but were not called to serve), Oliver and some of his brothers set up as millwrights.

Automatic mill

Labour shortage affected everything. Whatever might save labour was worth thinking about, but even machines to assist hand labour were still a new idea and anything which could actually replace men and horses was scarcely conceivable.

Evans' first invention was an automatic mill. Hitherto wind or water power had been applied only to the grinding stones themselves, all movement of grain and flour had been effected by hand. Evans applied water power to conveyors through simple gearing in order to replace all hand operations. His patent claims on milling are summarized by him thus:

> These improvements consisting in the application of power (originally restricted to the working of mill stones) to convey grain and meal up and down throughout the mill, to stir it and keep it in constant and equal motion, that the air might exercise its full influence to cool and dry it until completely manufactured without the aid of human labour.

There is some doubt how completely these improvements were put into practice, except in Evans' own mills. He was much ahead of his fellow millwrights. In the engraving of his mill, corn can be seen received by wagon or boat and elevated by leather or cloth belts equipped with cups. At the top it is cleaned and put into store; thence it flows into

* *Department of Chemical Engineering, University of Nottingham.*

hoppers and to the millstones; another vertical elevator takes it back to the top where it is spread out to cool, a rake turning it over and over. The next operation is bolting or sieving in rotary cloth screens which separate the bran and part-ground meal from the flour; finally, horizontal conveyors, made of metal augers on wooden shafts and running in a trough, carry the flour to the bagging store.

In all this we can easily see the beginnings of continuous flow automation. At the same time Ely Whitney, another American, was developing machines to simplify hand production to suit unskilled labour so that a musket could be assembled from interchangeable mass-produced parts. These two ideas—continuous flow and repetition work—were later combined, chiefly in America where labour remained scarce throughout the 19th century, into modern assembly line techniques and automation.

The idea did not catch on very quickly. One of the Evans brothers, Joseph, travelled about Delaware, Pennsylvania, Maryland and Virginia, offering without success a free installation for the first applicant in each county. No doubt this fruitless trip led to bitter feelings at home.

Nor did England prove any more receptive when Evans prevailed on a friend crossing the Atlantic to take with him plans and a model of his improved mill. This man wrote back to Evans:

> The milling business is carried on in a very different manner here from what it is in America, that is, all in a very small way, and almost entirely by windmills, each of which does but very little business . . . hardly a piece of high ground in England but is covered with them and all along the rivers they are so thick you may stand in many places and count more than thirty . . .
>
> I am afraid this country will not answer your purposes as well as we expected, small mills do not lend themselves to your machines; their owners would sooner jog on in the hard way. This surprises me, for this is the only thing that is not done better here than in America . . . Machinery here is in much greater perfection than you have any idea of.

When American millers began to get to know of Evans' improvements they began to imitate them without letting him know. The separate pieces of equipment being relatively simple, this was easy to do. For the rest of his life, he was increasingly involved in law suits to recover his royalties.

Steam propulsion ideas

We turn now to Evans' steam engine. He describes how his attention was first drawn to the power of steam at about eighteen, when he saw some young friends amusing themselves heating water in a stopped-up gunbarrel; but whereas they were content to make a loud bang, he "immediately thought there was a power here capable of propelling any wagon provided I could apply it." Later, after he had done some experimenting

> a book fell into my hands describing the old atmospheric engine, I was astonished to observe that they had so far erred as to the use of the steam only to form a vacuum to apply the mere pressure of the atmosphere instead of applying the elastic power of steam for original motion, a power which I supposed was irresistible.

When he had little more than a few ideas on paper Evans got to work to protect them by applying for a 'monopoly' to the States of Pennsylvania, Delaware and Maryland. This may indicate that he was already meeting trouble with payment for his mill improvements. He sought an overall cover for cotton card making, mill machines and steam wagons. At the early age of 22 he writes confidently

> Your petitioner hath long had in contemplation the very great power of steam and hath at great expense of study, time and labour invented an entirely new plan of applying the said powers to propelling land carriages to travel with heavy burdens up and down hills without the aid of animal power with such velocity as may be convenient and be guided by a person sitting therein and secure from the inclemency of the weather.

All the States except Maryland threw out the steam-carriage idea: and his Congressional Patent of 1790, was for mill improvements only.

For some years he made no real progress, probably because the family millwright business had few facilities for experiment. It seems likely that Evans made himself unpopular with his brothers by pursuing what they, no doubt, thought were time-wasting ideas that would lead nowhere. Whether he quarrelled with them or not, he sold his share in the business and set up for himself in 1793, when he ad-

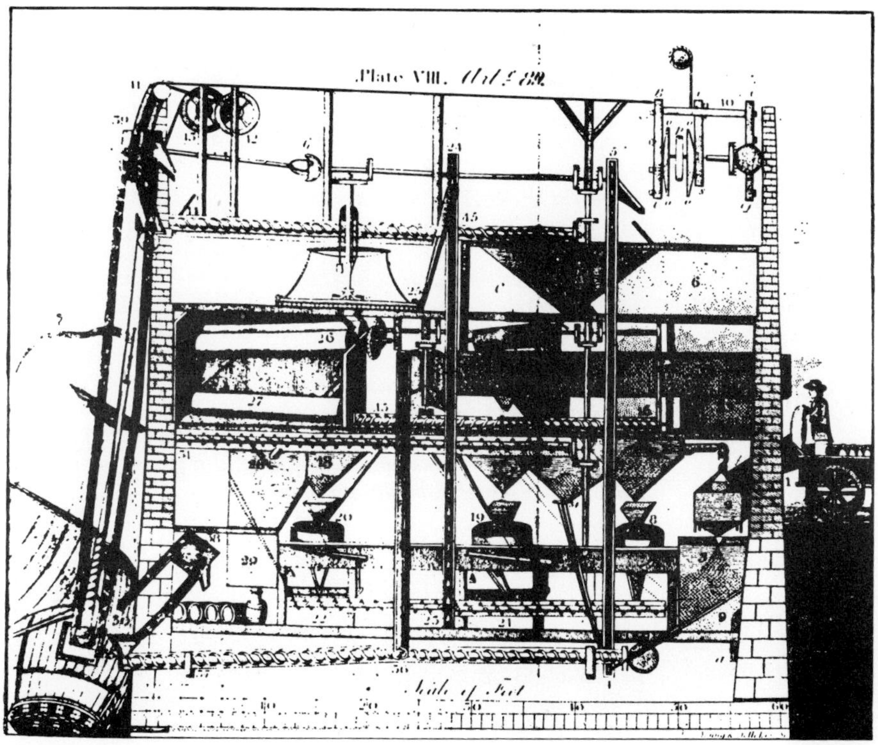

Evans' patented automatic flour mill, reproduced from 'The Young Mill-wright and Miller's Guide' of 1795

1—The wagoner emptying grain into scale pan.
2—Scale pan, to weigh the grain.
3—Small garner and wind chest for cleaning wheat.
4, 5—Elevator, to top floor of mill.
6—Main store for wheat.
7—Garner, feeding the shelling or rubbing stones.
8—The rubbing stones.
9—Grain is again elevated and deposited in garners 10, 11.
12—Rolling screen.
13—Fan for cleaning the grain.
15, 16—The conveyor to garners 7–17 and 18.
18, 19, 20—Millstones.
20, 21—The conveyor, collecting meal after it is ground.
23, 24—Elevator to hopper-boy 25.
25—Hopper-boy, which spreads and cools the meal and supplies the bolting chest.
26, 27—Bolting reels.
28—The chest containing the super-fine flour.
29—Spout for filling barrels.
35, 39—Elevator for unloading ships. This rises and falls in the curved slots and is driven by the universal coupling at G.
40—One view of the mechanism 42, 43 for hoisting the elevator clear of the ship.
38—A temporary elevator for short lifts.

vertised that he had removed to Philadelphia where he would receive applications for the use of his Improvements on Mills and described how he was able to produce the greatest possible quantity of fine flour from grain in one operation. He said further that he could build entire mills, supply cast iron gear-wheels to replace the easily-worn wooden gears then used, as well as mill-stones, bolting cloth (for sieving), plaster of paris, coal and sacks, in short, everything for the miller.

He had married in 1783 and had a growing family to provide for—seven children in all. For some years, therefore, he had to attend to building up this new business. But in 1792 he once again applied for a patent, covering this time some new ideas on boilers and engines. The opportunity again came to try for an English patent. A friend was going to England in search of a patent for himself and promised to do the same for Evans.

This man carried with him a set of Evans' drawings: what stage of development they showed we do not know. He tried for a patent and failed, then died in England. All trace of the drawings was lost but, according to some writers, the plans were shown to Trevithick, Vivian and other British steam engineers. This is quite possible (though there is no evidence) but it is only just to point out that Trevithick's high-pressure engine was different from Evans', the only similarities being the oscillating plug cock and the internal flue boiler; neither of these were new, even at that time. On the other hand, Evans was the first to realize the vast possibilities of high-pressure steam, including its use for locomotion.

Stationary engines

It now occurred to Evans, rather late in the day, that he could more usefully apply his steam engine to the driving of corn-mills where his chief money-making interest lay, rather than to steam road vehicles that nobody took seriously. He consulted the Professor of Mathematics at Philadelphia University, an Irishman, and two English engineers who were concerned with installing Boulton and Watt engines (made in America) in the local waterworks. They agreed that his principles were new and worth a more extended trial. Thus encouraged, Evans invested some of his growing resources of men and materials and soon found to his dismay that he had expended $4000 without tangible result.

Later, however, he had some success with what he called his little engine. This was a simple, vertical cylinder engine, the steam being admitted alternately above and below the piston via plug cocks driven from the flywheel.

Steam was supplied by a boiler on the left; the safety-valve, feed pump and feed water heater were above it. The illustration shows the stages by which Evans evolved his high-pressure boilers. His earliest, made between 1786 and 1801, were of thin sheet copper (iron plates not being available) bound round with 3 in thick wooden staves and iron hoops. This was the only way he could get sufficient strength for the pressures he wanted to use and the internal cast iron fireplace was also necessitated by the thin copper shell. Later he was able to make his boilers of riveted wrought iron plates, rolled from bar iron.

In his next design (1801-1805) he still kept the internal fireplace and circulated waste gas round the outside. Finally he gave up all internal tubes and had two interconnected plain cylinders, externally fired. This may have been the safest design, suited to the crude engineering and still cruder boiler firemen of the times

The Columbian grasshopper engine of 1811

The application Evans made in 1792 had claimed the use of oil, water or other dense fluid between the hot steam and the working parts of the engine to prevent steam from escaping; construction of the engine to produce immediate circular motion; and a boiler built so that the water surrounded the fire and the hot smoke surrounded the water.

The application of 1804, on the other hand, goes straight to high-pressure working as his main claim.

> I construct my boilers of strong metal in the form of cylindrical tubes, the best form to contain a great elastic power, and to enlarge their capacity I extend their length or increase their number which gives a large surface for the fire to act on, making them sufficiently strong to contain steam of elastic power of 1500 lbs to the inch area of piston but at the same time arranging the work that 50 lbs will be enough.

Other improvements were the return of condensate to keep the boiler clean; the use of forced draught for the fire and of waste steam for evaporation and for warming buildings.

In 1803 Evans made his third and last attempt to get a patent in England, again without success. In 1804, still dabbling in steam-propelled vehicles in spite of his resolution to forswear them and stick to corn-milling, Evans made a proposal to replace horse-drawn coaches between Philadelphia and Columbia with carriages powered by his newest engine. But unfortunately, this was turned down.

Steam dredger and boats

His next venture was a steam dredger for Philadelphia harbour which he called *Orukter Amphibolos*. Dredging was then done with hand-scoops or with a horse-driven chain of buckets and at one place in England at this time by some kind of steam engine; so the idea was not unknown.

Much has been made of Evans' machine because he claimed that he moved it by its own power one mile overland to the water. We have only Evans' word for it; could a fairly small engine move a boat on a temporarily wheeled carriage (a total weight of about 15 tons) on rough roads? The only picture of the machine we have was drawn thirty years later by an artist who, whatever his source of information, was no engineer.

The *Orukter* was the only locomotive Evans ever built although all his life he advocated their employment both on high roads and railroads. The dredger was used in the docks as intended but it was never a complete success. The authorities haggled about payment and Evans had difficulty in collecting his money.

Its engine and all his subsequent models were of the 'grasshopper' type; Evans probably evolved this indepen-

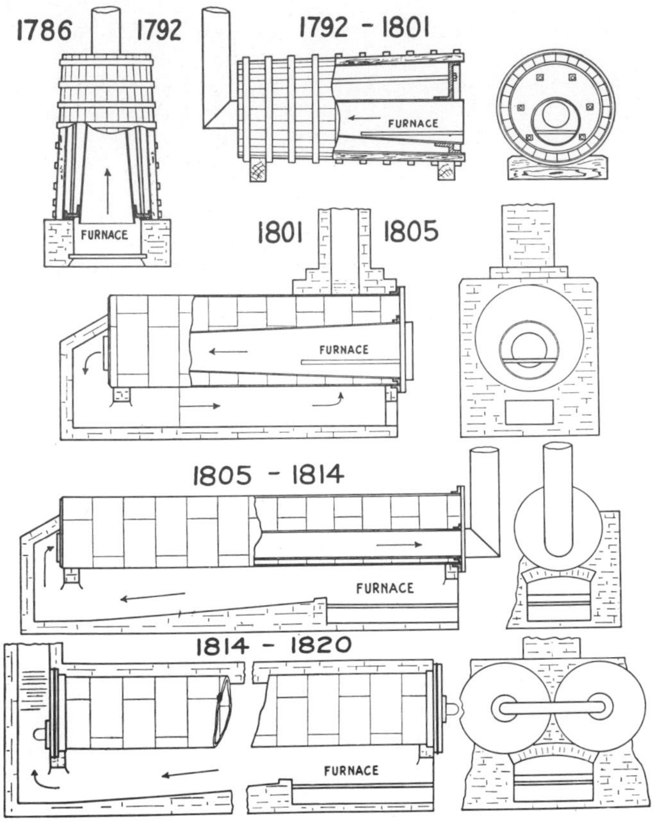

How Evans' boilers evolved over the years

dently but it was patented in England by Fremantle in 1803 and used on several early locomotives, including Hedley's *Puffing Billy*. It was a beam engine but, instead of having the beam pivoted at the centre like Watt's engine, it was pivoted at one end, the piston and crank being close together near the other end.

The improvements in this were a rotary valve for steam admission (patented by a fellow miller, Luther Stevens) the use of both pressure and vacuum and the return of condensate to the boiler.

About this time Evans set up a branch of his business in Pittsburgh under one of his sons. Having, as before, announced his milling activities in his advertisements he adds that he can supply "steam engines far superior to those on the principles of Boulton and Watt". His fatherly pride comes out in reporting back home after visiting this works

> it far excels ours, we had better travel a whole year to see such before we erect machinery for constructing steam engines. I therefore request Mr Cadwallader (another son) to come immediately, stay two months and make accurate drawings.

Evans now entered the steam navigation world where events had been catching up with him. The pioneer of mechanically propelled boats in America was John Fitch who, for a short time in 1788, ran a passenger-carrying vessel, moved by steam-operated oars, on the Delaware. Evans advised him to convert it to paddle wheels.

The first successful steamer service in the world was started by Fulton in 1808. The great attraction for prospective steamer operators was the Mississippi system that had just become U.S. territory. Fulton and his associates were granted a monopoly by the State of Louisiana but Evans and his sons were prepared to challenge this and began to build for the Mississippi service at their Pittsburgh works. Their first boat, the *Oliver Evans*, began running in 1816 but was later involved in one of the many steam-boat explosions which, though it was then out of his hands, caused Evans to write that this

> mortified and humbled me more than anything which has occurred affecting my interests or inventions . . . [the explosion] was due to the ignorance and inattention of the captain and engineers in letting the water in the boiler sink below the flue by which the fire passed through the centre.

The explosion may have been a contributing factor in persuading Evans to change from the internal-fireplace boiler to the externally heated cylinder. He wrote to the newspapers at a time when there was a public outcry about boiler explosions that people did not understand the difference between high and low pressures, blaming (rather unfairly) Boulton and Watt's low-pressure engines for using a degree of pressure dangerous to their boilers which were "constructed on wrong principles". He claimed the catastrophic failures were impossible with high pressure on the ingenuous grounds that "the elastic power of steam always rises gradually, giving time for the weakest part to yield".

Patent troubles

Like many other inventors, Evans, throughout his life had to contend with pirates who infringed his patents. He wrote plaintive letters about this to the legislature, he

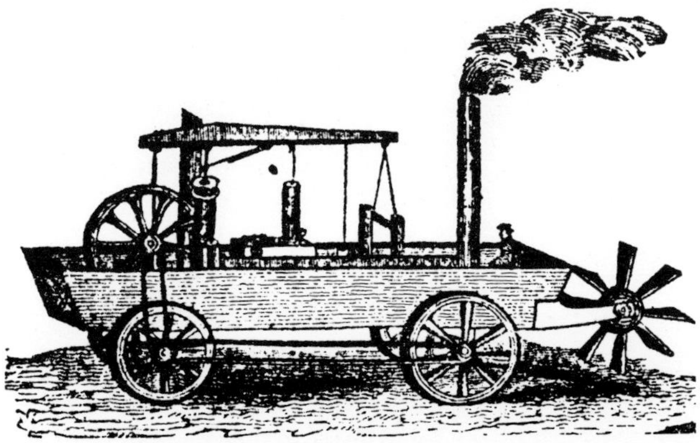

Two views of the 'Orukter Amphibolos': left, taken from 'The Mechanic' of 1834 and, right, steaming up the Delaware River

wrote to other inventors, including Whitney, proposing that they should "make up a purse" and engage a man to watch their interests. He also wrote to two Presidents (Washington and Jefferson had put in Evans machinery on their estates).

At one point, having heard that a judge had declared patent right to be an infringement of public right, Evans, with the approval of his family, burnt a collection of drawings and notes of projects.

> Such a doctrine from such an authority determined me that it was highly dangerous to leave my papers to lead my children or my grandchildren into the same road to ruin that had subjected me to insult, to abuse and robbery all my life. I was then in my 54th year and I had in the last three years done more to acquire permanent property by renouncing such pursuits and following regular business, than in the 30 years before.

It would have been well if he had held to this resolution, but it is scarcely necessary to add that he did not, at any rate as regarded dealings in patents. As soon as his factory prospered and he could safely leave much of the work in the hands of his sons, instead of relaxing, he redoubled his efforts to harry his patent-debtors and make them pay (it seemed to be chiefly his mill patents that gave trouble, no one infringed his steam engine ones).

Increasing prosperity gave him the means to indulge in more legal warfare. His faithful wife, who must often have wished he could leave inventing alone and attend to his proper business, died in 1816 and, from then until his own death in 1819, he was constantly on the move, on horseback or in a jolting coach, in all weathers. In the course of his travels he married a young wife and making provision for her was one of his last preoccupations.

He fell ill in New York and, shortly after receiving news of a disastrous fire at his Philadelphia works, he died.

Evans in America and Trevithick in this country did much to further the Industrial Revolution by showing, with their small high-pressure engines, that steam power was not a matter of ponderous beam engines tied to one place, doing fifteen strokes a minute. This opened the way for their application to threshing, railways, steam-boats and a dozen other uses. Having no condensers or air-pumps, they were simpler and cheaper to build; almost like the petrol engine later, they were sufficiently compact to go anywhere.

Again like other inventors, Evans suffered from lack of resources when he was young and could, perhaps, have pushed his ideas to a successful conclusion earlier. He had always to keep his mill work going to make a living. Since there was only one other engineering works in Philadelphia at that time, it is unlikely that any of his work could have been subcontracted and this must have held up development. Operating in a new country, he had to train all his men himself. He also had to be his own sales manager and advertising agent but he found time somehow to write two practical handbooks: *The Young Millwright and Miller's Guide* and *The Young Steam Engineer*.

He was certainly one of the great engineers of his age. Had he worked in more favourable circumstances, he might well rank as one of the world's great inventors.

REFERENCES FOR FURTHER READING

1. Bathe, G., and D., *Oliver Evans; a Chronicle of Early American Engineering*. 1935 Historical Society of Pennsylvania.
2. Wilson, Mitchell, *American Science and Invention*. 1960, New York.
3. Mirsky, J. and Nevins, A. *The World of Eli Whitney*. New York. 1952.
4. Thurston, R. H., *A History of the Growth of the Steam Engine*, 1878 and 1939 Oxford University Press and Cornell University Press, U.S.A.
5. Habbakuk, H. J., *American and British Technology in the 19th Century*. 1962 Cambridge University Press.

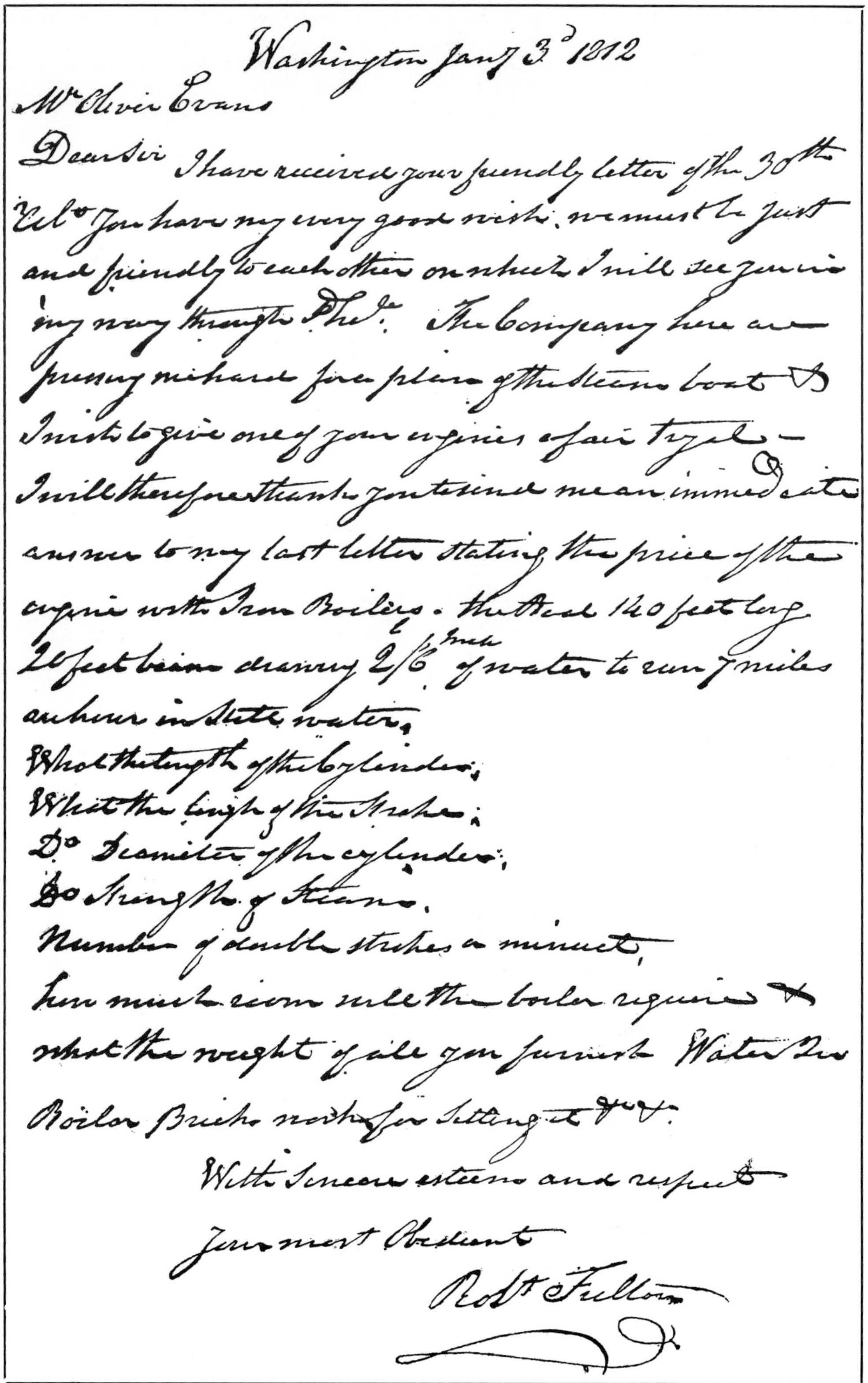

Washington Jany 3d 1812

Mr Oliver Evans

Dear Sir

I have received your friendly letter of the 30th Ult. You have my every good wish, we must be just and friendly to each other on which I will see you in my way through Phila. The Company here are pressing me hard for a plan of the Steam boat and I wish to give one of your engines a fair tryal—I will therefore thank you to send me an immediate answer to my last letter stating the price of the engine with Iron Boilers. The boat 140 feet long 20 feet beam drawing 2/6 inch of water to run 7 miles an hour in Still water. What the length of the Cylinders, What the length of the Stroke, Do Diameter of the cylinder, Do Strength of Steam, Number of double Strokes a minuet, how much room will the boiler require and what the weight of all you furnish Water In Boiler Brick work for setting it etc etc

With Sincere esteem and respect

Your most Obedient

Robt. Fulton

This letter shows that Evans and Fulton were not only rival steamboat operators but were at one time considering entering into business relations.
A chapter on Robert Fulton begins overleaf.

THE GREAT MASTERS

Starting in life as an artist, Fulton soon became interested in canals and navigation. Pursuing his aims with singleminded determination in three mutually hostile countries throughout a revolutionary age, he can lay claim to be the 'George Stephenson' of the steamboat.

Robert Fulton — Man of Vision

by I. McNeil, M.A. (Associate)*

If we are to take the word of John Rennie it is unlikely that Robert Fulton would find a place in this series on the Great Masters. In 1817, two years after Fulton's death, he wrote to a Mr Barrow of the Admiralty, "I consider Fulton, with whom I was personally acquainted, a man of very slender abilities, though possessing much self-confidence and consummate impudence." There is more than a grain of truth in the latter part of Rennie's assessment but, viewed through the perspective of a century and a half, his estimate of Fulton's abilities and achievement seems very wide of the mark.

Childhood

Robert Fulton was born in 1765 in Lancaster, Pennsylvania. Little is known of his parents except that they were Irish emigrés, probably of Ayrshire descent. His father appears to have been a respected citizen and held a number of responsible voluntary civic posts though in farming he was a notable failure. He died when Fulton, who had three elder sisters and a younger brother, was only three years old, leaving his wife and family poorly provided for. Robert was educated by his mother until he was eight when he went to a small local school. There the boy did little to distinguish himself scholastically although he displayed an undoubted talent for drawing and painting.

While little is known about Fulton's schooldays, it is clear from his subsequent behaviour and his letters after he had left home that he was more than a dutiful and affectionate son. As the elder of the two boys, he accepted the position of bread-winner quite early in life; whenever possible he contributed to the family's small income.

It was Robert who, before he left for England bought a home for his mother and sisters with his own small savings. To a young man setting out to seek his fortune on the other side of the Atlantic it must have been no small sacrifice.

Throughout his years in Europe, Fulton's affection and anxiety for his family is clearly shown in his letters home. His own financial state did not allow for any spare cash to be sent home but he was full of solicitation for the welfare of his mother and sisters.

Of his schooldays there are many tales bearing witness to his mechanical ingenuity and his prowess as an experimenter in both mechanics and chemistry. It is said that from his experiments with mercury he got the nickname 'Quicksilver Bob'. What these experiments were, however, we do not know.

The most notable event of Fulton's schooldays was the outbreak of the war of independence in 1776. Lancaster was well within the theatre of war and contained a large ordnance depôt and factory. It has been suggested that much of Fulton's later dislike of the monarchical form of government sprang from his boyhood impressions.

Painter of miniatures

At the age of seventeen Fulton left home for Philadelphia to earn his living. It has been said that he was at first apprenticed to a jeweller in that city but there is no evidence. Within two or three years he had established himself as a competent painter, particularly of portrait miniatures.

While there was a demand for paintings, particularly in an enlightened and cultured city such as Philadelphia, America provided little opportunity for a young painter to receive any tuition from more experienced artists. It is said that Benjamin Franklin was impressed with the young Fulton's work and advised him to go to England to get further training. Certainly, when he arrived in London early in 1787, he had forty guineas in his pocket besides a letter of introduction from Franklin to the American painter Benjamin West, then at the height of

* *For a photograph and biographical note of the author see p. 360, July 1963 issue.* Ed. CME.

his fame. West took Fulton into his studio. An introduction obtained there was to take him to Powderham Castle, near Exeter, to paint the portrait of Lord Courtenay and, indirectly, to lead to his career in engineering.

Fulton was a good portraitist, but without the essential touch of genius required to take him to the top of his profession. His pictures were however accepted by the Royal Academy. What induced Fulton to give up art for the more practical pursuit of engineering we do not know but it was a sudden decision taken during the months when he was painting in Devonshire. In giving up painting, he gave up what would certainly have been a comfortable if not affluent life.

It has been suggested that Fulton's first interest in things mechanical originated in his journey across the Atlantic when he was impressed by the slowness of the voyage and its utter dependence on the elements. He must have been taking an interest for some time, for in 1793 he designed a mill for sawing marble and other stone which was put to work at Torbay and for which he was awarded a silver medal the following year by the Society for the Encouragement of Arts.

Engineering projects

It was in 1793 also that he read a paper by Lord Stanhope on the design of the proposed Bude Canal to connect the Bristol and English channels. This canal presented considerable problems due to topography. Fulton wrote to Lord Stanhope with a proposal for using inclined planes instead of locks as a means of reducing the loss of water. His letter also mentioned a project for moving ships by steam.

Stanhope was an inventor of no mean distinction whose interest ranged from optics to printing machinery. At that time he had a 200 ton ship building for the Navy Board which was designed for steam propulsion, and he was therefore interested to hear more of Fulton's ideas. He informed him, however, that the idea of the inclined plane was nothing new. Fulton was a man to profit by experience and it was probably as a result of this incident that he later made himself familiar with previous work done in any field in which he was engaged. He has been accused of plagiarism but this procedure is one any prudent inventor would employ, then as now.

Fulton wrote to Stanhope, explaining his ideas for applying steam to ships and detailing experiments he had already made. Starting with a curious idea of a single paddle at the stern, driven by the reaction of a bow wound up by a steam engine, he had soon advanced to a paddle wheel, primitive but recognisable. His letters show clearly that, although he had little experience, he had a systematic and analytical approach to mechanical problems.

So far as the steamship was concerned, Fulton had nothing new to communicate to Lord Stanhope. He gave up the project, at least for the time being, and devoted himself to the study of canals which, in the 1790's, were in their heyday. During the next few years he travelled up and down the country and his own ideas on canal design, construction and operation became crystallised. These he published in 1796, advocating small canals, chiefly on the grounds of economy; and, of course, the inclined plane for which he obtained a patent in 1794. Over a slope of 20 to 90 degrees, two boats were to be carried in wheeled cisterns, so arranged as to balance each other. (Similar arrangements were in fact already in use, for example those built by William Reynolds on the Shropshire Canal in 1792.) He also proposed an integrated system of canal navigation, rather than haphazard individual constructions. Fulton's ideas were well ahead of his time.

In those early days of his mechanical career, Fulton's most important idea was probably a machine to speed up and reduce the cost of digging canals. He proposed a four-wheeled carriage, to be drawn by four horses and carrying a scoop-shaped cutter, four inches square, which dug into the ground somewhat in the manner of a ploughshare. Behind this a conveyor belt carried the soil upwards and backwards to where a four-bladed fly, driven through bevel gearing from the rear axle, was designed to strike it and so throw it sideways, clear of the excavation. Though the machine was crude and, in many details, impracticable, it displayed a high degree of ingenuity and is the first known attempt to produce a mechanical excavator. Such an idea was again well ahead of contemporary thought.

The lean years

Between 1793 and 1797 Fulton kept up a constant and lengthy correspondence with Lord Stanhope whose replies indicate clearly that he valued Fulton's opinions on canal problems. Much of these were concerned with the Bude Canal. Although this was not built until after Fulton's death, it incorporated seven of his inclined planes, rather than Stanhope's alternative, a kind of lift.

These were years of very great hardship for Fulton. He had no regular source of income and does not appear to have had much time, if any inclination, for painting. For his excavator he borrowed money from Robert Owen whom he met in Manchester. Three days after Christmas 1796 he wrote to Lord Stanhope a pathetic letter in which he asked for a loan of twenty guineas and hinted that he would like to obtain some sort of regular employment in his Lordship's canal schemes. Certainly he had for years been giving Stanhope all the help he could and does not appear to have received any financial help in return. It is not certain whether he obtained this loan or not but unfortunately he received no offer of employment. This may have contributed to his reasons for going to France.

Among the many public men to whom he had sent copies of his book was George Washington, President of the United States. Receipt was acknowledged and Fulton had replied with an enthusiastic letter on the advantages of developing a canal system in America. It is clear from this letter that in this, as in all his major projects, he was prompted largely by his strong humanitarian feelings. While he must have had a degree of self-interest, this was never the mainspring of his actions. His advocacy of canals was based on the firm conviction that, by providing a cheap and ready form of transport, they would expand trade and reduce the prices of basic commodities such as coal, grain, wool, timber and other building materials. This altruistic compulsion transcends other motives in all his work.

Moving to France

In April 1797 his luck at last turned for he had the good fortune to sell a quarter share in what he called his 'canal prospects' to a compatriot, for the sum of £1500; remarkable, when one considers that all he had to sell was his inclined plane patent and this was only valid in Britain. It is likely that the purchaser was Joshua Gilpin who was studying canals from 1795 to 1801, and who later built the Chesapeke and Delaware Canal. Now in a better financial position, Fulton took advantage of a temporary armistice to cross the channel for France.

His object in doing so was undoubtedly to try and put some of his canal ideas into practice before returning to America, and France would have seemed ideal for this purpose. The revolutionary government had taken over not only those canals that had been the responsibility of the various provinces, but also the privately owned ones. A considerable programme of public works was in hand.

Patent Law had also recently been established in France and to this Fulton had recourse. He was granted a patent for his inclined plane in February 1798 but by this time he had probably realised that France was too fully engaged in war to have much time for canal construction. However, Fulton had other ideas in mind and he set himself to develop these.

As a weapon of war the idea of the submarine was not new. William Bourne mentioned it in his *Inventions and Devices* published in 1578. About 1620 Van Drebbel is reported to have travelled from Westminster to Greenwich below the surface of the Thames. De Son's seventy-two foot vessel in 1653 failed mainly because its clockwork mechanism had not enough power while Mr Day, an experimental ship's carpenter, had been drowned off Plymouth in 1773 because the hull of his converted fifty ton sloop could not withstand the pressure at twenty-two fathoms.

In 1776, Fulton's compatriot David Bushnell had gone so far as to attack British ships that were blockading American ports. The assaults of his one-man-powered submarine were a failure but he proved that submarine warfare was a possibility and also that gunpowder could be exploded under water. Fulton had doubtless heard of Bushnell's experiments in which periods of thirty minute submersion had been achieved. The little vessel had vertical and horizontal screws and hand-pumps to control its depth.

The long story of his own attempts with the submarine in France is one of a constant struggle against bureaucratic inertia and of frequent disappointment. But he was finally defeated by the chivalrous sentiments of the French naval officers who refused to countenance such a barbarous weapon of war. Even to the Frenchman who almost daily decapitated his aristocratic compatriots, submarines were not cricket.

Within six months of his arrival in Paris, Fulton sent to the Revolutionary Directorate a remarkable letter in which he proposed the destruction of the English fleet by submarines. At first sight this would appear to be an act of treachery against the land which had harboured him for the last ten years and had given him what education he had in engineering and mechanics.

Having little success with the French, he approached the Dutch, again without response. His actions could be taken to be those of a man without principles and without loyalty or they could be interpreted as having an altogether wider significance.

As we have already seen, Fulton saw something of war as a boy and was not unaffected by it. In his work on the submarine and allied weapons he acted on the conviction that war was the greatest single evil mitigating against the happiness of the human race. He was the propounder of a scheme for universal disarmament and thought this would be aided by an 'ultimate deterrent'. There may have been errors in his logic but it was with this idea in mind that he acted. "I devoted my whole attention," he wrote, "to find out means of destroying such engines of oppression." The freedom of the seas was a human right which he intended to defend by his inventive efforts.

From 1797 until 1800 Fulton bombarded the French authorities with his plans but without success. His restless mind rebelled against inactivity and he engaged in other pursuits. He took out a French patent for a rope-making machine and another for a 'Panorama' that opened in Paris in 1800 and gave him a small income.

The 'Nautilus'

The turning point with his submarine scheme came in November 1799, with the overthrow of the Directorate and the appointment of Napoleon as First Consul. The atmosphere became much more favourable to his ideas and, by April 1800, he was well advanced with the construction of the hull of his *Nautilus*. By July the ship was complete and Fulton, with two others, first submerged for eight minutes, and then for seventeen minutes at Le Havre. The *Nautilus* had a copper hull, built on iron frames, an imperfect ellipsoid in form. It was 21 feet 4 inches long and 7 feet in diameter, with a heavy keel that could be detached in an emergency. Buoyancy was controlled by pumps and ballast tanks. The hull was designed to withstand a head of twenty-five feet of water. Its movement in the vertical plane was also assisted by a horizontal rudder of 30 degrees arc. At the fore-end was a hemispherical conning tower with glass look-out scuttles. Under water the ship was driven by a propeller, hand-operated through gearing, while on the surface a sail was mounted on a mast and framework, all of which could be retracted and laid flat along the top of the hull.

All of Fulton's tests were as successful as he could have hoped. His main difficulty at this stage was to provide enough air to allow practical periods of submersion. Candle lighting which had consumed too much oxygen was replaced by larger windows and later, in 1801, he experimented with a compressed air supply at 200 lb/in^2 which, it is said, allowed him to remain under water for up to five hours. When submerged just below the surface, he used what is now called a snorkel.

During these tests he not only completed a voyage of 70 miles but attempted to blow up two English brigs; this, although he had no commission or authority from the French.

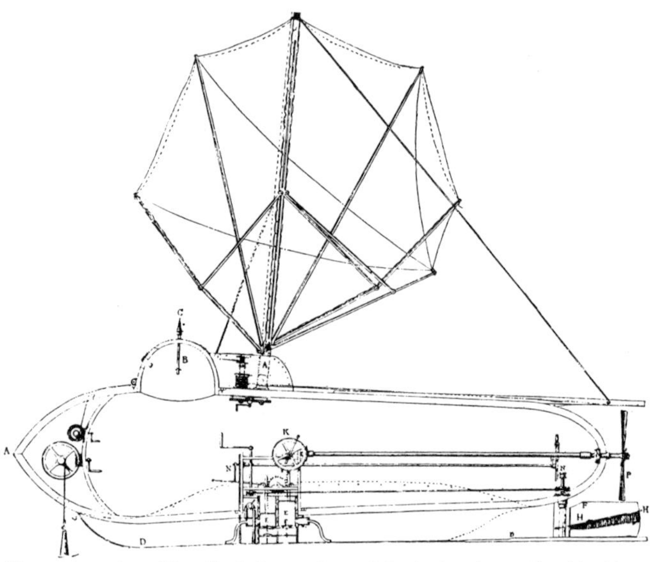

The submarine 'Nautilus' from the original sketch in the 'Archives Nationales' in Paris

A — mast
B — conning tower
C — spike for damaging enemy keels
D — keel
F — kingston valves
F — vertical rudder
G — anchor
H — diving rudder
K — handwheel drive
N — gears for F
P — propeller

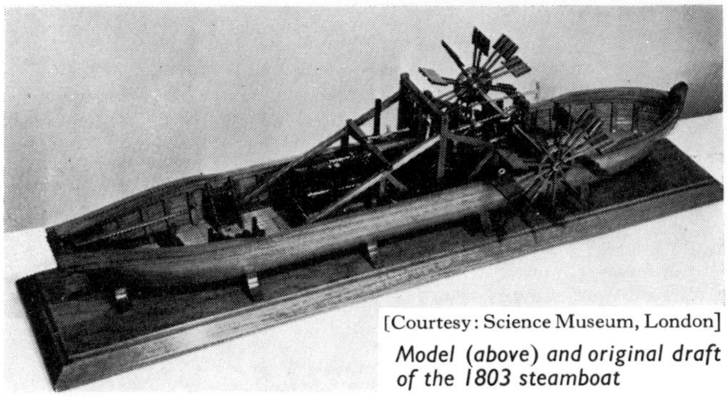

[Courtesy: Science Museum, London]

Model (above) and original draft of the 1803 steamboat

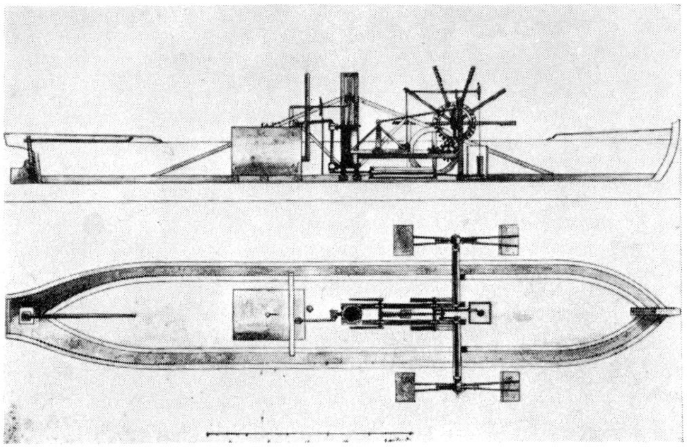

At least his courage cannot be criticised. Both the English ships moved away at the approach of the *Nautilus*, as they were forewarned of his activities and had no desire to join in his experiments.

The result of this was an increase of enthusiasm from the government, an immediate grant of ten thousand francs, and agreed terms of reward for the sinking of enemy vessels. The *Nautilus* was modified and refitted at Brest and arrangements were made there for the testing of another of Fulton's devices, the torpedo. This was in effect a submarine bomb containing up to two hundred pounds of gunpowder, which was carried by the submarine. With it he succeeded in blowing up an anchored 40-ft sloop from a distance of two hundred metres

Fulton sent the government a detailed report on his experiments. Within a year he had proved all that he had set out to do—the efficacy of the submarine, the efficacy of explosives under water and the fact that the compass could work reliably under water. So far as he was concerned, the experimental *Nautilus* had served it's purpose and, perhaps rashly, he broke her up. Unfortunately, the Minister of Marine resigned and was replaced by an older man who was not at all disposed in favour of Fulton's unorthodox methods of warfare and so the matter ended in 1801.

Fulton, however, had all the time been thinking about his early ideas on steam navigation and it was to these that he now turned. Doubtless he had reached the same conclusion as Napoleon, who had studied the progress of the submarine with interest. Submarine navigation could not become a practical form of warfare until the boat could be power-driven. The steam engine could never be suitable for such work and there was, at the time, no other independant prime mover.

On the other hand, Napoleon's projected invasion of England could not succeed without the mastery of the Channel. In the absence of naval superiority, this could only be achieved if the English fleet was prevented from sailing by lack of wind. Only a steamboat could operate under such conditions.

Early steamship projects

It is largely on his work in the practical development of steam navigation that Fulton's reputation must ultimately rest. His canal work was more methodical and more comprehensive than that of earlier workers but it contributed little that was new. His submarine work was on a grander scale and was more convincingly successful than that of Bushnell but was ahead of its time. His experiments could be little more than a basis on which later generations could build.

Where steam navigation is concerned the position is different. While Fulton's success in this field may not depend on any fundamentally new invention of his own, he is as worthy of the title of the Father of the Steamship as George Stephenson is of being called the Father of the Locomotive. Just as Fulton had precedent to work on, so did Stephenson follow the work of Trevithick, Blenkinsop and Hedley. Karl Benz and Gottlieb Daimler certainly did not produce the first petrol-driven car but they are generally held to be its true originators as they were the first to put it on a practical footing and to go into commercial production. Similarly, Wilbur and Orville Wright were the first to achieve powered heavier-than-air flight but they no more invented the aeroplane than Fulton invented the steamship. They were well aware of earlier trials such as those of Cayley, Maxim and Phillips.

Jonathan Hulls was the first to publish the idea of a steamship in 1736 but, had he attempted any practical experiment, it would have been doomed to failure as a Newcomen engine of sufficient power would certainly have sunk the hull by its own weight. The Marquis de Jouffroy d'Abbans achieved the first practical success but his reliance on a ratchet drive to obtain rotary motion would have stood little chance of commercial success.

From 1785 to 1796 an American, John Fitch, was a tireless experimenter with a variety of ingenious but mostly impracticable modes of propulsion such as banks of crank-operated oars and a continuous chain of paddle boards. His final design, a stern-wheeler, was put into commercial service on the Delaware. Financially it was a failure, however, as the weight of the machinery was too great to allow for an economic payload. His wooden cylinders, too, would have been a source of trouble if the service had been prolonged.

James Rumsey's water-jet boat certainly ran but with a very low efficiency. The inherent fault was that the water taken in at the bow was brought to rest relative to the boat, before being discharged in the stern jet. Rumsey unfortunately died of a stroke before he could make any improvements.

In Scotland Patrick Miller and William Symington built steamboats that had some success. However, their use of ratchets in the rotary drive not only occasioned a high rate of wear but was quite unsuitable for all but the smoothest of water.

Probably the most successful of Fulton's predecessors was Lord Dundas' little stern-wheeler, the *Charlotte Dundas*. Had Dundas had the vision to see it as anything more than a canal tug, he might well have gone ahead. As it was, the canal proprietors considered that the damage done by the paddle-wheel wash to the banks more than outweighed the tug's advantages.

Most, if not all, these experiments were known to Fulton, just as Stephenson was aware of earlier locomotive work. The early failures had made the public think that their proponents were either visionaries or madmen, but Fulton was not deterred. He had a practical enough turn of mind to analyse the causes of the failures and to benefit from them.

Even James Watt, who knew more about the steam engine than anyone, was not optimistic. He said to Henry Bell:

> How many noblemen, gentlemen and engineers have puzzled their brains and spent their thousands of pounds, and none of these, nor yourself, has been able to bring the power of steam in navigation to a successful issue.

If this was prophecy it was wrong.

In all invention and technical development, luck and timing play their part. Fulton had the good fortune to launch his steamboat into a world that was ready for it and also at a time when the rotary steam engine had achieved a sufficient state of development. But luck alone would by no means account for his success.

Practical outlook

Fulton approached the problem of steam navigation not only with a highly practical outlook and with adequate financial backing but with a study of theory that was quite unparalleled in his time. His letters and patents clearly show that he was the first to grasp the essential interdependence of such factors as engine and boiler power/weight and power/bulk ratios in relation to size, weight, form, and displacement of the hull and the economic payload. Fuel weight and fuel consumption too are frequent topics in his correspondence. His designs were based on his own numerous experimental observations, on the analysis of previous failures and on a broad theoretical appreciation of fundamentals. It was this combination of practice, theory and observation that was the basis of his success.

In his patent he describes the numerous methods of propulsion he has tried, stating that he is convinced of the superiority of side paddlewheels, and proceeds to what is the kernel of his patent and the key of his success.

> That any determined power of steam engine shall drive a given boat with the greatest speed which such engine can effect will depend on certain exact proportions which shall hereafter be explained, between the shape of the bow and stern of the boat, her length, draft of water and total resistance; and the diameter of her wheels, the velocity of their periphery, and the square feet of her propelling boards which continually act against the water . . .
>
> Be it therefore known that I found my claim to exclusive right for my invention or discovery not only to my particular mode of combining boats and machinery, but also to the necessary form of the boat or boats, to the size of the propelling boards, and the velocity they should run, proportioned to the shape of the bow and the stern of the boat, her draft of water, velocity, friction and total resistance, and to the calculations on the power of the steam engine to give the whole the necessary velocity.

In November 1801 a man arrived in Paris who was to play a large part in Fulton's future. This was Chancellor Robert R. Livingston, who came to France as Minister Plenipotentiary of the United States, an important post in the political climate of the time. Livingston was already interested in the development of steam navigation, so much so that in 1798 he had obtained a sole licence to operate steamships in the State of New York. Fulton and Livingston quickly came to a working agreement to exploit their ideas in partnership. Fulton had already made experiments on various forms of screws and paddles and on the relative weights and proportions of engine and hull. One of the partners' main problems was to obtain a suitable engine and Fulton, basing his calculation of the power required on Beaufoy's* experiments, wrote to Dr Edmund Cartwright† whom he had known in England in the hope of his being able to supply one. Cartwright's work had been purely theoretical and though this was a disappointment, Fulton set about more detailed tests on the hull design, using a 3 ft by 8 in clockwork model. These scale tests were, in fact, the first example of the now well-established procedure of tank testing and clearly demonstrate his practical approach.

The first trials

In 1803 Fulton started building his first steamship in Paris. Its flat-bottomed hull carried an 8 hp vertical, double-acting engine with a cylinder 17·7 in diameter by 31·5 in stroke, largely designed by Fulton himself and built by Jacques C. Perrier of Chaillot. The engine had a crosshead at the top of the piston rod working in vertical 'steeple' guides, whence the motion was taken down by return connecting rods to a pair of triangular bell-crank side-levers with equal three foot arms. From the upper part of these a horizontal connecting rod continued the drive to the paddle crankshaft. A pair of 5 ft flywheels, mounted on a further shaft low in the boat, was geared to this by 6:28 toothed gearing. The wheels were 12 ft diameter and 22 in wide, probably rotating at 15 rev/min to give some 4·5 mile/h. The low pressure wood-fired boiler (2 lb/in^2) was 7 ft long by 5 ft square. The hull of the ship was 66·5 ft. long, 10·6 ft wide and 3·3 ft deep, the draught being 1·5 ft.

No provision was made for sails, unlike the usual practice in steamships even much later. This was typical of Fulton.

After the ship was completed, she was moored in the Seine to await trials when a violent storm blew up. Under such circumstances the strength of the hull proved inadequate and it broke in half, depositing the heavy machinery in the mud of the river bed. Undaunted, Fulton built a stronger hull into which he fitted his salvaged boiler and engine. In his first trials on 9th August 1803, with a crew of three, he towed two other boats carrying many notable men of science who desired to witness the event. These included Citizens Bossut, Bougainville and Lazare Carnot who had been appointed by the Minister of Marine to report on Fulton's trials. Napoleon had asked for an urgent report.

Fulton, however, was not carrying out his work for the benefit of the French. He was realist enough to see that there was little that he could patent in the steamship and that, if he was to obtain any form of monopoly in France, it could only be by means of patent protection. Nor had he any desire to suffer again the sort of delays and frustration that had occurred in his negotiation with the French government over the submarine.

His intention was to return at once to America after his trials where his partner already had a steamboat-operating monopoly in New York State. The trials were meant to make sure that his designs were correct.

It is not certain if Napoleon's government intended to make an approach to him. His trials were a complete success, but no trace of the Commission's report has survived. Before they had any chance of action, however, Fulton had left

* Colonel Mark Beaufoy carried out experiments on floating solids of uniform section between 1793–1798 under the direction of the Society for the Improvement of Naval Architecture. His main conclusions were that skin friction is proportional to the wetted area and to the square of the velocity; and that bow and stern resistance are proportional to the square of the sine of the angle of obliquity of the bow and stern.

† *The Rev. Dr Edmund Cartwright, was the inventor of the power loom.*

France, not for America, but for England. By now the British government were showing increasing alarm about his activities and had sent a secret agent to meet him in Holland to start preliminary negotiations about his submarine inventions. He therefore postponed his voyage to America and went first to London.

Torpedoes for Britain

His first act on arriving in England was to call on Boulton and Watt to place his problems before them. (The letters he wrote to them at this time were addressed from the Storey's Gate Coffee House which stood on the site of the present Institution headquarters.) This correspondence is important as it clearly shows that, although he was dealing with the world's foremost steam engine builders, his grasp of the subject was such that he personally supervised all the details of the design and suggested many of the modifications that contributed to its ultimate success. His was the first bell-crank lever engine built by Boulton and Watt.

The British Government at once started investigations into Fulton's methods of submarine warfare and a special committee, consisting of Sir Joseph Banks, The Hon. Henry Cavendish, Major William Congreve, John Rennie and Capt. Sir Home Popham, decided that the torpedo was the weapon showing the greatest promise. Lack of a suitable engine virtually put the submarine out of court.

Fulton, as a result, went to Portsmouth in December 1804 to supervise the construction of a batch of torpedoes. These bore little resemblance to the torpedo of today but consisted of a carcass or bomb filled with explosives and fitted with floats, a grapnel, and a clockwork delay firing mechanism. The vessel was so weighted with shot as to float nearly flush with the water line and was towed by a black-clad seaman who, equipped with a paddle, sat low between the baulks of timber catamaran. Eventually the Admiralty ordered materials for a hundred of these, but only after Fulton had written direct to the Prime Minister, William Pitt, threatening that, if he did not receive more support from the Navy, he would have to resign from the project.

The torpedo was first tried out against the French fleet off Boulogne in October 1804 but only one of the five used was successful, sinking a pinnace and her crew of twenty-one. British newspaper reports of the attack brought a wave of public indignation against such an unsporting mode of warfare. Nevertheless, a further test attack was made against the French at Calais in December, but with no more success. The final test on the torpedoes was made at Deal when the brig *Dorothea*, now beyond her active service life, was moored and then blown in two by one of Fulton's submarine bombs.

His negotiations with the British Government had been begun at their own seeking, not at Fulton's. For this purpose he had delayed his journey to America, for two and a half years. Nelson's success at Trafalgar in 1805 had such an effect on the French fleet that the *raison-d'être* for Fulton's work in England disappeared. In January 1806 his principal supporter, William Pitt died; of the £15,000 promised to him by the Government, he received £1,653. After buying some paintings, he set sail for his native land in October that year.

Return to America

His return to America heralded the last, the busiest and the most rewarding period of his life. While his personal plans revolved entirely round his steamship projects, he was drawn away by the U.S. Government almost at once to demonstrate his torpedoes. Although their efficacy was proved by the successful demolition of a brig in New York Harbour in July 1809, further trials at the Navy Yard failed, due to the ingenuity of Commodore Rogers, the officer in charge, who protected his ship with the first torpedo netting. Fulton, not unnaturally, felt that this was unfair and the truth of the matter was that he wanted to get away from submarine work and get on with his new steamship.

In a final gesture, and mainly to influence Congress, he made public all his work in this field in a brochure *Torpedo War and Submarine Explosions* in 1810. He was also asked by the Government to undertake canal work but, with the vision of the steamship before him, he reluctantly refused. He had also to suppress his own inclinations to develop his ideas on railways for the transport of coal in Richmond, Virginia.

The years of success followed, greatly marred by the vast amount of litigation that he and Livingston had to engage in to protect their rights. As there was little that could be patented in their steamships, they could only hope for commercial success through obtaining and working operating monopolies.

Fulton's first ship in America commenced trials in 1807. The 146 ft long hull was powered by the 20 hp engine ordered from Boulton and Watt while the rest of the machinery was built by Fulton himself. The trials were a complete success but some work still remained to be done, such as boxing-in the paddle wheels and covering the engine.

Only ten days after the first trial, his steamboat, with some thirty or forty people aboard, completed the journey from New York to Albany, a distance of 150 miles, in 32 hours. The return journey was made in 30 hours. On 4th September, a regular service between the two points was started, the first commercial steamship project in the world. At first the ship was given no name, being known simply as 'The Steam Boat'. Later she was christened the *Clermont*.

In 1809 Fulton built *The Car of Neptune* for which he constructed his own engine. By the time he died in 1815, Fulton had designed no less than 17 ships, the series culminating in his 526 ton, 60 hp *Chancellor Livingston*. Under his initiative and guidance steam in navigation had been brought to a successful issue.

In the last two years of his life another project also occupied Robert Fulton, the construction of the world's first steam-driven warship. His design for the *Demologos* was typically original, consisting of twin hulls between which the paddle wheel was located for all round protection. A heavy upper deck sheltered the engine and boiler. A belt of solid timber 4 ft 10 in thick, protected each of the thirty 32-pounder guns. There were also two 'columbiads'—100-pounder guns fired below the water line. Trials took place in 1815, a 53 mile course being completed at an average of 6·4 mile/h. Cessation of hostilities between Great Britain and the United States in 1814 denied the *Demologos* its first taste of action but reports of it in the British press must have had some effect on morale.

Early in 1815 Fulton was delayed waiting for a boat when the Hudson was partly frozen. In inspecting the *Demologos* and other ships in process of repair he got wet through and caught a cold which turned to inflammation of the lungs. He died, in his 50th year on 23rd February.

Personal life

As a man, Robert Fulton was companionable and full of

lively conversation. He was able to make himself agreeable in any company but, in spite of this, there is no record of his having many friends. In England his poverty would hardly have allowed him to keep up with many who were on his own intellectual level or had the same interest and most of his time was taken up with his studies. In France he was more of a man of substance but in a country where even the language was not his own he tended inevitably to move among the relatively small circle of his compatriots. With Livingston, his partner, he was on the best of terms. Like many a bachelor, it was with a married couple that he had the closest bonds. When he first arrived in Paris he stayed in the same hotel as Joel Barlow and his wife and thereafter the closest of friendships existed between them. Barlow was a man of many talents and is variously described as statesman, philosopher and poet. His poem, *Columbiad*, which he published in 1807, contained a portrait of himself by Fulton who also arranged for the other plates and for the publication, as some return for the many kindnesses and financial help he had received from Barlow.

There was little time in Fulton's life for the pursuit of anything except his work. After he had taken up engineering he did little painting apart from an occasional portrait of one of his friends or their wives, a number of which still exist.

Even marriage was a thing he could not indulge in until late in life. He married two years after his return to America, when he was 43 years old. His wife was Harriet, daughter of Walter Livingston, a cousin of his partner. A son was born in 1809, followed by three daughters.

Perhaps Fulton achieved little in the field of canal work, the only one in which he came into direct competition with John Rennie. But he was a born engineer, a far-seeing inventor and an original administrator. In his twenty crowded years of mechanical engineering, he produced the first workable submarine, the first commercially practicable steamship, and the first steam-propelled warship. This was the achievement of the man whom Rennie considered to be of very slender abilities. Even if he had not had these successes his systematic combination of theory and experiment entitle him to a place among the Great Masters.

REFERENCES FOR FURTHER READING

COLDEN, C. D., *The Life of Robert Fulton.* 1817 New York.

KNOX, THOMAS, W., *Robert Fulton and Steam Navigation.* 1876.

SUTCLIFFE, ALICE, C., *Robert Fulton and the Clermont.* 1909 New York.

DICKINSON, H. W., *Robert Fulton, Engineer and Artist; his Life and Work.* 1913 London.

SUETER, MURRAY, F., *The Evolution of the Submarine Boat, Mine and Torpedo, from the Sixteenth Century to the Present Time.* 1907 Portsmouth.

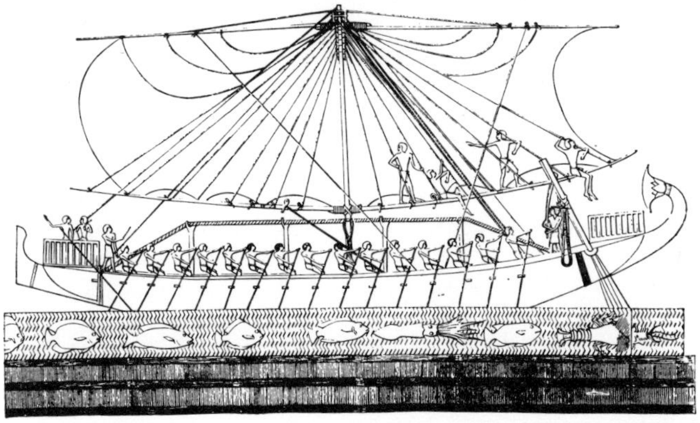

[Courtesy: Science Museum, London]

Fig. 1. A relief from the Temple of Deir el Bahari at Thebes, showing one of the sea-going vessels used by the Egyptian expedition to Punt in about 1500 BC

[Courtesy: Science Museum, London]

Fig. 2. The more sophisticated use of wind-power resulted in complicated rigging as shown in this illustration from a Dutch sailing manual of AD 1697

Marine Engineering

by C. C. Pounder, MICE (Member)

Steam had to serve a hard apprenticeship at sea, side by side and in competition with sail; and it was not until the end of the 19th century that the battle was finally won. This is the story of how the shipwrights turned into engineers and adapted the steam engine for marine purposes.

The oldest representations of boats came from Egypt. It is believed that Egyptians sailed over the Mediterranean to Syria and Crete before 3000 BC. Rock-carvings depict pre-dynastic boats as being paddled; but the earliest drawings represent boats as having both sails and oars, and being constructed of planks. Fig. 1 dates back to 1500 BC.

After Egypt, the mastery of the sea gradually passed to the Phoenicians, but not much seems to be known about their ships. Early Greek warships are said to have been dug-out tree trunks arranged with outriggers and carrying 24 oarsmen. Later Greek and Roman ships, according to surviving sculptures, were shallow draught structures suitable for beaching. Built with keel, ribs and planking, they were strengthened by transverse benches upon which rowers were seated.

Galleys

The first known representations of *biremes*—i.e., galleys with two banks of oarsmen—are embodied on two Assyrian reliefs and are believed to represent Phoenician warships.

The *trireme*, the Greek galley with three banks of oars, was first built in the seventh century BC. At the time of Alexander, galleys with four and five banks were apparently in vogue. The *trireme* was provided with a square sail and had a crew of 200. Over the centuries the galley became a long, narrow boat, carrying a lateen sail or two, but dependent upon oars for manoeuvring and safety. It remained the principal type of ship for at least 2000 years.

Between the two decks of a galley was the propelling power, some twenty-seven oars per side, manned by galley-slaves. chained close together on their narrow benches, who worked all the years of their miserable lives in this fearful occupation,

A galleass was a large, heavy galley, with three masts and a rudder. At Matapan, in 1717, the last battle in which galleys were used, the ships were 170 ft long and 27 ft wide. In each there were 427 slaves on 31 pairs of thwarts, i.e., 7 men per oar.

Progress is never steady, whether reckoned in terms of the development of a prime mover or the unfolding of the life of a nation. By way of example: during one term of six centuries no advance is known in shipbuilding. By contrast, between AD 1400 and 1500 sailing ships underwent more profound development than they had known during the whole of the preceding 5500 years. Nor was such relatively rapid progress achieved during the four centuries that followed.

The one-masted ship of AD 1400 became the two or three masted vessel of AD 1500. The ships had been given sprit-sails and topsails; a fourth mast was introduced and, later, top gallants. The effect was to make oars obsolete in large ships. It was these developments which enabled Europeans to venture across the oceans, thus making them masters of the world. Subsequent improvements to sailing ships, such as much more complicated rigging as shown in Fig. 2, were of secondary significance.

When the nineteenth century began, the only known means of crossing the ocean was by wooden sailing ship, the largest not exceeding 200 ft in length.

Up to this point in time, whatever might be the form and size of ship, its propulsion methed depended, inescapably, upon the application of human muscles and sinews, for large, crews were required to handle sailing ships. Then, not much more than a century and a quarter ago, came the power-operated ship. This was a technical break-through of unprecedented significance.

Steam propulsion

After the emergence of Western Europe from the Middle Ages, about 400 years ago, progress in material affairs was rapid, almost amounting to what biologists term a mutation. Parallel with the later phases of this upsurge there was the industrial application of the steam engine and the tremendous consequences stemming from its development.

Naturally, therefore, the steam engine had not long been in vogue for industrial purposes when the possibility of its application to ship propulsion occurred to many men, but only as a result of Watt's improvements did it become a practical proposition. Even then it took nearly another century before steam predominated over sail.

Jonathan Hulls, of Campden, Gloucestershire, secured a patent on 21st December 1736 and, in 1737, published *A Description and Draught of a new-invented Machine for carrying Vessels or Ships . . . against Wind and Tide or in a Calm.* He detailed his invention dealing with the principle of steam navigation.

There appears to be no evidence that Hulls actually constructed a steam boat. The only steam engine type available at the time was the clumsy slow-moving Newcomen atmospheric engine. Therefore it is highly improbable that he could have driven a boat by steam.

The work of James Rumsey on the River Potomac, that of John Fitch on the River Delaware in the 1780's, together with others in Britain and elsewhere, marks the beginning of the practical marine engine. Fitch's engine, harking back to the galley, worked twelve vertical oars; Rumsey, anticipating the future, invented a jet propulsion system.

Patrick Miller, an Edinburgh banker, constructed a vessel in 1786 which floated on three hulls with hand-rotated paddle-wheels placed between them. Since it proved too heavy for hand-propulsion William Symington in 1788 built a two-cylinder steam engine for a new vessel, with only two hulls; the engine was placed in one and the boiler in the other. Two paddle-wheels between the hulls propelled the little vessel at about four miles an hour. The cylinders were 4 in bore.

A second and larger engine was next installed, and Miller is on record as having said that Symington's engine was the most improper of all steam engines for giving motion to a vessel. In his method of condensing the steam, Symington infringed the patent of James Watt, who later obtained an injunction against him.

The first practical steamship was the *Charlotte Dundas*, a paddle vessel 56 ft long and 17 ft beam, driven by a Symington engine. In March 1802 it towed two 70-ton vessels 20 miles along the Forth and Clyde Canal, against a strong wind. The engine was a single-cylinder, double-acting, horizontal unit, 22 inches bore, 4 ft stroke, and with this simple and superior engine, Symington's prospects seemed bright. But, because of damage to the canal banks due to eddies from the paddle blades, this pioneer vessel was eventually laid-up and allowed to rot. Symington, who has well been named the father of marine engineering, died in poverty.

Robert Fulton and Colonel John Stevens were two notable USA protagonists of steam propulsion. Fulton's *Clermont*, built in 1807 (Fig. 3) and propelled by a Boulton, Watt and Co. engine, was employed on the River Hudson between New York and Albany. Stevens' vessel, *Phoenix*, was built during the following year for the River Delaware. The *Comet*, 1812, was the first merchant steamship in Europe. Built on the Clyde, the owner, Henry Bell—who was a carpenter—was the proprietor of an hotel at Helensburgh and the boat was built to convey his guests from Glasgow and Greenock to Helensburgh. It attained 6·7 knots. Bell reaped no financial reward from his efforts and was never able to pay the full price of the boiler. As early as the year 1801 he had suggested to the Government the application of steam engines to boats. The year 1814 is regarded in official records as the beginning of the steamship era in British merchant tonnage, with one steamer of 69 net tons.

Crossing the Atlantic

The *Savannah* is frequently mentioned as being the first steamship to cross the Atlantic. Launched as a full-rigged sailing ship of over 300 tons, a single cylinder steam engine was afterwards fitted, with paddle-wheels which could be unshipped. It crossed the Atlantic from Savannah to Liverpool during the summer of 1819 and, during a voyage of about 28 days, the engines were used for 80 hours. The decision of the owners to send the ship to Europe was not connected with any considered scheme for regular steam passages. Hence the voyage of *Savannah* and other ships did nothing to establish the commercial feasibility of plying the Atlantic with steamships.

What later became the Atlantic Ferry began in 1838, when three British companies sent steamships to New York. All the ships were made of wood; the engines were of side-lever type, driving paddle wheels. The first commercial steamer to make the Atlantic crossing was *Sirius*, tonnage 703, 178 ft long, 25·5 ft beam, with a 320 hp engine, cylinder bore 60 in, stroke 6 ft. The crossing occupied 19 days, in April 1838. Of these pioneer ships the most remarkable was Brunel's *Great Western*, a vessel 236 ft long, propelled by a two cylinder engine of 750 hp. The boiler pressure was 5 lb/in^2.

In 1843 the first screw-propelled all-iron merchant ship entered service on the Atlantic. This was the *Great Britain*, also designed by Brunel. The engines, comprising four inclined cylinders of 88 in bore, 6 ft stroke, developed 2000 hp. The six-bladed propeller was chain-driven from the crankshaft. A speed of 12 knots was obtained.

In attempts to establish regular steamship passages to Australia, several companies had sustained heavy losses. With the intention of steaming to Ceylon and back without

[Courtesy: Science Museum, London]

Fig. 3. The first commercial steamboat in the world: 'The Clermont' of 1807 on the Hudson river, U.S.A.

Fig. 4. Typical steamers of the early 1860's; the SS 'Sicilian' (above) was 272 ft long

re-coaling, the *Great Eastern* was built in 1854–8. It was an audacious experiment, behind which were Brunel and Scott Russell. The ship was 692 ft long, of 27 400 tons displacement, the coal bunkers accounting for 10 000 tons. The speed was 14·5 knots with 4900 hp developed at the 24 ft diameter propeller and 3400 hp at the 56 ft diameter paddles. The six masts could carry a total sail area of 58 000 ft^2. But as a commercial venture the ship was a failure.

It was said by Lord Pirrie, the great shipping magnate, at the turn of the present century, that the failure of the *Great Eastern* put back the era of the big ship by twenty-five years.

Engines

Many distinctive types of engine came into vogue. As well as side-lever engines there were diagonal, oscillating cylinder, and horizontal forms. Eventually the so-called 'steam hammer' designs prevailed in which the cylinder was above the crankshaft. Sometimes the engines were direct-coupled to paddles or propellers; sometimes they were geared.

Paddle wheel engines were of necessity slow-revolution types. When the propeller was introduced, the general practice was to use a paddle engine and drive the propeller shaft through gearing. This often took the form of a large spur-wheel on the engine shaft, with wooden teeth, and an iron-toothed pinion on the propeller shaft.

The progress made in marine machinery was due mainly to the efforts of men of unusual ability who had risen from the ranks of millwrights and mechanics. In this connection the names of Stephenson, Fairbairn, Telford, Laird, Whitworth, Field, Penn, the Napiers and the Maudslays, readily come to mind. But there were many others.

When the early steam reciprocators were installed in ships, the mechanics who tended them were denizens of workshops. That is, they went to sea simply because their engines went to sea. They were therefore treated as being nothing more than able seamen. It was the master of the ship who decided how the engine was to be run.

A century ago, the national background in Britain was, in high degree, favourable to progress in the mechanical arts. There was no dearth of great men in any department of life. Strong individualists were establishing industries which grew and prospered. It was the period of Darwinism, marking the beginning of a great intellectual upsurge which shattered the notions and myths of centuries. It was also the period of the Lenoir gas engine, which led to the Otto cycle, and so to later internal combustion engine practice. Fertility of body and inventiveness of mind were the characteristics of the times. There were, it is true, many grievous blemishes, too often smoke-screened by hypocrisy. But, despite its darker side, it was a great period, a period brimming with vitality.

Almost all the ideas which were developed in the second half of the century had been originated in the first half. Mechanisms and motions of all kinds were invented at a prolific rate, with the result that there is hardly any type of mechanism in use today which was not known a century ago.

Fig. 4 shows typical steamers of the 1860's. The *Sicilian* (above) had an inverted vertical engine. It was a four-masted iron barque of 2500 tons carrying capacity, 272 ft length of keel. The ship was provided with three steam winches, a capstan and self-reefing top-sails. The fore and main masts were of iron and the rigging of wire. The engine was not used when the vessel was sailing under canvas, with a fair wind. The propeller was then lifted out of the way by a special winch on deck, the line shaft being provided with two universal joints. Under sail alone, the speed of such a ship, fully-laden, might be ten or twelve knots; under steam alone it would be six or seven knots.

By the middle of the century the struggle of steam propulsion for recognition had been won, to the extent that an engine was accepted as a useful auxiliary in a sailing ship. From this period onwards, progress in marine machinery became an interwoven complex pattern of five constituents; boilers, engines, condensers, machinery room auxiliaries and pipe arrangements.

The marine engine was no longer a land engine installed on board ship, and the engine men had ceased to be mere mechanics. The marine engineer's status charter was the Act of 1862 which established examinations for them.

Propellers and iron hulls

By 1840, over thirty propeller types had been patented. As in many other marine engineering problems, the tussle between paddle-wheels and the screw propeller was long and intense, but in 1845, a tug-of-war between two Admiralty sloops, one fitted with paddles, the other with a propeller, showed the undoubted superiority of the latter. Thus the matter was settled.

Experience soon showed that screw-propulsion was not practicable for 300 ft long wooden vessels. When the power was applied at the extreme end of the ship, fine hull lines, which were needed for speed, were difficult to combine with the rigidity required to eliminate 'working' when the ship was wholly designed in wood. The solution was found in a composite system of an iron framework with the wood planking. This construction soon gave place to the all-iron ship, with a saving of about one-third of the hull weight. Although primarily introduced for screw-propelled steamers, the all-iron construction soon proved to be equally beneficial for sailing ships. A generation later, hull weights were reduced a further 15 per cent by the substitution of steel for iron.

It is not an overstatement that the work of the early marine engineers led to revolutionary advances in hull design and construction. Naval architects and shipbuilders thus had to follow their lead.

The annals of iron shipping included some spectacular losses. Thus one iron passenger vessel suddenly went to pieces in a gale, with heavy loss of life. Another went on the rocks and sank within ten minutes: 560 people were drowned. The iron shell plates were 6 ft long, 33 in wide, 0·75 in thick, double-riveted: all the rivets failed.

Boilers and condensers

In the 1860's, boiler problems were prominent and there were many explosions. The use of salt water as feed make-up was the source of many troubles; indeed, the prevention of serious saline incrustation was one of the engineer's chief duties. Boilers were blown-down every two hours. Neglect of this routine could lead to the water spaces becoming solid with salt. The water level in boilers was determined by test cocks, the water gauge having not yet been invented. Dead-weight safety valves were used until the 1870's; the salt-water evaporator was not to come into use for another 20 or 30 years, to be followed by the distiller.

The unsatisfactory state of boiler construction and maintenance is evidenced by the regularity with which boilers were replaced. Seldom did a set of boilers last for more than ten years. As soon as fresh water became available as feed make-up, boiler progress was accelerated.

Surface condensers were invented by Samuel Hall whose patent was dated 1834. Although this novelty was a factor in the success of *Sirius*, jet condensation continued to be general practice until the 1870's.

After being adopted on certain early transatlantic ships, surface condensers were shortly afterwards abandoned. This was partly on the score of expense and partly because they became clogged with the tallow so freely used in boilers and engines at that period. It was not until the 1860's that the surface condenser was applied on other than an inappreciable scale. The re-introduction was chiefly due to G. H. Spencer of the N.E. Marine Engineering Co. and Edward Humphrys.

It should not be forgotten that James Watt entertained the idea of a surface condenser with tubes and narrow cells made of flat plates.

In the application of the steam engine to ships engaged on long voyages, there were many difficulties to be overcome. For example, coaling stations had to be established at suitably spaced intervals on all sea-lanes. For this purpose the coal was carried in sailing ships. By 1883 the steamer with triple-expansion engines had begun the invasion of the world's longest trade routes, but it was the end of the century before the steamer really predominated there.

Sail versus steam

After 1870 engine development took the direction of compounding the cylinders, thus improving the fuel consumption. Fig. 5 shows an arrangement of cylinders favoured in high-powered ships of the period. The engine was of compound type, two HP cylinders in tandem with two LP cylinders. Coal consumption was about 3 lb per ihp hour.

In 1840 the total British steamer tonnage was 95 000, that of sailing vessels 3m. While the steam engine was struggling upwards towards increased reliability and better fuel economy, towards greater powers and improved manufacturing techniques, the sailing ship was advancing in size and speed. Unchallenged on the Far Eastern, Western American and Australian runs, the sailing ship reached its highest level of attainment and gracefulness in the swift sailing clippers of the period 1874–1883. That it was a very highly scientific level is clear from scrutiny of contemporary rigging arrangements. It was not until about 1880 that the steam engine could be said to predominate over sail because of its independence of the weather and because a much smaller crew was required to run a steam ship.

Multiple expansion

By the middle 1880's the triple-expansion, inverted vertical, double-acting marine engine had emerged due to a constant desire to economise in fuel costs. And, complementary to it, there was the multitubular cylindrical boiler, commonly called the Scotch boiler.

The development of the compound engine reached its greatest height in 1884, with the Cunard ships *Etruria* and *Umbria*. These were single screw vessels and the engines, which developed 14 500 ihp, comprised one cylinder of 71 inches bore side-by-side with two cylinders 105 inches bore, the stroke being 6 ft and the boiler pressure 110 lb/in^2.

The outstanding characteristic of the last quarter of the nineteenth century was the development and wider application of triple and quadruple expansion engines, working with steam pressures between 150 and 250 lb/in^2. These proved as superior to the compound engine as the latter had been to

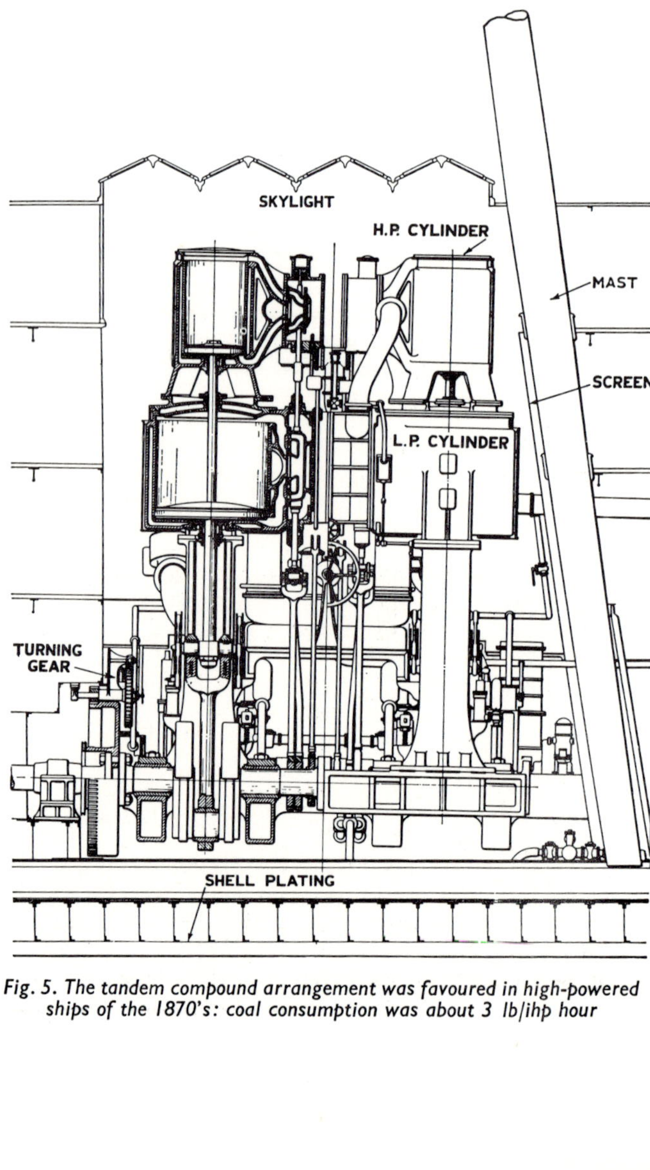

Fig. 5. The tandem compound arrangement was favoured in high-powered ships of the 1870's: coal consumption was about 3 lb/ihp hour

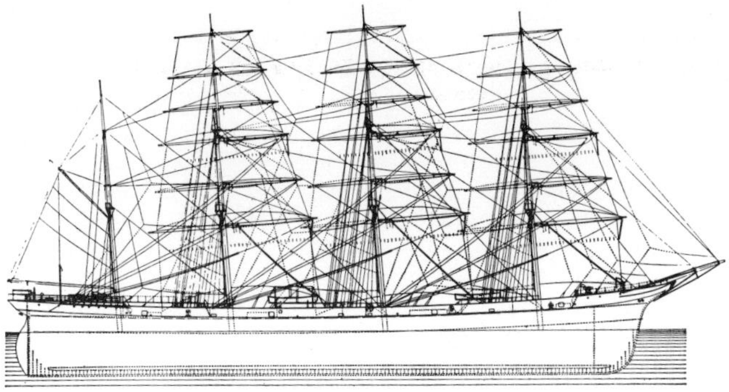

Fig. 7. The sailing ship was by no means finished in 1890: this iron barque was built in the same yard, at the same time and for the same owners as the 'Teutonic'

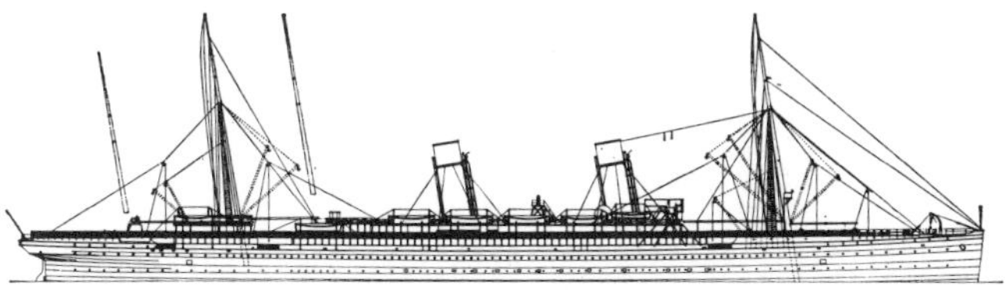

Fig. 6. The twin-screw liner 'Teutonic', completed in 1890, was then one of the two longest vessels afloat

the simple expansion in fuel economy, weight and space reductions.

Two passenger liners were outstanding, not so much for their machinery as for other features. They were *Teutonic* (Fig. 6) and *Majestic*, built in 1889–90, and at that time the longest vessels afloat. These ships were the first merchant vessels to comply with Admiralty requirements for cruisers, each having seatings for twelve guns. The port and starboard engines were isolated by a longitudinal bulkhead, which extended from the aft end of the engine room to the forward end of the coal bunkers, completely isolating the boilers, engines and pumps for each screw. There were 12 double-ended and four single-ended boilers

It is surprising to learn that when *Teutonic* was being built there was also being built in the same shipyard, for the same Owners, the sailing ship shown in Fig. 7.

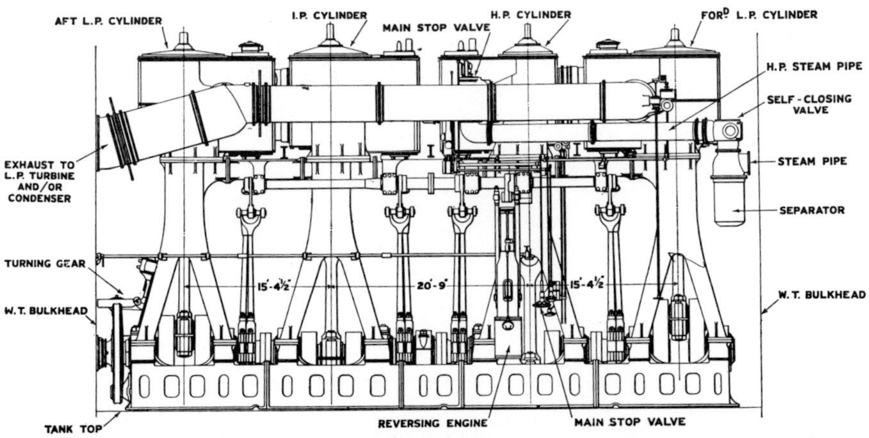

Fig. 8. The four-crank triple expansion engine of the 'Olympic', the most famous of the triple-screw passenger liners

Turbines

The opening years of the 20th century saw the reciprocating steam engine reach the peak of its development and its popularity. Thereafter its eclipse was rapid. The first shadows were thrown when Sir Charles Parsons' small ship *Turbinia* raced through the lines of warships assembled at Spithead for the Diamond Jubilee Review in 1897.

Fig. 8 shows a 4-crank triple expansion engine of *Olympic*, the most notable of the triple-screw passenger liners propelled by two reciprocating engines with turbines in the centre between them. The system utilised the toe of the engine indicator diagram by inserting a LP turbine between engine exhaust and condenser.

The direct-coupled steam turbine was not a serious commercial competitor with the reciprocating steam engine for other than special services since it was much too fast for efficient driving of the propeller. But the successful application of gearing by Parsons, in 1910, completely changed the perspective. By 1914 the geared turbine was well established. The impetus given by the war resulted in improved gear-cutting techniques.

Diesel versus steam

Sixty to fifty years ago, when Sir Charles Parsons and his staff were engaged assiduously and successfully in developing the steam turbine, other men at home and abroad were equally busy in developing the heavy oil engine.

About the beginning of the present century the internal combustion engine emerged as a marine propulsion engine. At first there were controversies regarding the merits of four-stroke and two-stroke cycles; later there were the claims of double-acting and single-acting types; the problems of heavy fuel burning; and so on. Ultimately the single-acting two-stroke pressure-charged engine became the widely accepted type.

In the decade following the first World War rapid advances were made in steam turbine propulsion, chiefly because of the shaking-up received from the diesel engine. In 1920 steam turbine sets of moderate power consumed about 0·9 lb of oil fuel per shp/hr, for all purposes. The diesel engine offered an all-purposes rate of 0·43 lb per shp/hr. Its greater first-cost was much more than counter-balanced by the saving in fuel and by the greater cargo-carrying capacity made possible by the reduced size of bunkers.

As steam was the engineering background of Britain, the new position created by the diesel engine was not complacently accepted. Within ten years boiler pressures were doubled and 450 lb/in^2, 750°F, became normal conditions. High-pressure-temperature marine turbine practice was founded upon land power station experience. This applied not only to the turbine itself, but extended to the boilers and the condensers as well as to the heat cycle.

Conclusion

Every system of ship propulsion hitherto devised has proceeded along its destined path to final stability. This evolutionary path has led through simple designs to more complicated schemes of greater initial cost but greater efficiency, always showing an advantage on balance. Then, to escape from the particular limitations of the propelling agent, a completely new step has become necessary.

Before the steam engine had run its course, naval architects had visualised a ship 1000 ft long; it was perforce a narrow ship, because of the power-weight-space limitations of the steam reciprocator and the coal-fired cylindrical boiler.

Then came the steam turbine, to override the limitations of the steam engine. With it, when aided later by the oil-fired water-tube boiler, much greater powers were readily obtainable; ships broader in the beam became practicable. The heavy oil engine with its high thermal efficiency had a profound effect upon ship propulsion economics.

Thus, in a hundred years of marine engine development, power per shaft had grown more than hundredfold; boiler pressures over fiftyfold; fuel consumptions a century ago were twelve times what they are today.

This brief review outlines some of the achievements of the past. What were the motives which led and, indeed, drove men to spend themselves and their substance in surmounting the problems inherent to those achievements? Probably the reasons were as diverse as the temperaments of the men themselves. But amongst them they made it possible for raw materials and peoples to be moved readily between Continents, leading to the world of today.

No branch of engineering has been more closely linked with great personalities than has marine engineering. Hence, to our young marine engineers is committed the custodianship of a great and living tradition. Adapting the words of Sir Isaac Newton: they stand upon the shoulders of giants.

REFERENCES FOR FURTHER READING

Landström, Björn, *The Ship*. Allen and Unwin.
Bonsor, N. R. P., *North Atlantic Seaway*. T. Stephenson and Sons Ltd.
Rolt, L. T. C., *I. K. Brunel: a Biography*. 1957 Longmans, Green and Co., London.
Hardy, A. C., *History of Motorshipping*. Whitehall Technical Press.

THE GREAT MASTERS

The large-scale supply of electricity —now doubling in quantity every ten years— was made possible by Parsons' development of the turbo-generator; his introduction of turbines and reduction gearing to ships' propulsion revolutionised water transport.

Charles Parsons— Purveyor of Power

by A. T. Bowden, Ph.D., B.Sc. (Member)*

Sir Charles Parsons will always be remembered as the engineer who developed the steam turbine-generator and gave it the performance, economy and construction which allowed it to become the successor to the condensing steam engine for the propulsion of ships at sea and the generation of electricity on land. Steam turbines of many types had, of course, been devised before Parsons applied himself to their design and construction, just as steam engines had been designed and produced before Watt; but the time was ripe for a comprehensive design, not only of the turbine itself but of the whole system of which it was a part.

The immediate success of his painstaking work stemmed from the fact that he considered the system as a whole and, in its application to power generation, he studied the electrical generator with the same devoted, almost fanatical, approach that he brought to the steam turbine itself. It is not generally appreciated that his contribution to the design of the generator parallels his work on the steam turbine and, in many ways, was much more important.

His genius is readily understood when one studies his original calculations which are still available in the works at Heaton, where he established the manufacture of the steam turbine. From these it is clear that, important as it was, the design of a multi-stage reaction turbine was in many ways overshadowed by his original work on gearing, critical speeds, bearings, electric rotor design and high voltage generation. A lesser man would have been content with mastery over the steam turbine: Parsons' greatness stems from his overall conception of the power systems to which he set his hand.

Childhood and education

Charles Algernon Parsons was born at 13 Connaught Place, Hyde Park, London, on 13th June 1854. He was the sixth and youngest son of the third Earl of Rosse whose ancestors had emigrated from England to Ireland at the end of the sixteenth century. Members of the family had made their names in politics, war and science. Parsons' grandfather had been a Vice-President of the Royal Society and his father, who was distinguished both as an engineer and as an astronomer, had been its President. The family seat at Birr Castle in Ireland was a rendezvous for the leading scientists of the day.

It was in this intellectual atmosphere that Parsons spent his boyhood. The brothers enjoyed advantages that most boys would envy. Attached to the castle were a forge and large workshops containing lathes and other equipment, including special machinery for polishing specula for telescopes. Charles and his brother Clere spent a great deal of time in the workshops, generally with their father. Most of the boys led an open-air life. Charles was attracted to rowing, fishing and shooting; his pleasure in these sports never ceased. Another sport they enjoyed was yachting: one of the three boats the family had in succession was the 188 ton *Titania* which had belonged to Robert Stephenson. Their trips took the boys across the Channel to Cherbourg and back to Ireland round Lands End; to Belgium, Holland and along the Spanish coast.

Charles Parsons never went to school but had, along with his brothers, the benefit of private tuition by men of such scientific calibre as Sir Robert Ball and Dr Johnstone Stoney. In July 1871, at the age of 17, he entered Trinity College, Dublin, and at the entrance examination secured a prize for proficiency in German. This was one of the results of his holiday jaunts on the Continent where he had made good use of his time. In the following year he won the prize for progress in mathematics.

After spending two years at Dublin, he proceeded to St John's College, Cambridge. At that time there was no

* *Director, C. A. Parsons and Co. Ltd., Nuclear Power Plant Co. Ltd., and International Research and Development Co. Ltd.*

Engineering School but Parsons attended such lectures as were given on Mechanisms and Applied Mechanics. He studied mathematics with such effect that he took his B.A. in the Mathematical Tripos of 1877, being placed as Eleventh Wrangler.

Parsons was a quiet and shy young man, no doubt as a result of his sheltered childhood. His shyness was, however, broken down by his close contact with his fellow members of the Lady Margaret Boat Club where, being very strong, he pulled an effective oar. His last appearance on the river was in the May races of 1877 when he rowed No. 3.

Parsons built his first turbo-dynamo in 1884. Output 7½ kilowatts at 18 000 rev/min

While at Cambridge, he schemed out a high-speed epicycloidal engine and made a model of it in paper, sealing-wax and steel wire. One of his fellow students has recorded that one day, when members of the boat crew visited Parsons' rooms, he said "look here you fellows, I have an engine which is going to run twenty times faster than any engine today". "Rot!" was the reply. Parsons picked up the paper model, blew into it and the wheels simply flew round. Still unimpressed, they put Parsons and the model engine under the table!

Apprenticeship and marriage

He began his engineering training in 1877 by joining W. G. Armstrong and Co. at Elswick, as a premium apprentice. His inventive faculties surprised the management and he was allowed time to construct an engine designed on the lines he had worked out at Cambridge. It was a compound engine with four opposed cylinders, revolving at half the speed of the crank-shaft. When running at 700 rev/min it developed 10 hp and, for a time, was coupled to a Siemens dynamo supplying current for a powerful arc lamp on the Elswick jetty. Armstrongs were not interested in its manufacture and, as Charles' brother Clere was on the staff of Easton and Anderson of Erith, who required a high-speed engine for driving pumps, the epicycloidal engine was sent there. The Erith firm subsequently made a number to the design, with good results. Clere later joined the firm of Kitson in Leeds who then began to make the epicycloidal engine, producing about 40, of 20 hp each, for driving dynamos. It was, however, ultimately eclipsed by the Willans engine and its manufacture discontinued.

At the end of his four years' apprenticeship at Elswick, Parsons decided to join his brother Clere at Leeds and entered into an arrangement with Sir James Kitson to set up a small experimental shop, each to pay half cost, for carrying out experiments on the rocket propulsion of torpedoes. This arrangement continued for two years but there were unexpected difficulties arising chiefly from the imperfections of the rockets, and the experiments were abandoned.

In 1882, Parsons had met Miss Katherine Bethell, at a time when he was said to be "a strange and weird young man". The courtship was marked by his extraordinary attention to the study of needlework. Soon he was able to embroider much better than Katherine could! His skill in all such matters was amazing.

They were married on 10th January, 1883, at Bramham, Yorkshire. The wedding tour was interrupted. At that time, Parsons was so absorbed in the design of a torpedo that he took his bride and a mechanic to Roundhay Lake for torpedo trials at 7 a.m. every day. On those winter mornings the mechanic was kept warm by work under the stimulating influence of his employer, but the bride was left, too often, shivering on the bank and ultimately contracted rheumatic fever.

Upon her recovery in the spring of 1883, the honeymoon was resumed for about five months. They visited the United States, including New Mexico and California, a great part of the way with the aid of a pair of horses and a hired buggy. They also visited Pittsburg, Chicago and other industrial cities, intent upon seeing engineering works.

On their return to England, Parsons decided to settle down and have a home of his own. There was a good opening with Clarke, Chapman and Co. as a junior partner, so he paid Kitson his share of the cost of the experimental work and left to go to Gateshead. In January 1884 he joined Clarke, Chapman and Co. and took charge of their newly organised electrical department. The firm then specialised in ship's deck machinery of all kinds and marine pumping machinery. They were anxious to produce a small, lightweight, steam-driven ship's lighting set, and Parsons was given the task of developing it.

At that time the optimum speed of the dynamo was much above that of any available reciprocating engine and many inventors were working on the problem of arriving at some efficient arrangement for coupling the two. Attempts were made to produce a low-speed dynamo for direct coupling to an engine: alternatively, belt-driven speed-up pulleys were used, but without great success.

Several other inventors produced high-speed reciprocating engines for this purpose, including Brotherhood, Willans and Bellis and Morcom.

The turbo-generator

Parsons thought the time opportune for an attempt to produce a practical steam turbine for driving a dynamo directly. From the outset he realised that, if the turbine were to be a success, it must rotate at what, in those days, were phenomenally high speeds. Preliminary experiments were begun with a view to ascertaining by trial the conditions of working equilibrium and steady motion of shafts and bearings at the very high speeds of rotation that appeared to be essential if the turbine were to be of moderate size.

Trials of shafts were run in bearings of different descriptions up to speeds of 40 000 rev/min; the shafts were $1\frac{1}{2}$ in diameter and 2 ft long, the bearings being about $\frac{3}{8}$ in diameter.

The next problem was the turbine itself. Various forms were considered and it appeared desirable to adopt a principle that had been successful in the water turbine and was easily adapted to a multiple or compound arrangement: the parallel flow principle was the answer. To obtain good efficiency with a moderate speed and wheel diameter it was necessary that the steam should pass through a large number of successive turbines or stages, with a small pressure difference across each. Under such conditions the velocity of the steam bore the right relationship to the peripheral velocity of the turbine blades.

Parsons had many practical problems to solve before his ideas could be embodied in an actual machine. Not only had a suitable form of blading to be invented and appropriate manufacturing methods to be devised, but the design generally had to conform to conditions quite outside the range of ordinary engineering practice. A rotational speed of 18 000 rev/min was adopted, which was over fifty times that of the fastest reciprocating engine of the day. Means had to be provided for the continuous lubrication of the bearings, and a totally new method of controlling the speed of the machine had to be devised.

In addition to designing a revolutionary form of high-speed prime mover—the turbine—Parsons had also to produce an almost equally revolutionary d.c. generator. He required a dynamo that could be driven directly by his turbine at a speed of 18 000 rev/min and faced the question with the same boldness that he showed in the design of his turbine. He achieved an equally striking success.

Both electrical and mechanical problems had to be solved, for the changes in magnetisation of the core were vastly more rapid than in any other machine then built. The dynamo was of the bipolar type with an output of 75 amp at 100 V. The armature spindle was of high carbon steel, bored through from end to end to permit the circulation of oil for cooling. The diameter of the armature over the windings, which were of the external drum type, secured by binding wire, was $2\frac{3}{8}$ in, the diameter of the core $2\frac{1}{2}$ in. The latter was composed of thin iron sheets, built up to form a length of 8 in, with notched driving discs at each end. The armature had fifty conductors and the commutator twenty-five segments.

On 23rd April 1884, Parsons took out two patents, describing the generator in detail and also its turbine drive. A picture of the machine which was completed that year is shown in the illustration. It developed $7\frac{1}{2}$ kW when running at 18 000 rev/min and had a steam consumption of about 150 lb/kWh. This plant worked regularly for some years, supplying current for lighting experimental shops at Gateshead.

During Parsons' partnership with Clarke, Chapman about 360 turbo-generators, ranging in output from 1 to 75 kW, were constructed, mainly for shipboard electric lighting. The 75 kW sets, built in 1888 for the Forth Banks Power Station, included the first turbines to be installed in any public power station.

His own firm

Parsons dissolved his partnership with Clarke, Chapman and Co. in 1888 and decided to found his own works. He established this in June 1889, on a site about two miles east of Newcastle's centre. It covered an area of only two acres and the total staff numbered forty-eight. Unfortunately, his previous patents remained the property of Clarke, Chapman and Co. so that Parsons was precluded from building the axial flow type of turbine. He therefore produced a modified form in which the flow of the steam was radial. Although this restriction was a serious drawback, he had nevertheless constructed a 100 kW unit by 1891. This machine was the first condensing turbine and, when tested by Sir Alfred Ewing, was found to have a steam economy equal to the best reciprocating engine of the day. The largest radial flow turbine built for land purposes was a 150 kW turbo-generator supplied to the Portsmouth Municipal Electric Light Station in 1893. In the following year, Parsons bought back his original patents and was then able to return to the manufacture of the axial flow type which he preferred.

Progress from then on was rapid. One of the first machines to be built was a 350 kW turbo-generator for the Manchester Square Power Station, London: in 1900 the first 1000 kW set, with twin cylinders in tandem, was built for Elberfeld, Germany. By 1912 the unit output had reached 25 000 kW; by 1923, 50 000 kW.

The growth of electricity supply, consequent upon the invention of the turbine, created a demand, not only for larger generating units, but also for higher transmission voltages in order that more extensive areas might be served economically. In the early days the practice had been to generate at about 2000 V and to step-up the voltage when required. By 1905 Parsons had constructed turbo-machines to generate at 11 000 V and this became the standard generating voltage for many years.

In 1928 he again attacked the problem and produced a 25 000 kW turbo-alternator, generating directly at 36 000 V.

The first Parsons condensing turbo-alternator in 1891; it had an output of 100 kW at 4800 rev/min

The 'Turbinia'

The problems and disappointments arising from the rapid development of

A 50 000 kW Parsons turbo-alternator installed in Chicago in 1923

an entirely new prime mover, together with the establishment of a new works, would have been quite enough for most men, yet it was during this period that Parsons turned his attention to ship propulsion. The use of the steam turbine for ships was among the claims made by Parsons in his original patent of 1884. Ten years later the design and construction of a small vessel, *Turbinia*, was under way.

Parsons himself designed the hull and, in order to arrive at the most suitable form, made a number of models, about 2 ft long, which he towed by a fishing rod and line in a small pond at Ryton-on-Tyne, to study their respective wave formation and other characteristics. He next experimented with a 6 ft model, driven by twisted rubber cord through single-reduction spiral gearing on a propeller shaft. He decided to use the radial flow principle for the *Turbinia* as he had then built machines of this type up to 150 kW, against only 75 kW of the axial flow type. The power developed by the turbine was 2000 shp when running at 2000 rev/min.

The first trials of the *Turbinia* were undertaken late in 1894. Due to cavitation, of which little was known at that time, they were quite disappointing. Experiments were made with various arrangements of propellers of different diameters, running at different speeds. The best performance obtained was only 19¾ knots with a turbine speed of 1780 rev/min. Clearly, either the turbine was not developing its rated output or the efficiency of the propellers was extremely low. To settle the matter, Parsons devised special apparatus to measure the torque exerted on the propeller shaft. This instrument was the prototype of the modern torque-meter; by its use he assured himself that the fault was in the propellers and not in the turbine.

It was concluded that the trouble arose from the inability of the water to follow the rapidly moving propeller blades and he constructed a tank, with glass sides, in which a stroboscopically illuminated propeller could run at high speeds.

It was recognised that cavitation would be favoured by working with water near its boiling point, so the first experiments were made with hot water. However, it was more convenient to attain the same results by maintaining a vacuum above the water in the tank, and the nature of cavitation was exhaustively studied in this manner. The knowledge gained led to great improvements in the design of high-speed propellers and the methods of study initiated by Parsons have since been generally adopted.

In the meantime further experience had been obtained on some 350 kW axial flow turbo-generators built for land purposes and it was decided to replace the single radial flow turbine by an axial flow turbine with three separate cylinders. Each cylinder was coupled to a shaft on which three propellers were fitted, making nine propellers in all. In April 1897, trials were resumed which eventually culminated in the *Turbinia* attaining the remarkable speed of 34½ knots.

Later in the same year Parsons demonstrated her performance at the great Naval Review held to celebrate the Diamond Jubilee of Queen Victoria. Following up this success, he formed the Parsons Marine Steam Turbine Company with a registered capital of £500,000 and proceeded to build the Turbinia Works at Wallsend, for the manufacture of marine steam turbines.

The extraordinary speed attained by the *Turbinia* had impressed the Admiralty and, in 1898, after much deliberation, they placed an order with Parsons for a 30-knot turbine-driven destroyer, H.M.S. *Viper*. Parsons and his associates were, however, required to deposit a sum of no less than £100,000 as a security against the vessel meeting the specified performance. It achieved remarkable results, but was, unfortunately, lost at sea during a fog.

In 1901, the first turbine-driven passenger vessel, the *King Edward*, was built for service on the Clyde and others followed. In 1907, the 38 000 ton Atlantic liner *Mauretania* went into service. It attained 26·04 knots with her turbines developing 70 000 hp, and continued in service until 1935.

In 1905 a committee on naval design, appointed by the Admiralty, advised that in future turbine machinery should be used exclusively in all classes of warships. H.M.S. *Dreadnought*, fitted with Parsons turbines, was accordingly commissioned in 1906.

Reduction gearing

Up to this time practically all the turbines constructed had been direct-coupled, either to the generator or the propeller shaft and a compromise had been necessary with regard to the most suitable speed. Parsons realised that, if further progress were to be made, the only solution would be reduction gearing between the turbine and propeller shaft.

He began by carrying out exhaustive experiments to determine what tooth-speeds could be employed and what power could be transmitted, consistent with safety and durability. In 1909 he purchased an old cargo vessel, the *Vespasian*, of 4350 tons, and replaced her 750 hp triple expansion engines by geared turbines. With the same boilers and steam conditions the geared turbines used 15 per cent less fuel. The gearing worked perfectly and after the *Vespasian* was broken up, the turbines and gearing were transferred to another vessel.

Parsons also supplied a 750 bhp turbine, running at 2000 rev/min, for driving a rolling mill: the rolls had to run at 70 rev/min, which was successfully achieved by means of double reduction gearing.

In order that gear wheels should work quietly and without deterioration under the conditions of speed and power imposed by their new duties, it was essential that their teeth should be extremely accurate, both in form and pitch. Parsons, therefore, turned his attention to the production of better gears.

In 1912 he devised what is known as the Parsons Creep Mechanism. The work table of the machine was caused to rotate slightly faster than the master wheel, with the result that any errors existing in the latter were distributed spirally round the wheel being cut, instead of being concentrated at one part of the circumference. In this manner the unavoidable defects of the master wheel were, for all practical purposes, completely eliminated in the work.

The method of 'creep-cutting' created an entirely new standard of accuracy for mechanical gearing and facilitated the use of turbines in steam ships ranging from slow cargo vessels to the fastest warships and liners.

On land there was not the same scope for geared turbines as in marine practice, but the introduction of gearing nevertheless widened the field to, for example, continuous current dynamos, centrifugal pumps and small alternators.

Optical work

Although the name of Parsons is best known in connection with the turbo-generator and the marine turbine, he devoted much time and energy to other important activities. While he was with Clarke, Chapman he took an active interest in the newly formed Sunbeam Lamp Co., of which he was one of the founders. The company manufactured electric lamps, both arc and incandescent, and Parsons devised many ingenious devices for this purpose. But after he left Gateshead in 1888, he took little further interest in the subject.

Optical work held a great interest for him, as it had for his father. In 1886 Charles began the manufacture of parabolic reflectors for searchlights and three years later set up a special department for their production in his Heaton works. The manufacture of glass reflectors remained one of the standard products of the Company for many years. In 1921 he acquired the controlling interest in Ross Ltd of London, the well-known makers of binoculars and other small optical apparatus. Here he introduced various improvements in the methods of grinding but soon turned his mind to the larger question of the manufacture of the glass itself. In the same year he purchased an optical glass factory near Derby and made many improvements to the annealing furnaces. By these methods he was able to make exceptionally good optical glass discs of any required size.

The first turbine driven vessel, the 'Turbinia', attained 34·5 knots in 1897

In 1925 the old established firm of Sir Howard Grubb and Sons Ltd, makers of astronomical instruments, was threatened with closure. Parsons had made the glass for the lenses of some of their telescopes and the construction of these instruments appealed to both his scientific and mechanical instincts. He purchased the assets, formed a new company under the name of Sir Howard Grubb, Parsons and Co., and established a works at Walkergate.

Parsons—the man

This, then, was Parsons the engineer and scientist, whose name will always be associated with the development of the steam turbine and its applications on land and sea. He it was, above all others, who made possible today's abundant supply of comparatively cheap electricity and in his day he revolutionised marine transport. What was he like in his private life?

When he and his young wife left Leeds to set up house, they thought how delightful it would be to dwell in the country and their first choice fell on a house at Corbridge-on-Tyne. This was a beautiful place but it meant that Parsons had to leave for Gateshead at 7.30 every morning and did not return until 8 p.m. Within less than a year, therefore, they made their home at Elvaston Hall, Ryton-on-Tyne, in Durham.

In 1885 their daughter Rachael Mary was born and in 1886 a son, Algernon George. They remained at Elvaston Hall for about ten years—the most strenuous and anxious in Parsons life. Later he took a grouse-moor by the North Tyne, where happy autumn months were passed.

Although pressed by important work, Parsons still found time to amuse the children. He produced toys never before imagined, or permitted, in a nursery. Among these was *The Spider*, a three-wheeled contraption boasting a steam engine and boiler heated by methylated spirit. This toy travelled at high speed round the lawn with Parsons and his children, chased by their three dogs, rushing and shouting after it. Miniature locomotives were also constructed and great consternation was caused when these propelled themselves along the library carpet, leaking methylated spirit all the way and blazing a path through occasional pools of flame.

He also took an interest in his kitchen and decided to make strawberry jam, notwithstanding the indignation of the cook; but fortunately the result was a great success. Not so successful, however, was his attempt to fix the kitchen boiler which resulted in much smoke and more acid comment from the cook.

In 1894 the family moved to Holeyn Hall, Wylam-on-Tyne, and here Parsons provided himself with a complete experimental workshop. About this time he also purchased a country house, *Ray*, at Kirkwhelpington, some twenty miles northwest of Newcastle. He was very fond of fishing and also enjoyed shooting, though his pleasure seemed to be mainly in the enjoyment derived by his guests.

A less easy-going side of his character was shown in the episode of a motor lorry which, travelling in the opposite direction, had forced him on to the grass verge. This annoyed Sir Charles so much that, having collected the nearest policeman, he turned round and drove 30 miles until he caught the unfortunate lorry driver and charged him with dangerous driving. Wherever he told this story, he would add that more people ought to do what he had done in order to stop dangerous driving.

Another pet aversion of his was the liberty people took with the apostrophe after his name. In the end he issued instructions to leave it off altogether, preferring, as he said, to offend those who knew rather than suffer indignities from the ignorant. From that day to this the apostrophe has been banished in the organisations he founded.

Throughout his life he took his pleasure in tinkering at small jobs with his own hands. In the early days of motoring, he was ready with expedients which, to most people, would seem fantastic. On one occasion the family were to meet some friends for a picnic. They arrived late and Parsons' explanation was "the car would not go and we had to undress the parlour maid". It transpired that a piece of steel spring was required to repair the distributor and the fastening on a woman's stays afforded just what he wanted. The parlour maid was accordingly sent upstairs to remove her stays.

As an employer, Parsons was as advanced in his treatment of staff and workpeople as in his technical outlook. In 1919 he set aside a large sum to create a fund for the benefit of his employees. Any workman could receive assistance from this fund, either by way of weekly payments or a lump sum. The employees were encouraged to form their own sick benefit fund to which the Company contributed. In a trust he set aside shares for selected employees, the dividends being paid to Parsons until they accumulated to the share par value. Then, on application, the transfer of the share to the employee was completed. He believed that employees who had given long, faithful, or outstanding, service to him, should have a direct interest in the business but he felt that shares would be valued all the more if they had to be paid for by accrued dividends.

He was a terrific worker and, in the early stages of turbine development, would frequently work far into the night, using his great manual skill in tackling the many problems that arose. After being all day at the Works, he would settle down after dinner to several hours' reading of reports or patent specifications.

In his way of life Parsons was extremely simple, and appeared to have no expensive habits. Nothing pleased him more than to join in the social events of the Company and he always made a point of being present at annual dinners, sports days, etc., at which he thoroughly enjoyed himself. One of the most delightful features of his character was his extraordinary modesty. Even at the zenith of his fame he seemed unable to realise that his achievements had been exceptional or that his ability was anything out of the ordinary. The natural humility of his disposition led him to expect other people to possess an insight equal to his own, and he was always ready to listen to any reasonable argument and to discuss it on terms of equality.

He was equally capable, however, of an abrupt explosion of impatience with what he considered stupidity in thought or action. But normally his manner was kindly, courteous and considerate.

His public and private benefactions were many. He contributed freely of his knowledge to the proceedings of scientific and technical societies and many of them, in return, conferred distinctions on him. His honours included a knighthood and nine honorary degrees. In 1927 he received the supreme distinction of the Order of Merit, being the first engineer ever to receive it.

Parsons died on 11th February 1931, while on a cruise in the West Indies.

The 'Turbinia' at the Naval Review at Spithead in 1897

[Courtesy: Science Museum]

Detail of the calculating wheels on the Scheutz difference engine: the development of calculators is described overleaf

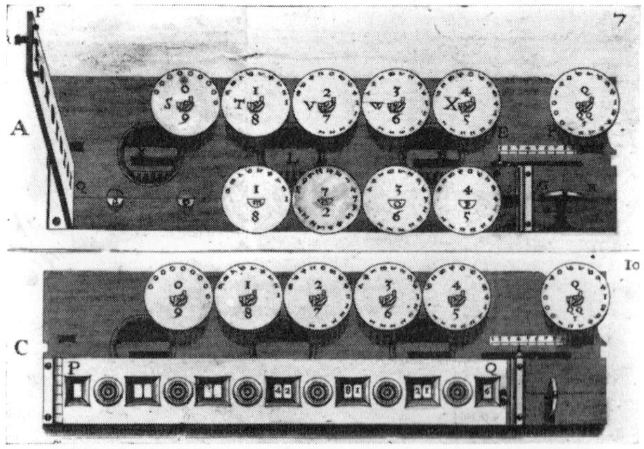

Fig. 2. Sir Samuel Morland mechanised Napier's 'bones'

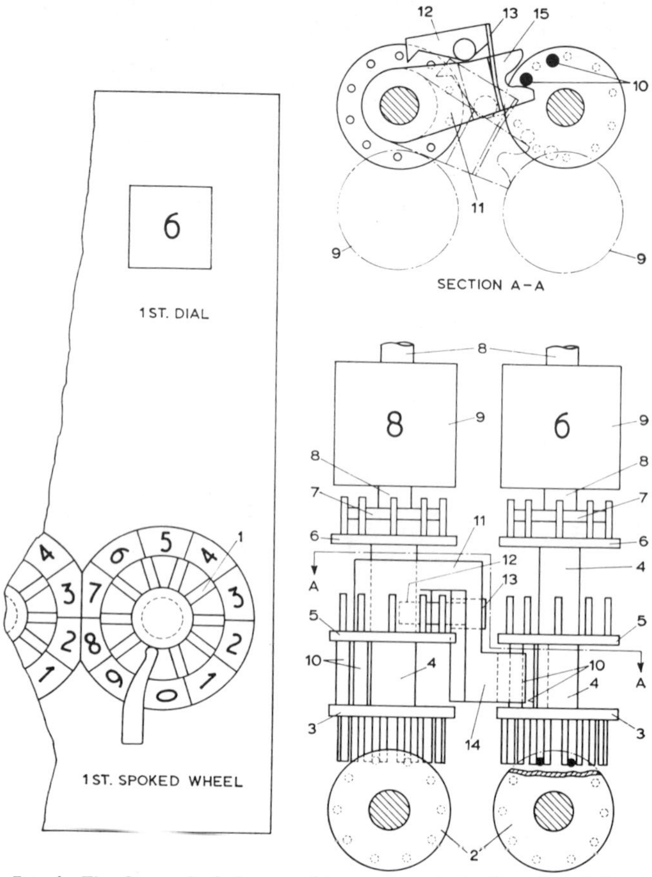

Fig. 1. The first calculating machine was made by Pascal to help his father, a tax gatherer, with his arithmetic

Calculating machines were not built by clockmakers but by country gentlemen, mathematicians and, later, bank clerks. The mechanisms employed, all based on gearing until quite recently, became increasingly ingenious, culminating in the printing tabulators of this century. Though the calculating is now largely done by electronics, the mechanical engineer's contributions to computing techniques are by no means finished.

Digital Calculating Machines

by J. Last (Member)

In view of the fact that a clock is essentially a counting instrument, it is a little surprising that the earliest steps towards an adding machine should have been taken by philosophers, mathematicians, a printer, at least two bank clerks and, latterly, machinists and motor-car-engineers. Perhaps this diversity of background, which, added to the early hazards attending intercommunication, is responsible for the absence, in the main, of any connected line of development or even similarity of approach in the first 170 years.

Adding by dials

For instance, in 1642 Blaise Pascal, a philosopher in several senses, provided for his father, a tax gatherer, a box having on its lid a row of eight ten-spoked wheels, each surrounded by a fixed outer ring, bearing the digits 0 to 9, arranged from right to left and rotated like a modern telephone dial. The internal arrangement is shown in Fig. 1, in which each selection wheel (1) drives a pin-type crown wheel (2), engaging a similar wheel (3). Wheel (5) on the same shaft (4) receives a 'carry' from the neighbouring mechanism to the right and (6) is engaged with a star wheel (7).

Thus the rotation of (1) is repeated, in the opposite sense, by (9) the unit digit of the result of an addition being shown as a figure on the dial.

There is a difference between the sequence of adding by the rotative movements of a wheel in a machine and that employed in manual calculations. In the latter case, each column is added by itself, probably necessitating a carry to the next column which may even run into two figures. A machine adds each row to the total of the rows already accumulated and a carry can never be greater than unity.

Pascal solved the carrying problem very simply by what is known as a 'stored-energy carry'. Between the wheels (3) and (5) and coupling them together are two pillars (10). Freely pivoted on each shaft (4), except the right-hand one, is a weighted lever (11) having pivoted on it a pawl (12), biased by the flat spring (13). The lever (11) terminates in weight (14) furnished with two approximately epicycloidal teeth (15), the whole forming a gravity actuator for the pawl.

As a dial reaches the position showing the figure '6' or thereabouts the pillars (10) begin to engage the teeth and, by the time number '9' is exhibited, the second pillar is supporting the weight as shown in the chain-dotted profile. At the same time the pawl (12) will have passed, as shown, the next pin in the wheel (5). As the dial moves to the '0' position, the weight is free to fall and to advance the wheel (5) with which the pawl is engaged by a one-pitch movement.

If suceeding dials to the left are also in the '9' position the advancing movement releases their associated weights successively, the work of raising these having already been done.

By this arrangement, Pascal avoided the possibility of 'over-carrying' due to jerky operation of the selector wheels. Gravity-loaded detents engaged the latter to define the resting positions of the dials. Each pin of wheel (2) is engaged between a pair of pins on (3), to minimise backlash, the crown wheel having ten such pairs for this purpose.

Fig. 3. A model of Leibnitz's machine, the first to incorporate stepped gears

Multiplication

The art of mechanical calculation then received a posthumous prod from John Napier, Baron Merchiston who, before Pascal, had not only invented the logarithm but embodied in his 'bones' the square multiplication table. These bones were mechanised in 1666 by Sir Samuel Morland, a landed gentleman, who proposed the machine shown in Fig. 2, which carried a store of factor digit wheels in the upper row from which wheels may be taken according to the digits of one factor and arranged in the appropriate juxtapositions in the lower row, where their centres engage pinions permantly in gear with a rack carrying a pointer. A digit of the second factor is now registered on the scale by moving the rack with the pointer, thus rotating each wheel through the same number of pitches.

As Napier's bones are numbered from 1 to 9 and each bears a succession of the products of its number and the digits 1 to 9, so each wheel bears near its rim the products of its number. But the 'tens' figure of each two-digit product is placed on the wheel diametrically opposite to the units figure.

The movement of the rack brings the partial products to the horizontal line through the centres of the wheels but the 'tens' figures appear adjacent to the units, to which they must be added. A plate over the wheels has windows to expose the product figures.

This notion of multiplication by the summation of partial products waited two and a quarter centuries before receiving a more complete mechanical embodiment.

After Morland appears Gottfried Wilhelm Leibnitz, another prolific philosopher, who designed, in 1672, a machine to effect multiplication by repeated addition. This machine, shown in Fig. 3, is in the Landesbibliothek in Hanover. Mechanically it is much more mature than that of Pascal.

Sixteen shafts, bearing dials for registering a product, are carried in a horizontal, fixed, frame and each carries a ten-toothed product pinion. In a lower frame are eight shafts, each bearing a nine-toothed gear, approximately as wide as nine product pinions, but its teeth are successively shortened, axially, from the rear end. The longest is nine pinion-widths long and the shortest one pinion width. The nine teeth only occupy a portion of the whole circumference and leave a blank zone at root radius. This is the much used Leibnitz stepped gear.

The eight stepped gears can engage with any eight adjacent pinions out of the sixteen product pinions because the lower frame can be slid transversely on the product shafts by means of the quick-thread screw. This provides for column-shift in order to multiply by ten before applying the next higher order digit.

Each stepped gear is adjustable to control the number of tooth pitches through which it will rotate a product pinion. Each shaft bearing a stepped gear protrudes in front of, and is engaged by, a lug which can be moved by a rack and pinion mechanism, ending in a flat dial. The eight setting dials are each engraved with the digits 0 to 9, one at a time visible through a window. They indicate multiplicand digits.

Each stepped gear shaft also carries at its front end a pinion permanently engaged with one or two of seven long, but not stepped, gears whose shafts are mounted in the lower frame. This forms the operating gear train of the machine.

When the multiplicand digits have been set on the flat dials, each multiplier digit is built up by rotating the operating handle to the left, the value of the digit being first registered by a peg placed in the divided plate at the right of the lower frame. The pointer is moved one hole pitch for each turn of the handle, and the peg limits the number of turns to the value of the required multiplier digit.

Each product shaft, in addition to the 10-toothed product pinion, has a single-tooth arm and a smaller, 10-toothed carry-pinion. Every two successive product shafts have a carry spindle between them. Each carry spindle has a five-pointed rear star wheel, serving the purpose of a 20-toothed gear with three teeth out of every four left out. This star wheel has a different axial position on successive carry spindles.

A rear star wheel can be engaged by a single tooth arm and

Fig. 4. The Earl of Stanhope used a spider mechanism for carrying on this machine which was able to divide as well as multiply

Fig. 6. The massive construction of Babbages' 'Engine' must have increased the cost

rotated one-tenth of a turn to the left when the product dial related to that arm is approaching or passing zero. This movement stores a carry and brings a left-hand point to the horizontal and to an engaging relation with the 10-toothed carry pinion to the left.

The shaft of the long gear under the carry spindle has a carry arm in an axial position to engage the front star wheel and when the stepped gear has completed its sweep past the product pinion, the carry arm turns the carry spindle through one-tenth of a turn, by engaging the front star wheel, if a carry has been stored. This movement causes the left-hand point of the rear star wheel to rotate its related carry pinion through one tooth pitch, advancing the product dial by one unit.

The machine is imperfect because the action just traced, if applied to a product shaft having its dial already standing at 9, will cause that shaft to store a carry but there is then no carry arm to pass it on and complete the carry. Recourse was had to pentagonal plates on the rear ends of the carry spindles, in such a relationship that when a carry had not been stored, an upper side of the pentagon could be horizontal. A stored but incomplete carry would cause a corner of a plate to be uppermost and the operator was expected to rotate the plate to the left by hand, so as to complete the carry. The necessity of completing carries resulting from carries was overlooked.

The important new features of this machine were the ability to add multi-digit numbers repeatedly without re-setting and to give column shift to the multiplicand when multiplying by ten.

One can only guess at the method of division. If performed by reverse rotation, the dividend being set by hand on the product dials and the divisor on the multiplicand dials, then the peg on the right-hand turn-counter or multiplier dial would be discarded and the extreme left-hand pentagon would be watched for the carry-stored position, indicating over-subtraction, the quotient digit being read on the multiplier dial. Leibnitz hinted at subtraction by complementary addition.

In 1703 Leibnitz wrote that 'in place of the progression by ten and ten I have used for several years the most simple of all progressions, which is by two and two'. He himself gave a Chinaman the credit for its invention but it was for the purpose of coding and not, as in the case today, to simplify mechanised computation.

However imperfect, Leibnitz's was the first multiplying machine actually made.

Carrying by spider

A century was to pass before another significant attack was made on the problem.

In 1775, Viscount Mahon, afterwards Earl of Stanhope, a most practically minded gentleman, devised a beautiful compact mechanism, made by James Bullock, which could multiply and divide correctly.

This machine was of the push-pull type. An upper frame could slide sideways in a guide frame, itself able to slide forwards and backwards.

Fig. 5. Thomas of Colmar's 'Arithmometer', a tidied up version of Leibnitz's machine, gained him the Legion of Honour

Fig. 7. This printing device was added to Babbage's engine at the suggestion of his son

The upper frame carried a number of parallel shafts in a fore and aft direction. Each shaft had a ten-toothed multiplicand setting wheel, latched by a spring detent, and a dial numbered from 0 to 9.

Behind the setting wheel on each shaft were nine racks arranged axially around a cylinder, the tenth position being vacant. These racks each had the number of teeth corresponding to the digit diametrically opposite to them on the dial. Thus, when '3' appeared at the top of the dial, a rack having that number of teeth would be undermost.

Extending laterally, lower than the upper frame and supported on the base, was a row of ten-toothed gears, also latched by detents. When the guide frame was pulled forward, these gears were rotated through as many pitches as there were teeth in the selected racks which engaged them. The guide frame moved forward far enough for the racks to clear the gears by some margin and, on being pushed to the back, the frame was lifted by rocking latches in the end guides and the racks moved back clear of the gears.

Each 10-toothed gear had at its left a product or dividend dial numbered from 0 to 9 and a single-tooth arm on the left of the dial. As the dial passed the 0 position, the arm wiped a toothed wheel below it through one pitch. This carry wheel bore a three-legged spider and, in the carry-stored position, one leg would be about to engage the next product gear to the left. Below each spider was an arm on a transverse shaft bearing a pinion at its right-hand end, the arms being arranged helically round the shaft.

A rack was bracketed to the guide frame below the pinion, pointing from front to rear. On the forward stroke of the frame in multiplying, the rack was clear of the pinion but the backward stroke brought the rack into mesh with the pinion and the carry shaft was rotated, successively engaging the spiders and rotating them through a quarter of a turn, completing all carries, whether stored, or themselves resulting from carries.

The movement of the latches was reversed for the purpose of division. The rotation of the carry shaft was also reversed and, in order that carries could progress in proper sequence, arms on the carry shaft were in two series, arranged in helices of opposite sense.

A further row of dials in front of the product dials served as stroke counters. In multiplication these were pre-set to the multiplier digits and the strokes counted back to zero. In division the quotient digits were added on by the back strokes. The upper frame slid sideways for multiplication.

Two years later, in 1777, Viscount Mahon produced another well thought-out design with a rotary action. The product dials, gears and carrying arrangements were much as before but the cylindrical rack groups had now been replaced by an array of relatively large multiplicand dials below the product gears on a common shaft. Fig. 4 shows this with top covering bars removed. The product wheels, counter wheels and carry spiders are all most effectively detent-controlled. The machine is a pleasure to operate, if rather slow to set up. The ingenuity of the sideways engagement of the multiplicand dials, in conjunction with their angular setting, to vary the digit transmitted to the product wheels, was used, in principle, for many later machines.

In 1810 Thomas, of Colmar in Alsace, thoroughly tidied up the Leibnitz machine and put the result, the *Arithmometer*, on the market. He became a Chevalier of the Legion of Honour for his trouble.

He seems to have decided that most things ought to be the other way round: the stepped gears do not slide, the multiplicand group stays still and the product group is given the column shift to the right, to multiply by ten.

A replica of this machine is shown in Fig. 5. When all dials have been set to zero, multiplier digits may be pre-set by small knobs above the front row of dials and will then, one by one, have their settings reduced to zero by successive operating cycles. In division, after setting all dials to zero, the rear dials are pre-set with the dividend digits and the quotient digits are built up from zero on the front dials.

To sum up progress so far, Pascal began the march of progress with his stored energy carry, Leibniz gave us his stepped gear and addition in parallel, Morland proposed multiplication by the summation of partial products, Mahon staggered his carries and reset his multipliers to zero: and Thomas de Colmar did the same for the Leibnitz machine and avoided reverse rotation in division.

The printing difference engine

At the time when the Chevalier Thomas was placing his *Arithmometer* before the world, a student had been entered at Trinity College, Cambridge, to further his knowledge of mathematics, among other things. The powerful method of differences engaged his attention and, before leaving Cambridge for Paris, Charles Babbage had realised the importance of a machine which could calculate tables, based on the method, by the simple process of addition. He seems to have displayed a curious blindness to what had already been achieved in adding and carrying. Apparently he started from scratch although, at one time, he actually owned the two Mahon machines.

The first Difference Engine was made between 1820 and 1822 by Babbage himself, with the help of a 'wheel cutter of Lambeth'. He then addressed himself to the problem of printing the output. A later Engine was begun in 1823 with the support of the Government.

In both these machines, carrying involved storage and progressive completion, as in the Mahon and the Thomas machines. The relatively massive construction must have aggravated the crippling cost. Fig. 6 shows how far the machine was built.

Babbage also visualised simultaneous carrying. This is a time-saver when successive additions are to be made.

Work on the Difference Engine continued until April 1833, after which the engineer, Clement, discharged the men employed, thereby releasing one J. Whitworth, who brought away with him the notion of fixed numbers of T.P.I.,

Fig. 8. Inspired by Babbage's work, Scheutz in Sweden developed the first practical printing calculator

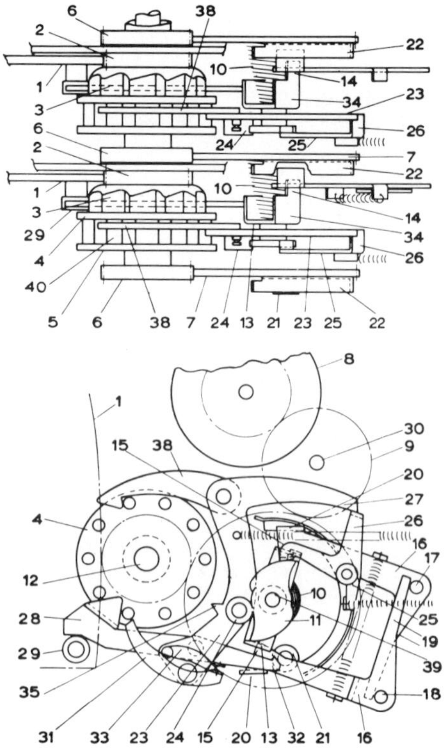

Fig. 10. The principal parts of a machine for plural column operation. For key see text

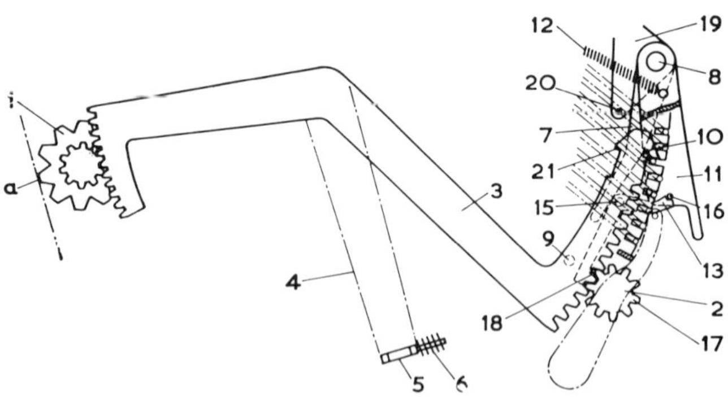

Fig. 9. The basic Burroughs mechanism. For key refer to the text

established about 1827, as well as the practice of gauging. A section of the Analytical Engine was constructed by R. W. Munroe, London, after 1880, at the instance of Major-General Henry P. Babbage, second son of Charles. This included a printing device, shown in Fig. 7.

Babbage seems to have added no mechanical device to those already known. He certainly did plan most of the essential functions in the modern computer, distinguishing the devices for operation control, variable control and number control, later to be applied through Jacquard card chains. He died in 1817, having produced no successful machine.

A description of this work inspired a printer of Stockholm to build the first printing difference engine. Mr Scheutz was not an engineer but obviously a clear thinker and, with his son, a graduate of the Stockholm Technological School, developed the machine shown in Fig. 8, after many setbacks and difficulties of finance, familar to all pioneers.

This machine provides for four orders of differences, expressed in fifteen digits. It was completed in 1854 and received the Gold Medal of the Great Paris Exposition, Napoleon III acquiring it for the Imperial Observatory. The dials exhibit the digits of the five quantities in horizontal rows and the function, as calculated, is set up in digit punches against which a plate of lead or sheet of papier-mâché is pressed to form a stereotype mould.

The carrying is stored by a small lever at each dial and is completed, sweepingly, by the pillars which are dragged by an endless chain. The embossing punches are selected by stepped cylindrical sectors, arrested by snail cams, mounted above the eight higher order function dials, the remaining seven being relied upon to derive an adequate 8th place digit. The snail cam method of selection was, of course, borrowed from the well established striking clock practice.

Accounting machines

The 1880's may fairly be credited with the birth of modern accounting machine practice, beginning with the struggles of William Seward Burroughs. Being much concerned with the errors of manual accounting, and having given up banking for engineering because of his tuberculosis, he decided to apply his new skills to making a machine which would not only add but would print what it added, and the totals arrived at. Joseph Boyer, of pneumatic hammer fame, furnished him with bench room.

Burroughs was pioneer of several devices which are commonplace today, such as full-bank latching keyboard, automatic noughts, over-run carry and dashpot control of the operating lever. He adopted a cycle in which swinging members, moving to an extent controlled by selected and latched digital keys, presented at the end of their outstroke a row of type which could print the digits selected by the keys.

The members carried gear sectors which could be engaged by 10-tooth pinions, at the completion of the outstroke (the machine was for decimal notation) so that, on the return of the sectors to their home position, the wheels could each be turned through a number of pitches corresponding to the value of the digit printed.

Fig. 9 shows the essentials of the basic mechanism. The outstroke has been completed, the type-wheel (1) presents the proper digit at *a* and the adding wheel (2) has been engaged with the sector (3). The swinging frame (4) has a crossbar (5), bridging all the sectors and a compression spring (6) for each one. On the return stroke, the frame acts through the spring to return each sector but the upper position of the sector is determined by the stop (7) which engages the stud (9). The arm (11) is loaded by the spring (12) but is retained in position by the latch (13), pivoted on the frame by the pivot stud (15).

The latch (13) is of bell-crank form, the upper arm stepped to engage the stud (16) in the arm (11) and the lower extending into the path of a stud (17) in the side of the adding wheel (2) of the next column to the right for unlatching the stop.

The stroke is sufficient to return the sectors to their home position and to compress the springs (6) through a further movement corresponding to one tooth pitch of the sectors.

The arrangement is self-staggering, as each sector can release its neighbour to the left in moving to the carry position. It is, in fact, the first stored-energy carry since Pascal's in 1642, the storage element being a spring. Many things have been modified since this first model but the principles have remained the same.

A little later two machinists, Dorr Eugene Felt and Robert Tarrant, set out to produce an adding machine in which the depression of a key would immediately complete an adding operation. Their first patent is dated 1888 and the machine again features stored-energy carrying.

The keys actuate sectors through angles proportionate to the values of the digits so that the '9' key produces the greatest throw. Fig. 10 gives the arrangement of the principal parts of the first machine designed for plural column operation, several auxiliary details being omitted. On the downward stroke of a key its related sector (1) rotates the wheel (2) and the internal ratchet (3) trails over a pawl on the disc (4) which is one wall of a lantern gear (40), freely pivoted at (12) having ten bars. The other wall (5) carries a 20-toothed wheel (6) which engages a 40-toothed wheel (7), freely pivoted at (39). Thus the depression of a key and its release upwards causes first the idle rotation of the wheel (3) and then its reversal, to rotate the other wheels, and finally the dial, through the number of pitches which corresponds to the digit on the key. The return of the sector ends with its roller (29) raising the lock pawl (28) to engage the lantern wheel and prevent its overspin.

This model appeared about 1918 and is still familiar. It is produced by the Bell Punch Co.

A machine which is in a class by itself is the *Logabox*, in which relatively large numbers of balances can be stored on trays of registering racks. It was originally invented in 1923 by another bank clerk, Francisco Campos. Since that day much development has taken place and even the original adding wheels have disappeared, the adding being performed on the registering racks themselves.

Partial products again

The embodiment in practice of a principle first incorporated more than two centuries earlier by Samuell Morland was achieved in 1889 by Leon Bollée, whose name was later associated with a three-wheeled motor car. Effecting multiplication by the summation of partial products, he used, in place of Morland's little dials, a plate having pegs of which the heights were proportional to the value of all the digits in the table. The partial product stores, one for each multiplicand digit, are adjusted by handles which engage notched plates. The entire calculator is moveable sideways, by means of a crank handle for selecting multiplier digit racks in the stores and, by means of whole rotations, to shift from column to column of the product. A single rotation of the main shaft effects complete multiplication of the multiplicand by one multiplier digit.

After the stores have been adjusted by the handles according to the digits of the multiplicand, the multiplier digits are pre-set on dials by means of knobs. The machine cycle is in four phases, each of 90 degrees.

This machine was exhibited at the Paris Exposition of 1889 and provided for twenty product digits.

In 1893 Otto Steiger of Munich offered a choice of four alternatives to Bollée's calculating plate. These were: rotatable stepped discs, slidable stepped blocks, stop pins in racks and profile plates attached to racks; each type in two separate sets, one for tens and one for units digits of the partial products. His machine reached the market, through Hans Egli of Zurich, as the *Millionaire* calculating machine, beginning with a simple hand-driven model. By that time, however, the Bollée store had replaced the Steiger devices and effected much simplification. Steiger's real contribution was in getting the result with only one store instead of one for each digit of the multiplicand. The Bollée store was used to control the strokes of the adding racks in accordance with the multiplier digit selected by adjusting the level of the store, this being the equivalent of Bollée selection.

The racks were applied to the pegs of the store by the movement of the frame, twice in each machine cycle, the store being shifted lengthways between strokes of the racks so as to add the tens digits in the first stroke and then the units digits in the second stroke. The occurrence of the same multiplicand digit in several columns (the setting of the store being common to them all) transfers to all the related shafts the indication of the single store for the particular factors concerned. Only one store is therefore required.

The movements of the shafts are transferred to the register through pin-and-disc clutches, the dials being turned by one or other of a pair of bevel gears, just as in the Thomas machine, but the mode of carrying is much more elaborate.

The sequence of the arithmetical operations is as follows: In multiplying 23 by 67 the products by the left-hand multiplier digit are $6 \times 2 = 12$, $6 \times 3 = 18$.

Set in the multiplicand keyboard	23	
enter tens digits	11	from the store, same columns.
column shift	110	
add units	28	from the store.
	138	

The products by the next multiplier digit are $7 \times 2 = 14$, $7 \times 3 = 21$

enter tens digits	12	from the store.
	150	
column shift	1500	
add units	41	from the store.
Product	1541	

The machine was convenient and fast, the later models having more fully developed keyboards and being motor-driven. It has not been available for some years.

Combination with typewriter

Towards the end of 1903, Halcom Ellis, of Newark, N.J., proceeded to put an adding system behind a typewriter; whereas Hubert Hopkins of St Louis, put a typewriter in front of an adding and listing machine which he had designed and which eventually became the Dalton adding machine with a 10-key keyboard and sectors for adding and printing.

Ellis preferred a full-bank keyboard and his machine became the National accounting machine in 1928. The adding wheels are driven by straight racks, the stroke of which is limited by the engagement of studs on the keys with steps in the upper edge of each rack. As in the Burroughs machine, adding is done on the return of the rack to the front of the machine but carrying is not left to the rack. This is a printing or listing machine, so each rack has a slot which engages a roller on a stud in a sector and this sector is coupled to the typehead. The printing blow is given by swinging the typehead bodily up to the platen on the arm.

The notches on the rim of the sector are engaged by a bar which swings on a shaft after the rack is arrested, in order to align the type selected accurately. The pin-and-slot connection between the sector and the rack enables the bar to restore both, after moving the rack rearwardly, as far as the selected key allows, through the springs, coupled to the sector and the bar.

In adding, the wheels are engaged with the rack at the end of the rearward stroke, the forward stroke, giving the proper rotation to the wheels.

Each wheel has 20 teeth, in the decimal field, and has two carry toes, one of which will displace a pawl in reaching or passing zero. Each pawl has a lug which, when undisturbed, retains the heel of a carry cam. The displacement of a pawl releases the cam, the spring on which causes a small movement, bringing the heel up to the bar which bridges the accumulator. The cam slot engages a stud extending from a carry sector in the next column to the left.

When the return stroke is completed, the wheels are disengaged from the racks and engaged with the sectors. The bar is then moved away from the heel, the resulting swing of any released cam causing the remainder of its slot to swing the next sector to the left through one-pitch. The action may cause further releases of cams providing the essential carry sweep.

Where an accumulator is required for balance purposes it must subtract as well as add. The wheels are then engaged before the rearward stroke and are reversed in rotation. The slot, in this case, is Y-shaped and the sector is biased by a spring hooked to a movable bail to enter the arm of the Y appropriate to the necessary direction of swing of the sector. Two carry pawls, are employed and a shift bar takes one or the other of them out of action.

Negative balance

An accounting machine in which subtraction may be done is sure to register a total in complementary form sooner or later, thus:

	£237	15	10
subtract	345	9	7
	9999892	6	3

To be useful, this should be converted to the true negative balance form. In the National machine it was customary to add this complementary sum into a vacant accumulator, clearing the subtraction accumulator, and then to transfer it back, subtracting it from zero, thus:

	0000000	0	0
—	9999892	6	3
	0000107	13	9

The four noughts would, of course, be suppressed, automatically.

The operator was responsible for this conversion but it will be noted that, in the first operation, subtraction carries passed to the left-hand column. A dummy carry cam is provided here and causes the raising of a two-colour inking ribbon and the subsequent printing of the balance in red. For several years now the machine has taken care of the whole operation.

A very important provision is the shifting carriage with a variable stop bar, automatically controlling operations according to the field presented to the numerical printing head. A normal typewriter keyboard is situated in front of the full bank keyboard, the type bar array being much as usual, but gathered to one side to make room for the numerical head. With its many accumulators and flexible cross-adding provision, the machine has been used very successfully as a printing difference engine.

In 1908 Hopkins devised a multiplying machine incorporating a typewriter but using straight adding racks carrying typeheads and operating two accumulators, rather similar to Ellis's. These racks could be directly key-controlled through the shifting stop basket of the former machine or by set of stepped cams which were lifted to various heights through plungers shifted bodily to register multiplicand digits. The plungers were lifted by crossbars, raised to product digit heights by profile cams in pairs, used successively. Each pair represented the tens and units value from a Bollée store and was coupled to a separate row of multiplier keys which raised them through a constant height.

As usual, two adding cycles are required for each multiplier digit, one of the accumulators, devoted to adding partial products, having provision for column shift. This machine became the well-known Moon-Hopkins billing machine and is the lone survivor incorporating Samuel Morland's principle.

The more generally used multiplying machines today are those based either on the Chevalier Thomas' machine of 1810 or on that of Viscount Mahon of 1777, this latter type finding revival in 1890 in the machine of Willgodt Odhner of St Petersburg, Russia, which also came to be known as the Brunsviga and the Muldivo. The counting teeth A are caused to protrude by a disc shaped plate in each column having a circular cam slot in two parts of different radii, with a sloping step between them. The plate is adjusted angularly by a projecting lug and is held in angular relation to the counting shaft by a series of detent notches and a spring plunger. A set of laterally displaceable carry teeth in the zone clear of the counting teeth to give effect to carries which are stored by setting a cam lever into the path of the carry teeth.

Both the Thomas and the Odhner type have many representatives today, differing mainly in the auxiliary control for such purposes as automatic long division, square root, storage of quotients and various registering devices.

Punched cards

We now have to go back in history, to the mechanical device invented by Joseph Marie Jacquard early in the 19th Century to help him weave multi-coloured textiles. Charles

Babbage thought of using it for calculating machines but it remained for Dr Herman Hollerith of New York to apply it in the compilation of the U.S. census of 1890. He adopted Jacquard's card but kept it separate, instead of linked in a chain. The actual punching seems to have had little attention.

His first apparatus contained spring-loaded sensing plungers, any one of which, on passing through a hole in the card, dipped into an individual cup of mercury. The cups were connected, variably to a prototype of the present plugboard, coupled to an array of unit counters and to a group of 22 sorting boxes, each having a lid which would open on receiving an impulse from a chosen field in the sensing press. After being sensed, the card was removed by hand and placed in whichever box had its lid open. In the simplicity of this sorting technique there is a touch of irony for those of us who have had to deal with modern high-speed card sorters.

In 1894 Hollerith designed the digital card with '0' to '9' in columns and in 1908 the accumulator which is still in use. This was designed to be controlled by punched cards in which the significant holes are sensed by contact brushes while the card is in motion. This method makes the card a part of the machine for the sensing period, a type of commutator, in fact.

The accumulator dials are driven through gears from a continuosly running shaft carrying sliding dog clutches. Each of these can be engaged by means of a magnet when the passage of a hole past a brush allows an existing current to pass. The clutches are all disengaged mechanically on the passage of the zero point of the card by the sensing line. Thus the value added to the reading on each wheel depends on the distance of the hole in the card in advance of zero.

The dials have a ratchet wheel on their left and this has a long tooth. When the wheel passes, from the '9' to the '0' position, a pawl is lifted by the long tooth beyond the usual movement given in passing the other ratchet teeth and a stud lifts the curved end of the latch of the adjacent lever to the left, allowing the pawl on that lever to be drawn back through one pitch and thus storing a carry.

Hollerith thought and claimed that he had invented a simultaneous carry mechanism but all he had done was to provide for applying simultaneously the carries stored in a sweep. Babbage already understood and explained the necessity for detecting when the '9' position was arrived at, in order to provide for simultaneous carrying.

In about 1925 a type of commutator was geared to each dial for the purpose of reading out electrically. The Hollerith accumulator has given very good service for many years and was the calculating element in Dr Aiken's first Harvard digital computer which produced some very important tables.

Card feed or sorting

It is difficult to say when cards were first fed from a magazine by a picker knife but it was an established practice when James Powers broke into the punched card field by becoming involved in the U.S. Census of 1910. At this time, in addition to providing a magazine-fed punching machine, he put three Dalton adding machine units, minus keyboards, on a base plate and, having fixed stop baskets under the Dalton sectors, then put an inverted form of Hollerith's plunger sensing head underneath a pair of pin plates between which a card could lie while being sensed. The sensing head on being raised with any plungers that came through a card, lifted those selected from a set of push wires, controlling the Dalton units directly from the card. The cards were delivered to the pin plates by a pair of feed rolls from a magazine.

In the first machine the wires were supported in the frame which carried the Dalton units above the pinplates, but soon they were held in cast aluminium frames detachable from the machine. These units were then treated as changeable members and the connection box frequently had its wires interlaced in various ways to provide special layout programmes. Powers' method was to punch the entire card in one blow, following set-up on a keyboard in which corrections could be made before punching. His tabulator has been developed enormously and has printed alphabetic characters since 1925. The Hopkins type of sector is still in use.

To complete the story of calculating mechanisms, reference may be made to several circular designs such as those by Braun 1727, Muller 1746 and Hahn 1789, that of Braun being made for Charles VI. It took a modern designer to roll up the Chevalier Thomas' product so that it would go into a half-pint tumbler. This machine is the little *Curtor*. The stepped gear drum has both direct and complementary gear segments, the rotation not being reversible. As there is only one stepped gear, there could only be one equivalent for Thomas' carry arms so the obvious alternative was to store carries by shifting the carry pinions and this was done.

Conclusion

After a while, the development of calculating machines began to progress in step with the advance in mechanical engineering techniques. More competent machining, better gear cutting, the growth of press-tool practice, all tended, to make construction of mechanical calculators easier, and permitted greater freedom in design.

The pioneers did their best with what they had, their sole machine tool being probably a treadle lathe. What they produced were mechanical statements of the same basic functions which, today, we design into our punched card tabulators, computers and other modern equipment.

By means of electronics we have now learnt to make parts operate in milliseconds or less, to keep in step with electronics, but the logic behind the operations has not greatly changed, and mechanical devices of one sort or another are still necessary. The last has not been heard of mechanical calculation, by any means.

Acknowledgment

I am very grateful to Dr H. R. Calvert of the Science Museum, London, for information about, and access to, many of the early machines. The photographs are all reproduced by courtesy of the Director of the Science Museum.

THE GREAT MASTERS

Designer and maker of the first British motor car and one of the early scientists of flight, Lanchester was a portent of the modern world. His discoveries and inventions were a foundation of the motor and aircraft industries of today.

Photos taken in 1920—aged 52;

F. W. Lanchester—Automobile Engineer

by P. W. Kingsford*

When Lanchester died in 1946, the Royal Society paid its customary tribute to a Fellow which included this passage:

> In him we had one of the very rare combinations of a great scientist, a great engineer, a mathematician, an inventor and a true artist in mechanical design . . . He was also a poet and a philosopher.

Education

Frederick William Lanchester was born on 28th October 1868 in the village of Lewisham, the fourth of eight children and the second of five sons of Henry Lanchester, an architect, and Octavia Ward. On his father's side his ancestry went back to the well-known Devis family of painters, and his mother's people were London silversmiths. His childhood, spent at Hove where his father was designing a new housing estate, seems to have been a free and a happy one in the security of a large and lively family and a comfortable, though never luxurious, standard of living.

It was a happy, lively family. The children had a good deal of liberty but were taught that, unless individuality was coupled with a disciplined mind, it resulted in lack of purpose or eccentricity.

His elder brother followed his father's profession and became a national and international figure; his elder sister was a teacher, painter and writer.

Frederick's education, begun at his mother's knee (she herself had been a teacher of Latin and Mathematics), continued at a kindergarten school at the age of six and to a preparatory boarding school; thence he went to the Southampton Hartley Institution and finally the Royal College of Science at South Kensington.

The kindergarten was an enlightened, progressive school where freedom of development was emphasised. He was a boarder for four years at the preparatory school but, as it was in Brighton, he was never far from home. At this school also the teaching was good, the classes small, and Lanchester quickly achieved facility in mathematics, though English was a burden to him, then and later. There he also acquired a life-long distaste for games.

As he had shown a distinct aptitude for science, his parents agreed to send him, at the age of 14, to the Hartley Institution, then the foremost place of its kind in the South of England. Much later, after many changes, it became the University of Southampton. Again he was fortunate, since pupils were given much time for private study, there was a good library and the studies ranged widely from metallurgy to physiology. He worked hard for the two years and won a national scholarship to South Kensington. In Southampton he made friends in a shipbuilding firm and at the gas works, and gained his first experience of workshops.

The years 1885 to 1888 at the Normal College of Science (later the Royal College of Science) and the Royal School of Mines were tremendously formative in different ways. He came under the influence of outstanding men such as Thomas Goodeve, Frederick Guthrie and C. V. Boys, the physicist. From Goodeve he had the lectures on the gas engine which led him to go into that field of work. But the College course he found inadequate because it contained no practical engineering and this had two important consequences. He supplemented the course by attending evening classes in workshop practice at Finsbury Technical College and became, in his own words, "an average good tool maker." But he also became so discontented with his studies that he gave them up.

Why was he an academic failure? In the final year he had chosen the wrong subject, mining rather than mechanical engineering. On the one hand, he thought he saw a lucrative career in mining (and money was scarce) while, on the other, the mechanical engineering course did not contain the practical work which he wanted. But even the mining course turned out to be too academic.

Early inventions

Instead of taking the finals, he occupied his time in other ways. He spent long hours in private study, absorbing Newton's *Principia* and the works of W. J. Macquorn Rankine. It was these which led him to study flight. On the other hand, while still a student, he made his first two inventions, precursors of many to follow. These were the radial cursor, a calculating device on the slide rule; and the pen-

* *Head of Dept of Business and Social Studies, Hatfield College of Technology.*

and about 1945

dulum accelerometer, which he later developed for use on the motor car.

It is impossible to say what influences really led Lanchester to become an engineer; as a child, one never-to-be-forgotten day, he had looked out of his nursery window at Hove and seen the *Great Eastern* steaming down the Channel. He had certainly intended to be an engineer at the early age of 12 or 13. But the most likely reason is that his father had several inventions to his credit, including a hollow brick for house-building and a safety bicycle.

At first he had difficulty in finding a job, and was obliged to take work at 6d an hour for a hack draughtsman who drew up specifications for would-be inventors. But this turned out to have advantages, for he gained a knowledge of patent procedures and, when he designed a new draughtsman's instrument, he was able to draw up his own specification, make samples himself, and sell the patent. However, his first real opportunity came when, through an uncle, he obtained introduction to the owner of the Forward Gas Engine Company, a small but progressive firm in Birmingham, employing some 60 men. He was appointed assistant works manager at a salary of £1 per week. It was typical that, although he accepted this wage because of the experience to be gained, he insisted on deleting from the agreement a clause that any of his improvements would become the Company's property.

Gas engines and aerodynamics

Lanchester became works manager at the age of 21 and soon after had his younger brother George apprenticed to the Company, paying the premium himself. George became his eminent assistant and colleague throughout. Much of the Company's work was servicing and modifying gas engines and much of Lanchester's efforts were to obtain gas economy; eventually he designed an engine which created an economy record.

He invented two other improvements in 1889 and 1890. His pendulum governor earned him royalties from his company, and a development of it was later used in his cars. His engine starter not only removed the hazard of starting by hand but also earned him royalties of at least £3000. In negotiation with Crossley Brothers he showed he did not underrate his work by asking £10 000 for the patent.

As works manager he showed his innovating zeal by removing the customary secrecy which surrounded production. Charles Linford, his predecessor, had kept the essential details in his private notebooks for fear of information leaking to competitors, but Lanchester made new drawings of all the engines in production, and these showed the full details needed in the shop. Confident in his ability as designer, he set his face against the mystery and secrecy which was prevalent at that time.

At this point in his life Lanchester's future was poised between the motor car and the aeroplane. He had a vision of air transport as well as road transport. He designed and built both the engine and the hull of the first British motor boat, with the purpose of obtaining financial backing for a motor car. The flat-bottomed river craft with stern paddle wheel was driven by a vertical single cylinder engine with his wick carburetter.

But his prime interest at this point was the flying machine, and from 1892 onwards he carried out extensive experiments with model gliders in his garden to investigate the problems of lift and drag. Mathematical theories of fluid motion had already been far developed in the 19th century but they had not explained the lift of a wing and its dependence on the relative motion of air over it.

Lanchester now took this step and, with characteristic boldness, imagination and insight into physics, associated the lift of an aerofoil with the circulation of air. He postulated the existence of vortices at the wing tips which continued downstream as trailing vortices. By this he laid the foundations of modern aerodynamic theory.

He explained his vortex theory of lift in a lecture to the Birmingham Natural History and Philosophical Society in 1894 but, when a paper was offered to the Physical Society in 1897, it was rejected. Lanchester, deeply disappointed, filed it away, though taking out a patent as protection.

It was clear that the flying machine would need an engine far lighter than anything yet made and he felt confident of being able to design one. But his friend Dugald Clerk told him:

> Lanchester, you may be right, but you are a young man and must consider your reputation. If you were to put forward such a proposition seriously you would be regarded as a crazy inventor and your reputation as a sane engineer would be ruined.

Lanchester therefore turned to the motor car, though he regarded it merely as a preliminary to tackling the flying machine.

His first cars

A legacy of £1000, his royalties, and money from a small syndicate of friends enabled him to make his experimental

Aeronautics was Lanchester's first love and he wrote one of the early papers on the subject; here he is with an elastic-propelled model designed to test his theories

cars. The first was made in a workshop adjoining the premises of the Forward Gas Engine Co. and the second and third in somewhat larger accommodation, a galvanised iron building in Ladywood Road.

Two younger brothers joined him. George came as personal assistant after a year or two and remained in partnership with him until the demise of the Lanchester Motor Co. in 1931. Frank joined the first company, formed in 1899, and was invaluable in sales and promotion. When, after five years, the experimental cars had been completed, Lanchester had spent about £1400 of his own money, out of a total expenditure of over £4000.

Birmingham offered skilled labour, industrial know-how and progressive government, so that he had good prospects for producing what was to be a new type of vehicle, not merely a horseless carriage. In Germany and France cars were being made to the designs of Daimler and Benz but in a Britain hampered by unfavourable road legislation, there had been practically no development of the motor car. In 1894, when Lanchester was designing his first car, the French were well ahead and even in the Automobile Club's reliability trials of 1899 Lanchester's seems to have been the only British entry.

Thorough in everything, he inspected all he could find abroad and was not much impressed. He meant to eliminate the appalling vibration of all the cars then in production and he thought out his design almost from scratch. This meant making almost every part himself.

The first experimental car, which took the road in February 1896, was not only a complete innovation but also introduced and combined, for the first time, many essential features of the modern car. Outstanding were its torsionally rigid frame, a perfectly balanced and almost noiseless engine, dynamically stable steering, and epicyclic gear for low speed and reverse, with direct drive for high speed. Brilliantly designed though it was, this car was underpowered and Lanchester immediately designed not only a new engine but a new transmission system.

In the reconstructed model, changes affecting the whole future of the car were made. The two-cylinder, 8-hp flat twin engine retained the principle of balancing by reverse rotation; a new Lanchester valve gear and a new Lanchester ignition method, based on a magneto generator, were introduced. Equally new was his own worm and wheel gear, used for the first time instead of chain transmission or bevel gear, and the gear cutting machines he had to design to be able to make it.

This car, which had a maximum speed of about 18 mile/h, was the prototype of all future Lanchester cars. By 1898 he had completed a second, similar in most respects, but with nearly twice the maximum speed. Christened the *Gold Medal Phaeton*, it was a brilliant success in the official trials, and eventually found its proper home in the Science Museum.

The next stage in Lanchester's work, from 1899 to 1904, was production of cars for the market, and here again he developed new methods which foreshadowed the future.

Failure in management

At the same time his unwillingness or inability to compromise or collaborate with business or business men became apparent. When the Lanchester Engine Co. was formed in 1899 he was for some time not only general manager but also chief designer and works manager. Added to this, the Company was under-capitalised for the task of producing a quality car and harsh disagreements between the finance and design functions soon arose.

Lanchester himself was not always the easiest of men to deal with; although kind and generous, he could never suffer fools gladly, and his humour is exemplified by a remark about his directors: "They seem to change their minds pretty frequently, but if I had a mind like theirs I should change it as soon as I could."

He insisted on production being based on a system of fully interchangeable standard parts which included even the bodywork, an extraordinary achievement in those days of coach builders. The system, which was well ahead—perhaps too far ahead—of its time, involved him in devising completely new gauges and tolerances. At the same time, after trouble with Whitworth nuts and bolts, he evolved his 'M' thread which anticipated the British Standard thread by nearly ten years.

> The difficulties of management, *he wrote*, were very great, partly owing to the fact that no ancillary trades had then developed and we had to make everything ourselves, chassis, magneto, wheels, bodywork, etc, everything except the tyres.

Although the 10 hp and 12 hp cars, engineering masterpieces of their kind, had been sold in considerable numbers, the Company went bankrupt and Lanchester relinquished all responsibility except that of chief designer which he retained until 1914. He lost a good deal of money but he could now return to his true *métier* of engineer-artist and also return to aeronautical science; the time for this was ripe.

[Courtesy: City Museum, Birmingham]

The Lanchester 10/12 hp car of 1900 was designed as a motor vehicle, rather than a horseless carriage

Lanchester in 1898 (left), driving an 8 hp Lanchester Gold Medal Phaeton, with his brother George as passenger

'Dyott End', Moseley, was designed by Lanchester himself

Aerial flight'

A great deal had happened in the world of flight during the 10 years since Lanchester had discovered the vortex theory. Kutta and Jowkowski had made important theoretical contributions but actual flight had so far been essentially empirical. Now was the time for scientific aeronautics.

In 1907 and 1908 Lanchester published his first theoretical treatise *Aerial Flight* in two massive volumes, the fruit of exhaustive experiment and analysis over a period of more than thirteen years. In the first volume he proved the importance of skin friction and arrived at the elements of boundary layer theory, as well as re-stating more fully the vortex theory. In Volume 2, he established the basis of a theory of stability.

Although these books went into several editions, so novel were his theories that they were not accepted or even understood in England until after 1920, when the more comprehensive studies carried out in Germany by Ludwig Prandtl became known. No doubt this was mainly due to opposition at the National Physical Laboratory which prevented research in the desired direction, but Lanchester's own impatience with minds slower than his own must have had its effect. His deeply felt resentment of this neglect was somewhat lessened by the ready understanding which his work received in Germany and while his books were being translated he became a close friend of the physicist Carl Runge at Göttingen University. Runge's daughter described him as he was at the height of his powers, about 40 years old:

> Lanchester stood in odd contrast with the English scholars whom Runge had met hitherto. In the first place, he was certainly not an academic type. Heavily and strongly built, with a large head, yet he was quick of movement and also had something of boyish ungainliness in his appearance. There was a boyish freshness and vigour in his attitude to everything in his own particular field of interest. He thought clearly and incisively and, when he felt he was understood, he used a humorous manner of speech, often accompanied by paradoxical remarks and flashes of wit.

Consultant

His main work was now done but many years of achievement lay ahead. He had done important pioneer work for the two principal modes of transport of the future; and in 1909 he became a member of the Government's new Advisory Committee for Aeronautics and also consultant to the Daimler Company; thus he continued his work in both fields. To the Committee he contributed reports and memoranda; he enjoyed this work among his peers for 12 years. In the other field he made two important inventions. His damper for crankshaft vibration (1910) proved completely successful, according to Sir Harry Ricardo, in eliminating torsional vibration and thus laid the ghost which hitherto had haunted the 6-cylinder engine. His harmonic balancer in the next year, his last direct contribution to Lanchester cars, dealt with vibration in the 4-cylinder car by an extension of his old principle of balancing by reverse rotation. Both these inventions brought him substantial sums.

Lanchester's double pre-eminence during these years was exemplified by two other events. As president of the young and vigorous Institution of Automobile Engineers* he gave his address on factors that had contributed to the advance of the automobile and his James Forrest lecture to the Institution of Civil Engineers was a masterly survey of the states both of scientific knowledge and of aeroplane construction.

During the 1914–18 war Lanchester devoted all his notable energy to helping the national effort and had a considerable effect on air policy, then in a painful process of formulation. His influence was exerted, both as an expert and, in a new role for him, as a populariser.

He became embroiled in the controversy about defence against air attack and, to some extent, in politics. He even spoke for the Government candidate at a sensational by-election. Conservative at heart, he regarded politics as a dirty game but felt too strongly about the need for a new air arm to keep away.

Ten years earlier he had prophesied the importance of air forces, and now in his book, *Aircraft in Warfare*, he explored the possible tactical and strategic uses of the aeroplane. In it he put forward his N-square law that, under certain conditions, the fighting strength of a force is proportional to the square of its numerical strength, multiplied by the fighting value of its individual units. This application of scientific method to military problems became recognised only during the second world war as a contribution to operational research.

Golden afternoon

The war over, Lanchester enjoyed the golden afternoon of his life. Now living in London, he was able to indulge his life-long passion for music by opera going and musical evenings. Many years before he had had his voice trained and singing had become one of his main pleasures, particularly the *Lieder* of Schubert, Brahms and even Hugo Wolf. Music expressed his gusto in living.

Visitors would find him hard at work on the pianola at a Beethoven sonata and, characteristically, he invented an improved control for this instrument.

His other great pleasure was deep sea sailing, not for the thrill of racing, but for the skill of navigation and the relaxation from care. He made many trips across the Channel in his cutters *Iseult* and *Troubadour*. He preferred casual holidays and the old clothes which went with them; though he was inclined to take a drawing board away with him!

Added to these pleasures was the long-delayed recognition in aeronautics; the Lanchester–Prandtl theory of lift and drag was now an accepted term. The Royal Society elected him a Fellow (although a doctorate from Birmingham gave him more pleasure) and the Royal Aeronautical Society gave its gold medal. To crown all, the bachelor of 51 married.

He had first met Dorothea Cooper ten years earlier when her uncle was teaching him to sing. They became engaged

* *merged with this Institution in 1947,* Ed. CME

early in 1919, while Lanchester was in the Isle of Wight, fitting out his yacht, and were married in the same year at Field Broughton, where the bride's father was vicar.

They made their first home at 41 Bedford Square, and Lanchester took much pleasure in furnishing the spacious Georgian house, which was much to his liking and which made an elegant home for his collections of Chinese porcelain and Japanese prints; as well as for his two grand pianos.

Home, family and friends always meant much to him. Although, or perhaps because, they had no children of their own, the Lanchesters found much joy in their nephews and nieces and, when these had grown up, in their children.

The rest of his life was spent in Birmingham where he returned in 1924, having designed himself a new house which was to be his last home. For several years he continued as consultant to Daimler and he also undertook work for Wolseley and Beardmore.

However, relations with Daimler became difficult; Lanchester's desire for research and experiment conflicted with Daimler's interest in production and financial economy; and he was never discreet in controversy. In 1929 he found himself without employment, and faced with the task of launching what was to be his last venture, Lanchester's Laboratories Ltd.

Music and poetry

This was a venture in quite a new direction for, linking with his interest in music, it specialised in musical reproduction and quality radio products, such as moving-coil loud speakers, based on his own patents. Now over 60 years old, he struggled for success for four more years but the severe economic depression, shortage of capital and overwork defeated him. When he was struck down by a serious illness, the little company closed down.

In 1919, at the age of 51, he became engaged to his future wife. Here they are, shortly before their wedding, in her father's garden at Field Broughton, examining a wedding present: a complete set of Shakespeare

[Courtesy: Mrs Lanchester]

During his enforced idleness he wrote, and published under a pseudonym, a book of verse, an art which had always interested him. It was a long ballad, in a strict metrical and rhyming scheme, called *The Centenarian*, a Lakeland story in which some of his own experiences may be read.

After his illness, though physically weakened, his intellect remained strong as ever. Between the two world wars he wrote several more books: one on relativity and another on the theory of dimensions; he hoped to make some money with them, but without success. He continued to contribute important papers to the professional bodies and in his 70th year gave the Thomas Hawksley Lecture to this Institution. He also attempted more poems, one of which, *A King's Prayer*, attacked the prevailing mood of appeasement. The following two extracts are from the introduction to a reprint:

> Versatility is not regarded with favour by the British public ('Jack of all Trades', etc.). So, when writing my Lakeland Story, I attributed the authorship to a 'ghost', Paul Netherton Herries, a fitting nom-de-plume, and one that has proved useful on other occasions.

> 'A King's Prayer' originated as a protest against the suicidal policy which prevailed prior to the year 1936, and which culminated in the Abyssinian imbroglio and the shameful Hoare-Laval Treaty. Two thousand copies were printed and distributed broadcast to Members of both Houses of Parliament and others. The effect (if any) on our National Policy is, and must remain, an unknown quantity, but certain it is that a change came about, followed by a period of re-armament.

But money worries pressed, he could barely make ends meet and he was afraid that the bank would foreclose on his house. The grand old man of the motor car and of aeronautics could not himself afford a car or an air fare. "It is an amazing experience", he wrote, "to find that, without any loss of my faculties, the world has no use for me."

In quiet retirement during the 1939–45 war, struck now by Parkinson's disease and increasingly afflicted by blindness, he spent his remaining time in writing his *Musical Scale*, a work of great technical complexity: and in gardening, in aid of which he designed his own two-wheeled barrow.

The last honour he received, the James Watt Medal, gave him more pleasure than all the others.

He died on March 8th, 1946. In spite of disappointments, had enjoyed life and his great achievements. In one of his poems occur the lines:

> His soul was in his work which, living on, bears fruit:
> So doth a sou attain to immortality.

[Courtesy: Imperial War Museum]

The S.E.5a fighter first saw action in 1917. Made by the Royal Aircraft Factory, Farnborough, it was powered by a 150 hp Hispano-Suiza engine and carried one Vickers machine gun, synchronised to fire through the airscrew. *An article on the evolution of the aeroplane begins on page 160.*

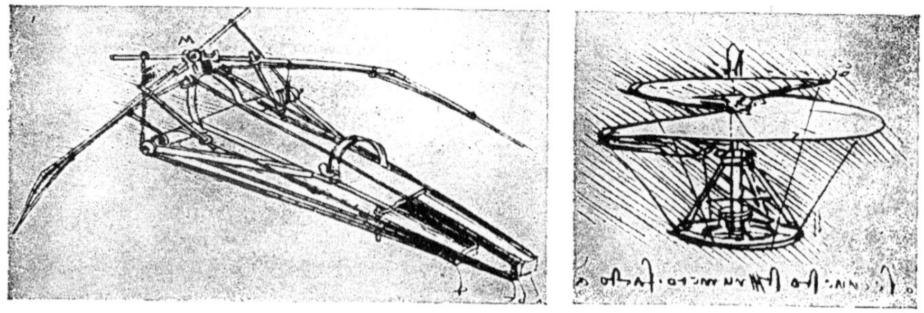

Fig. 1. Two of Leonardo's many aircraft sketches: orniothopter (left) and a rudimentry helicopter

The Development of the Aeroplane

by P. W. Brooks, B.Sc. (Eng.)

The Chinese seem to have discovered many of the fundamental concepts of aviation between 600 and 2400 years ago. These included: the kite (5th century B.C.)—which was used for man-lifting some time between the 6th and 13th centuries A.D.; the parachute (date uncertain); the ornithopter (1st century A.D.); the glider, the helicopter and the airscrew (all in the 4th century); the rocket (12th century) and, possibly, the idea of separating lift from propulsion (14th century).

Leonardo's basic concepts

Like other ancient Chinese inventions, such as gunpowder and printing, these aeronautical ideas appear gradually to have spread to Europe. The first of them may have reached the West in the 12th century or earlier but several had certainly done so by the fifteenth. The first comprehensive record we have of many of these concepts is contained in the notebooks of that great and versatile genius, Leonardo da Vinci (1452–1519). Working in northern Italy and France, Leonardo drew many remarkable sketches of aeronautical devices and also made an extensive study of the flight of birds. His drawings—of which more than 400 concerned with flying survive—and his notes show that he conceived man-carrying flying machines working on the ornithopter, helicopter (see Fig. 1), parachute and fixed-wing principles. He may also have given thought to the possibility of separating the functions of lift and propulsion. One significant innovation by Leonardo was the addition of horizontal and vertical tail surfaces for the control of one of his ornithopter designs. He also made what may have been the first proposal for mechanically powering a flying machine (with a bow-string mechanism). Leonardo apparently gave no attention to kites but other records survive of European kites in the fifteenth century.

Unfortunately, Leonardo da Vinci's remarkable notebooks were lost to posterity for nearly 300 years after his death and made no contribution to subsequent progress. However, it is likely that his sources of information, and possibly even unknown contemporaries with similar interests, contributed to the aeronautical tradition of medieval Europe and thus inspired and informed later workers.

The ensuing 270 years saw many abortive attempts at human flight, mainly on the ornithopter principle but these were of little significance. However, a Frenchman, A. J. P. Paucton did propose a crude form of man-powered helicopter in 1768 and the lighter-than-air principle of flight—which was probably first adumbrated by Roger Bacon in the 13th century but is outside the scope of this study–had meanwhile, been incorporated in the design of an aerial ship by Francesco de Lana-Terzi in 1670. A model hot-air balloon may have been flown by Bartholomeu Lourenço de Gusmão in Portugal in 1709 (and possibly much earlier by the Chinese). A successful model hot-air balloon was certainly demonstrated by the Montgolfier brothers in France in 1782, followed by the first manned balloon flights in 1783.

Fig. 2. Sir George Cayley's design already incorporated the major features of modern aeroplanes, including separate tail, control surfaces and propulsion

Cayley develops the theory

These developments and a successful helicopter model made by Launoy and Bienvenu in France in 1784 (itself apparently deriving directly from China), rather than anything handed down from Leonardo, seem to have inspired an English country squire, Sir George Cayley (1773–1857), to take an interest in the problems of human flight; he was to be the dominant figure in the early history of the aeroplane.

His first aviation experiment (in 1792) was with a helicopter model copied from that of Launoy and Bienvenu. During the next 18 years, he devoted a lot of time to reasoning out the basic principles of fixed-wing flight. By 1799 he had formulated a clear concept of an aeroplane with most of the fundamental features which eventually proved successful. Cayley's proposals, sketched in Fig. 2, included: a fixed wing for lift with a cambered surface; separate systems of lift and propulsion; a car to carry the pilot mounted below the wing to give lateral pendulum stability; separate horizontal and vertical surfaces for stability and control; and propulsion by two pilot-operated horizontally-flapping paddles.

Few, if any, developments in engineering have made a greater impact on human affairs than mechanical flight. Just 60 years ago the first successful man-carrying flights were made with a powered, heavier than-air machine. Since then progress has been spectacular and its consequences have altered the course of history. But the principles of modern flying machines were foreshadowed by Leonardo and much of modern aeronautical theory was developed by Cayley at the turn of the 18th century.

During the next ten years he gradually clarified and extended these ideas. In particular, he reached the fundamental conclusions that lift and drag vary as the square of the airspeed, multiplied by the density of the air; that drag relates to lift as the tangent of the incidence on an inclined plane; that the centre of pressure is much nearer the leading than the trailing edge; and that, for plane surfaces, it moves rearward as the angle of incidence increases. He came to realise that a region of reduced pressure above an aerofoil contributes to lift.

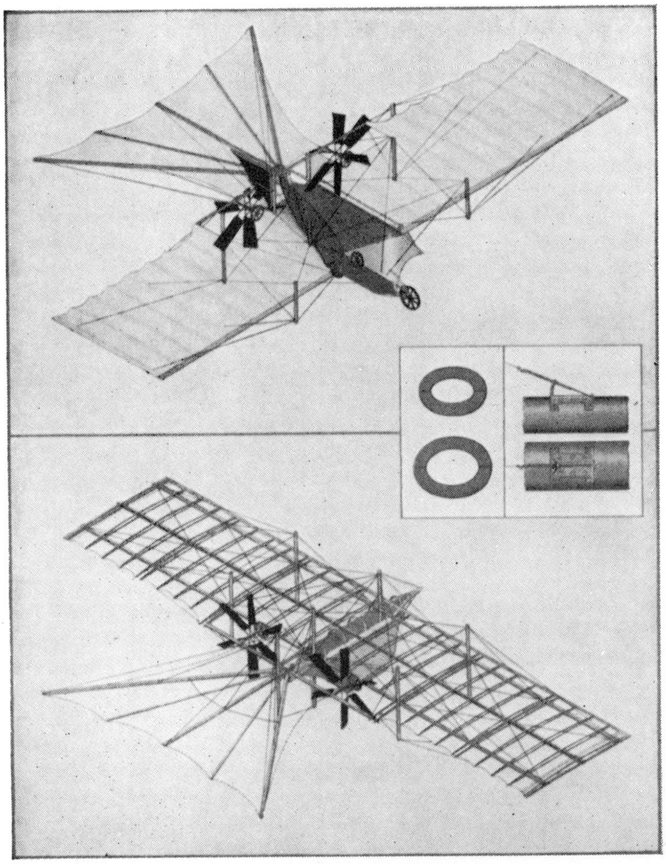

Fig. 3. Henson's aerial steam carriage was to be built out of bamboo and braced with wires. The inset shows sections of tapering outer spars and turnbuckle for tightening the bracing

His conclusions were based on observation of the flight of birds (particularly while gliding), on experiments with whirling arms (to measure the forces on inclined planes) and from successful flights in 1804 of a model glider with a flat, kite-shaped wing of low aspect ratio, over distances of up to 90 ft at gliding angles of about 1 in 3. From his study of birds, Cayley established that a cambered wing, at an estimated incidence of six degrees, would carry a load of a pound per square foot at a speed of about 24 mile/h.

During this period, Cayley also suggested wheeled undercarriages for flying machines, so that they could gain flying speed during a take-off run along the ground (assisted, if necessary, by running down a slope) and he designed the light tension bicycle-type wheel for this purpose. In 1809 he built and tested a full-size glider. Unfortunately, we do not know whether this historic first successful flying machine had a rectangular, sail-like wing of low aspect ratio with leading and trailing edge spars (a type which Cayley may have already envisaged) or whether it employed a kite-shaped wing, as did the 1804 model. The machine, which weighed 56 lb and had a wing area of 300 ft^2, apparently made many glides and achieved gliding angles of up to 1 in 8 while carrying 84 lb of ballast. Later it was flown for short hops carrying a man (possibly Cayley himself) and finally crashed with a boy on board during experiments in extending the glide with flapper propulsion.

As a result of his experiments, Cayley established that man-power was inadequate for sustained flight (at least in the then state of knowledge) but he thought it might be used to extend the glide of his machines. He foresaw the use in aircraft of one of the then new prime movers and pointed out the likely advantages of the still-to-be-invented internal combustion engine over heavy steam engines.

Cayley also noted at this time that slow-flying birds have long narrow wings (high aspect ratio), that all birds have wings of cambered section and that they usually hold them with a lateral dihedral angle while gliding. He applied the two latter observations to his gliders but apparently did not fully understand the advantages of high aspect ratio—almost all his designs had aspect ratios of around unity. Another of his discoveries was that streamlined shapes (derived from fish and birds) reduce drag. He established at this time that birds propel themselves in flight by varying the incidence of their outer wings (and sometimes of individual main feathers) as they flap up and down, while the greater part of the wings act primarily as fixed lifting surfaces. Consequently, Cayley advocated flapper propulsion in preference to the more efficient airscrews (which had first been tried unsuccessfully for balloon propulsion by J. P. Blanchard in France in 1784).

Probably inspired by false reports of successful man-powered flight by J. Degen in Vienna, Cayley published the results of his researches in 1809–10. He gave all his arguments and conclusions but no precise details of his gliders.

Only 40 years later did he resume full-scale experiments: for many years, his only interest in aeronautical affairs was in lighter-than-air and rotary wing developments.

Henson's steam carriage

Cayley's published results in 1809–10 provided the inspiration for the next important step: the remarkable design for an 'Aerial Steam Carriage' patented in 1842 by another Englishman, W. S. Henson (1812–88), shown in Fig. 3. This was of a monoplane of 150 ft span powered by a 25/30 hp steam engine driving twin pusher propellers.

Fig. 4. Hargrave experimenting with his box kites: note the two-cell, in-line lay-out

Although impracticable in the size proposed, the Aerial Steam Carriage was widely publicised and established the basic configuration of the modern aeroplane. It incorporated all the design features advocated by Cayley with the important additions of a rectangular wing of relatively high aspect ratio (5) which was double-surfaced over cambered ribs, threaded on to span-wise spars which were most effectively braced by wires and king-posts. The use of propellers was also an advance on Cayley's flapper propulsion—though the latter had himself advocated propellers for airships in 1817 and 1837.

Henson failed to give his wings dihedral angle as had been sensibly suggested by Cayley. Instead, he mounted a large fin above the centre section to provide lateral stability. His tail control surfaces were rather inadequate in area and had too small a moment arm.

Cayley published his comments on Henson's design in 1843. He quite unjustifiably criticised the strength of the wing high aspect ratio and, taking up a multiplane proposal by Thomas Walker made some time before 1827, suggested a 'three-decker' instead, which would be stronger for a given area.

Henson followed his full-size monoplane proposal with experiments with models to prove the various features of the design. Powered by small steam engines, two of these—the second designed by an associate, J. Stringfellow—achieved free flight, although it is doubtful whether either made sustained flights.

Structural configuration

Cayley resumed full-scale tests with gliders in 1848–49 at the age of 75. Following his multiplane suggestion, his 1849 glider seems to have been a triplane with roughly square, closely-spaced, sail-like surfaces, mounted parasol-fashion above the pilot's wheeled car. There were two separate tail units, each with horizontal and vertical surfaces, one fixed for stability, the other movable by the pilot for control. This glider may have made short flights carrying a boy. Take-offs were made after gaining speed by rolling down a hill and after being pulled into the air against a breeze with a rope.

A large model glider, with square, sail-like monoplane wings between leading and trailing edge spars achieved glides of about 200 ft at a gliding angle of 1 in 8, at 23 mile/h. Next came a second full-size glider in 1853. This glider is reported to have made a manned flight of possibly as much as 1500 ft.

Although Cayley published details of one full-size glider in 1852, unfortunately no precise details of the gliders actually flown were publicly identified at the time. Had they been, they would probably have inspired immediate emulation, possibly with far-reaching effects on aircraft development. As it is, we still do not know precisely what Cayley's successful gliders looked like. All we do know is that his full-size designs almost certainly had lifting surfaces which were barely satisfactory, either structurally or aerodynamically. Cayley adhered throughout his long life to kite or sail-like surfaces of low aspect ratio, even after Henson had demonstrated the correct next step.

Cayley's biggest contribution to progress was undoubtedly his monoplane-with-tail configuration. This inspired Henson and was confirmed by the work of the Frenchman, A. Penaud (1850–80), who invented the twisted-rubber-driven model aeroplane in 1871 and thus showed how a stable model of this configuration could make sustained flights. Cayley also gave much attention to the tandem wing layout, first suggested by Walker in 1831. This led to the work of the American, S. P. Langley (1834–1906) who flew successful steam and petrol-engined monoplane models of this type between 1891 and 1901 and then very nearly achieved powered manned flight with a similar full-size design, shortly before the Wrights. However, this configuration, although tried at intervals ever since, has not survived.

Cayley's suggestion of a multiplane in 1843 was endorsed by F. H. Wenham (1824–1908) who demonstrated the aerodynamic advantages in 1858. The first multiplane of the modern, high-aspect-ratio, rectangular configuration was Stringfellow's model triplane of 1868. His son, F. J. Stringfellow, built a model biplane of similar layout in 1886. In parallel with these developments, Wenham in 1858, and H. F. Phillips in 1884, showed the improved lift to be obtained from aerofoil sections with their greatest thickness near the leading edge, and the importance of the nose shape of wing sections.

Octave Chanute (1832–1910) in America finally combined these concepts into a truly efficient structure. In 1888 he built and tested a box-kite type of model glider. In 1893, the Australian, L. Hargrave, invented the box-kite proper, thus applying the multiplane concept to that ancient device which had traditionally been a monoplane. At the same time, he invented the arrangement shown in Fig. 4 of two linked tandem biplane cells from which the first European biplanes of the early 20th century were to derive. In 1896 Chanute tested the full-size biplane glider shown in Fig. 5, (after first trying it as a triplane). This machine had a monoplane tail but the biplane wings were braced in bays (employing the principle of the Pratt truss) and thus set the definitive pattern for the standard biplanes of the first 25 years of aviation.

With the addition of a sufficiently light and powerful internal combustion engine and an effective control system, the structure evolved could now form the basis of the successful aeroplanes of the first decade of this century. Before turning to this crucial next step, however, the dominant influence must be noted of the pioneers of practical gliding who took up where Cayley had left off in 1853.

The key figure of this period was the German, Otto Lilienthal (1848–96). Like Cayley, Lilienthal and those who imitated him ignored the rectangular form of braced monoplane or biplane wing which had crystallised between Henson and Chanute. Instead, they adopted the bird (or perhaps, more accurately, bat) wing form with radiating structural members over which fabric was stretched. Using this crude basic design, Lilienthal produced a whole series of highly-successful monoplane and biplane gliders between 1891 and his death in a flying accident in 1896. During this time, he made more than 1000 glides, covering distances of up to about 1300 ft. Lilienthal controlled his gliders by swinging his hanging body but he also experimented ineffectively with a movable elevator and a system of wing-warping for lateral control. He tried unsuccessfully a carbonic acid gas engine to power one of his gliders by means of flapping wing tips.

The effect of Lilienthal's practical demonstrations was far-reaching. He sold his gliders to other experimenters. One of his disciples, the Englishman P. S. Pilcher (before he too was killed in a gliding accident) became interested in the box-kites of Hargrave and apparently adapted this configuration to a flying machine, thus introducing this new and more practical form of multiplane to Europe in 1899.

The Wright brothers' success

However, Lilienthal's greatest contribution to progress was his stimulation of the interest of the American brothers Wilbur (1867–1912) and Orville Wright (1871–1948). These two men were to be the first to solve the problem of sustained and controlled powered human flight. Starting in 1899 with a controllable kite, and continuing during 1900–02 with gliding experiments on the Lilienthal pattern, they used a trussed rectangular-wing biplane of relatively high aspect ratio, deriving directly from Chanute's successful 1896 glider but without the latter's tail.

The Wrights' decisive contribution to all that had already been learnt was their realisation of the importance of control. They made other, less important, contributions—in particular, they developed more satisfactory wing sections, a light petrol engine and efficient slow-turning propellers—but it was their highly effective control system, comprising movable forward elevator, rear rudder and wing warping, which led to their success. In 1903, after more than 1000 glides of up to 600 ft, they built their first powered aeroplane and, towards the end of that year, made the first successful aeroplane flights in history: four brief straight hops, the longest lasting just under one minute and covering 850 ft. The great thing about these flights was not their length but the fact that they were made close to the ground in a 25 mile/h wind, thus demonstrating a degree of control never previously approached.

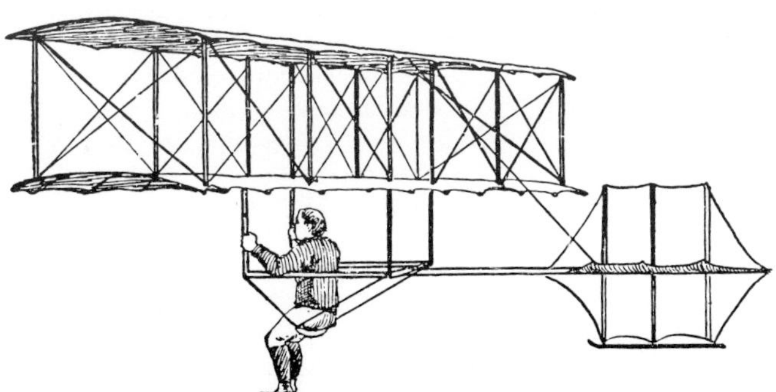

Fig. 5. Chanute's first full-size glider, tested in 1896, set the standard pattern for the biplanes of the next 25 years

In 1904 the Wrights flew their second aeroplane, a slightly improved development of the first, and with it achieved a circular flight of over five minutes, during which $2\frac{3}{4}$ miles were covered. The following year, a further improved biplane made nearly fifty flights, the longest covering 24 miles in 38 minutes. The aeroplane had arrived.

In the first aeroplane which flew successfully the Wrights had deliberately adopted the unstable tail-first configuration, believing that stability was incompatible with adequate control. In this they were to be proved wrong and, because they were slow to recognise this, their design was soon out-dated. However, in 1905 they were ahead of all others. They therefore decided to stop their experiments, keep their work secret and find a customer for their invention. Unfortunately, as so often happens, they proved to be less successful in business than in engineering. For $2\frac{1}{2}$ years they negotiated, first with the U.S. government and then with interests abroad. Not until 1908, did they conclude contracts with the U.S. Army Department (for $25,000) and with a private French syndicate (for $100,000). Although they did no flying in the interval, they had by then completed several new machines (two-seaters with improved engines) one of which had been shipped to France in 1907.

Development in Europe

The Wrights' success attracted surprisingly little attention even in the United States and the brothers were, in any case, soon doing all they could to keep details secret. However, all the more important features of the Wright biplane had, in fact, already been revealed in Europe by Chanute in a lecture in Paris, early in 1903. This information and unconfirmed reports of the first successful powered flights stimulated development in Europe. Although several experimenters attempted to copy the Wrights' design directly, they were—fortunately as it turned out—unsuccessful, so that attention was diverted to other and more promising configurations.

A Brazilian, A. Santos-Dumont is credited with the first powered flights in Europe. These were made in France in 1906 with a tail-first biplane of unsatisfactory design. However, in 1904 a Frenchman, F. Ferber, who had been experimenting since 1899, built a glider with Wright-type biplane wings and a tailplane on the Cayley/Lilienthal/Chanute pattern which set the Europeans on the right track.

This led to two Hargrave-derived box-kite biplane gliders, with their tail cells at the rear, which were built by the brothers Gabriel and Charles Voisin in 1905. They set up in business in Paris as aircraft manufacturers. Though building most of their early aircraft at least partly to other peoples' designs, they gradually developed a box-kite biplane which emerged as the most successful European aeroplane in 1907, with a flight of $\frac{2}{3}$ mile in $1\frac{1}{4}$ minutes. By May, 1908 this had been increased to $7\frac{3}{4}$ miles in $15\frac{1}{2}$ minutes. In 1909 a standard Voisin made a flight of nearly $2\frac{3}{4}$ hours.

The Voisin biplane had its engine driving a pusher propeller, rotating between booms connecting the wing and tail cells. It had no lateral control and relied on the stability imparted by 'side-curtains' which joined the upper and lower surfaces, box-kite fashion. Pitch control was provided by a forward elevator deriving from the Wrights, and there was a rudder in the rear box-kite cell.

An Englishman living in France, Henri Farman, perhaps the greatest of the pioneer pilots, used Voisin-built biplanes. In October 1908 he made the first cross-country flight in history (16½ miles) after fitting his biplane with ailerons. In 1909, Farman produced the final form of pusher biplane: this dispensed with the side curtains (and, much later, with the front elevator also) and was equipped with ailerons. The type remained in widespread service, until the end of the 1914–18 war.

The tractor monoplane

The other important type to be developed in France during the period prior to the Wright brothers' first public flights in August 1908, was the tractor monoplane. Some of the first (unsuccessful) powered aeroplanes built were of this layout. They included designs by the Frenchmen, F. du Temple and C. Ader.

A Hungarian in France, T. Vuia, produced a tractor monoplane in 1906 which had a bat-shaped wing reminiscent of Lilienthal. This made only a few short hops but showed the way to straight-winged monoplanes designed by the Frenchmen L. Bleriot and R. Esnault-Pelterie and by Santos-Dumont in 1907 and 1908. These designs incorporated ailerons or wing-warping for lateral control.

Bleriot's No. VIII monoplane flew ⅔ mile in July, 1908. In the same month the *June Bug* biplane, designed by the Aerial Experiment Association in the U.S.A., was flown a mile by G. H. Curtiss. Bleriot's No. XI (1909) is shown in Fig. 6.

Fig. 6. The Bleriot XI which first flew the English Channel incorporated the external bracing which, for some time, prevented the monoplane eclipsing the biplane

By the Spring of 1908, the Wrights realised that their virtual monopoly of a practical flying machine was about to be overtaken by developments in both Europe and North America. By then, they had fortunately been able to conclude satisfactory commercial arrangements. Therefore they resumed their flying in secret in May 1908. Their first public flights were made by Wilbur in France in August and by Orville in America in September. Their demonstrations startled the world, particularly because of the high degree of pilot control which they revealed. Moreover, they showed conclusively that the design which had been perfected in 1905 was still superior to those developed by all other experimenters up to the end of 1908 (by which time Wilbur had made a flight of 2 hours 20 minutes, covering 77 miles). The Wright 1907–09 biplane is shown in Fig. 7.

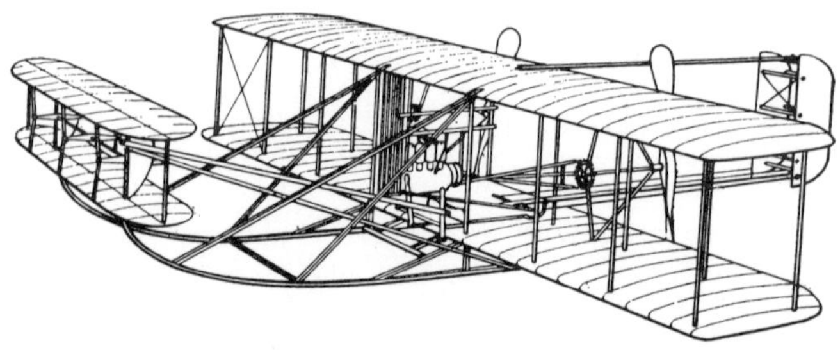

Fig. 7. One of the 1907–09 biplanes built by the Wright brothers

However, the ascendancy of the unstable, tail-first Wright biplane was to be short-lived: within a year it began to be eclipsed by the more stable European monoplanes and biplanes and the Curtiss biplanes in America. The stability was, of course, by no means satisfactory in many early designs. Work undertaken just before the 1914–18 war—particularly by G. H. Bryan, L. Bairstow and E. T. Busk in England—led to important improvements in stability. Much of this was based on basic aerodynamic theory first formulated by F. W. Lanchester in the 1890's.

For some years after 1908 the tractor monoplane, as one of the two successful European configurations, looked a likely short term rival of the biplane. In the event, this did not happen. With the structures then practicable, the monoplane had to be externally-braced, which meant that it had only a small advantage in drag while it was often weaker than the trussed biplane. For these reasons the monoplane declined in popularity before and during the 1914–18 war and was favoured by only a minority of designers.

The tractor biplane

The first tractor biplane was tentatively flown by Ferber in 1905. A. de Pischoff tested another in 1907–08. The first which really established the type appeared in 1909 and were due to the Frenchmen, A. Goupy and L. Breguet. Not until 1910, however, did Breguet's design begin to show its real promise, followed by the equally-successful British Avro biplanes in 1911. The type rapidly gained in popularity, combining the best features of the pusher biplane and tractor monoplane. By 1914 the tractor biplane had established itself as the most important type and was to hold this position for the next 15–20 years.

Something like 5000 aeroplanes were built up to 1914 while nearly 200 000 were produced during the 1914–18 war. Large-scale manufacture and use had a profound effect on development. The aeroplane was converted from an experimental device into a proven production article designed on the basis of widely accumulated operational experience. However, in spite of this, real progress in aircraft design was—with a few notable exceptions—surprisingly limited between 1914 and 1918. Substantial advances were made in the size of aircraft and in the power and reliability of aero-engines but the methods of construction remained largely unchanged.

The first aero-engines derived from motor-car practice. Air or water-cooled, they were usually of in-line or vee configuration and some developed up to about 150 hp by 1914. In 1908, the specialised air-cooled rotary appeared and provided a notable advance in power/weight ratio over previous stationary engines. This type, of which the French *Gnome* design of 1907 (Fig. 8) by L. Seguin was the first practical example, made a notable contribution to progress and was widely used up to about 1920 when it gradually lost favour because of its high fuel and oil consumption and because it could not efficiently develop more than about 240 hp.

By 1920 the stationary water-cooled engine was developing up to about 500 hp and engines of up to 350 hp were in widespread use in military aircraft. Such powers enabled single-seat fighters to achieve speeds of up to 150 mph and multi-engined aircraft to have loaded weights of up to 15 000 lb in general service (30 000 lb for a few types), with correspondingly increased ranges and military loads.

The multi-engined designs developed from the Russian Sikorsky *Russkii Vitiaz* of 1913, which was the first successful four-engine aeroplane.

Fig. 9. The Junkers stressed and corrugated skin construction set one pattern between the wars; thick wings were used, designed on the cantilever principle

Metal structures and cantilever wings

The lack of progress in the structural design and aerodynamic efficiency of the majority of aircraft used in the 1914–18 war continued well in to the 1920's. Wooden wire-braced tractor biplanes covered with fabric remained the most popular type until metal gradually replaced wood in the framework of the biplanes and externally-braced monoplanes of the later 1920's. At the same time the monoplanes began to regain their popularity, particularly in the United States. Metal—both steel tubing and light alloy strip—was used inefficiently to replace wood directly. Used in France during the war, it was encouraged, notably in British military aircraft during the 1920's, because of the difficulty in obtaining suitable timber in sufficient quantity. Metal had been used experimentally in this way in a few of the earliest practical aircraft and it was widely adopted in the fabric-covered steel-tube fuselages of the highly-successful aircraft produced for the Germans during the war by the Dutchman A. H. G. Fokker. He also used wooden cantilever wings which derived an important part of their strength from their ply-wood covering and continued to use this type of structure in a most successful series of mainly civil monoplanes flown by the airlines from 1919 until the early 1930's.

The cantilever monoplane with a wing sufficiently thick to dispense with bracing which Fokker, and later many others, achieved in wood, derived from the designs of a German, Hugo Junkers who patented an all-wing aeroplane in 1910. Two of Bleriot's early, only partly-successful, monoplanes in 1907 had cantilever wings but J. Weiss in England was the first to demonstrate the possibilities of thick wings (10–14 per cent thickness/chord ratio) with model gliders (from 1905), full-size gliders (from 1907) and aeroplanes (from 1909). Weiss's designs were strut-braced but Junkers proposed cantilever structures.

Fig. 8. The 'Gnome' was the first notable example of a power unit specially designed for aircraft propulsion. This air-cooled rotary engine had a lower power/weight ratio than the previous types, developed from car engines

The first cantilever thick-wing monoplane built was the unsuccessful Antoinette *Monobloc* of 1911, designed by the Frenchman L. Levavasseur* who was also responsible for the wire and king-post braced Antoinette monoplanes which were Bleriot's chief competitors after 1908.

Junkers built his first cantilever monoplane in 1915. He employed welded steel sheet with the skin carrying a major part of the loads. Although flown successfully, the structure proved to be too heavy to be practical. After a second (also unsuccessful) machine of similar construction in 1916—this time a low-wing monoplane fighter—Junkers turned to duralumin, the aluminium alloy which had been developed by the German Wilm in the first decade of the century. Junkers used the new material in corrugated sheet form over a load-bearing tubular structure. He used it for both monoplane and biplane wings which were up to 20 per cent thick. Bi-plane wings were usually 3–7 per cent thick.

The efficiency of such thick-wing sections had by then been demonstrated in wind-tunnels by Junkers himself and by L. Prandtl at Göttingen. The latter was subsequently responsible for the series of Göttingen high-camber wing sections extensively used between the wars. The Junkers corrugated-skin metal designs were used in limited numbers during the 1914–1918 war and remained in production until well into the 1939–45 war. An early model is shown in Fig. 9.

Fuselages which derived their strength entirely from their skins—so-called monocoques—were built in 1912. Deperdussin used wood, Morane-Saulnier steel and Huth aluminium. During the first world war, wooden monocoques came into limited use in France and Germany and, at the end of it, Short Brothers in Britain pioneered duralumin monocoque fuselages and flying boat hulls.

Stressed skins and flutter

The most significant structural development during and just after the 1914–18 war, however, was due to the Germans C. Dornier and A. Rohrbach, who produced complete airframes with smooth, stressed skins. The Dornier designs of this period were partly fabric-covered but those of Rohrbach, which first flew in 1920, were all-metal, with a central torsion box to the wing structure. Another German, W. Messer-

* *See also his work on cars, p. 163 Volume 1.*

schmitt, produced a modified type of metal wing with a D-nose torsion box in 1924. These, and similar, stressed-skin methods of construction were used on a limited scale in the 1920s but they led directly to the development, primarily in the U.S., of the modern type of metal monoplane in the 1930s.

Fokker, Rohrbach and Messerschmitt also pioneered the monoplane wing which tapers in chord and thickness from root to tip, thus improving structural and aerodynamic efficiency. Other developments of the 1920s which contributed to the aerodynamically-clean modern type of monoplane of the 1930s were: radial air-cooled engines (developed first in Britain and then in the U.S.) and efficient methods of cowling them (due to Townend in Britain and to the U.S. National Advisory Committee for Aeronautics); reduced drag from, notably, retractable undercarriages (first proposed in 1911 and demonstrated in 1920); more efficient, thinner wing sections (due, primarily, to the American NACA); wing flaps to increase lift for take-off and landing (first tried in Britain in 1914); and increased wing loading, made possible by the use of the better wing sections with flaps and by the use of larger aerodromes. The leading edge wing slot to delay the stall was developed by F. Handley Page in 1920.

Important in the development of clean unbraced monoplanes was the overcoming of flutter (aerodynamically-excited structural oscillations) as a result of work by, notably, W. J. Duncan and R. A. Frazer in Britain. The improved stressed-skin structures themselves owed much to the studies of a German, H. A. Wagner, who produced his tension field beam theory in 1929, and to the American, J. K. Northrop, who showed the way to lighter and more durable multicellular structures which reduced the stress concentrations.

These developments led to the establishment, after the middle 1930s, of the metal stressed-skin, low-wing, monoplane as the standard type in place of the braced biplanes of the previous era of which the one shown in Fig. 10 is a representative. This monoplane has been much improved in detail during the past 30 years but it remains substantially unchanged, being still built up of aluminium alloy sheet, extrusions and pressings riveted together. A famous example of the metal monoplane of the 1930's was the *Spitfire*, shown in Fig. 11.

Advances in manufacturing technique have included flush riveting, bonding and spot welding; much improved light alloys, honeycomb structures and integrally-stiffened skins, machined from the solid. Steel and titanium have come into limited use alongside the generally-adopted aluminium alloys. They are likely to replace the latter, however, only in those future aircraft in which the effects of kinetic heating create problems of strength at high skin temperatures.

In the early 'Thirties, multi-engined monoplanes usually had engines driving propellers which were mounted along the leading edges of the wings. Jet engines, introduced in the 1940s, have however been installed in various positions which are still far from standardised. Engines were buried in the wings or hung in pods below the wings in most of the first generation of multi-engined jets as can be seen in Fig. 12. However, pods mounted on the rear fuselage are becoming increasingly popular.

The jet engine and vertical take-off

As important as the aerodynamic and structural advances incorporated in the new all-metal monoplanes since the early 1930s was the refinement of the petrol piston-engine. The advent of the air-cooled radial in the 1920s was followed by its progressive development, combined with the use of improved fuels and, after 1934, constant-speed, variable-pitch propellers, until maximum powers rose in the largest sizes, to about 3500 hp. Such outputs were often maintained at heights of 20 000 ft and more, by gear-driven or turbo-supercharging. At the same time, specific weights were reduced to rather less than a pound per horsepower, reliability was much improved and lives between overhaul, in civil use, were increased in some cases to as much as 1500 flying hours. The liquid-cooled engine of Vee configuration kept pace with the air-cooled radial during these years—although less widely used—and was particularly popular for fighter aircraft up to the end of the last war. The low-drag, ducted radiator, developed notably by Meredith in Britain, made a particular contribution to the success of this type of engine.

In the late 1920's A. A. Griffith and F. Whittle in Britain made the first practical proposals for the use of gas turbines for aircraft propulsion. This was to prove the most important development since 1903. Whittle's patent for a turbo-jet engine was published in 1932 and inspired similar developments in Germany by H. von Ohain and M. Mueller, from 1936. The first German jet engine ran in 1937 at about the same time as the first British engine. Early British development concentrated on engines with centrifugal compressors while the Germans paid more attention to axial compressors which have since been generally adopted. The first jet aeroplane, the German Heinkel He 178, flew in 1939 and the first

Fig. 10. The de Havilland 'Moth' of 1925 is a good example of the wire-braced biplane which maintained its lead over the monoplane until the late 'Twenties

Fig. 11. Victor in the Battle of Britain, the Supermarine 'Spitfire' is an excellent example of the smooth stressed skin, cantilevered monoplane which has remained basically unchanged since the early 'Thirties

military jet fighters—German and British—entered service in 1944. Since then, the jet engine has gradually taken over from the piston engine in all but the smallest aircraft.

The gas turbine geared to a propeller—the so-called turbo-prop—has had a limited vogue since it was first introduced for certain civil and military purposes. Although some turbo-props now develop up to 15 000 hp, the type seems unlikely to survive in competition with the pure jet.

The most important developments in the jet engine so far have been the adoption of the twin-spool layout and, particularly, of the dual-flow principle, i.e: only part of the air flowing through the compressors is burnt with the fuel and drives the power turbines. These developments have roughly halved the penalty in higher specific fuel consumption suffered by the early simple jets compared with piston engines or with comparably-developed turbo-props. The largest jet engines now develop static thrusts of up to 30 000 lb for a weight of about 6000 lb.

The high power-to-weight ratio of the jet engine has created new possibilities for the integration of lift and propulsion which had been kept separate since Cayley. The helicopter has provided vertical take-off and landing since 1936, at the cost, compared with the fixed-wing aeroplane, of much inferior performance. The recent developments, however, make it possible to retain higher speeds in V.T.O.L. aircraft but carrying capacity and economy remain low. Griffith made the first suggestion for special light-weight jet-lift engines to be used only for vertical take-off and landing, shortly after the end of the last war. The first test rig, the Rolls-Royce *Flying Bedstead*, flew on this principle in 1953. But already in 1946 a Frenchman, M.Wibault, had patented a design for jets which could be directed either downwards for lift, or horizontally for propulsion, by means of swivelling nozzles. The first aeroplane to fly on this principle, the British Hawker P.1127, did so in 1960. Other methods of using the jet engine to provide vertical or short take-off and landing are under development.

Conclusion

During the past 50 years aeronautics has unquestionably been one of the most sophisticated of the technologies, often setting the pace on the frontiers of applied knowledge. This has meant that advances in aviation have frequently influenced the rate of progress in other technologies and, indeed, in many aspects of industrial activity. Mechanical engineering, in particular, has benefited from aeronautical progress.

In the field of materials, the development and application to aircraft of, notably, light alloys, high-strength steels, heat-resistant materials, synthetic rubbers and plastics has subsequently contributed to progress in the design of road, rail, sea transport and power units. Between 1907 and the late 1940's, the design and development of petrol engines and improved fuels owed much to research and development directed primarily towards aeronautical ends. The modern gas turbine, of course, derives very largely from the even greater R. and D. effort which has gone into the jet engine and propeller turbine since the early 1930s.

Fig. 12. The General Dynamics Convair CV-990 jet transport shows one popular post-war development, that of multi-jet aircraft with 'podded' engine nacelles

Advances in manufacturing techniques and processes made in the aircraft and associated industries have often benefited other industries—including the design, development and use of machine tools, machining control equipment, synthetic adhesives and bonding processes, paints and protective finishes and the manufacture of light-weight structures. The same is true in methods of structural, installation and systems design, lofting, stressing, weight control and the application of computers; in production, inspection, quality control and interchangeability; in production planning, estimating and progressing; in methods of design, office and factory organisation, management and perhaps even in the handling of labour relations. Practices initiated by military and civil air operators have also benefited the operation and maintenance of power plants and vehicles.

Progress in aircraft design during the past 60 years has resulted primarily from the development of power units of ever higher power/weight ratio, with secondary emphasis on small over-all engine dimensions, improved reliability, reduced specific fuel consumption and longer life between overhauls. At the same time, as far as airframes are concerned, progress has resulted from lower drag and from holding down the ratio of structure to all-up weight, while size and performance have increased and thus demanded greater structural strength, stiffness and durability.

The objectives of this development effort have been higher operating speeds—associated with not-too-high minimum speeds for take-off and landing—and greater disposable loads in relation to installed power.

Mechanical engineering in general has benefited most obviously in the field of power plants. However, other by-products achieved by advances in the structural and aerodynamic fields have also been considerable.

INDEX TO VOLUME 2